Short Pulse Laser Interactions with Matter

AN INTRODUCTION

Short Pulse Laser Interactions with Matter

AN INTRODUCTION

Paul Gibbon

Research Centre Jülich, Germany

Imperial College Press

ICP

Published by

Imperial College Press
57 Shelton Street
Covent Garden
London WC2H 9HE

Distributed by

World Scientific Publishing Co. Pte. Ltd.
5 Toh Tuck Link, Singapore 596224
USA office: 27 Warren Street, Suite 401-402, Hackensack, NJ 07601
UK office: 57 Shelton Street, Covent Garden, London WC2H 9HE

British Library Cataloguing-in-Publication Data
A catalogue record for this book is available from the British Library.

First published 2005
Reprinted 2007

SHORT PULSE LASER INTERACTIONS WITH MATTER
An Introduction

ISBN-13 978-1-86094-135-1
ISBN-10 1-86094-135-4

Printed in Singapore

To
Susanne,
Theresa and Claire

Preface

One of the more delicate tasks in putting this book together was deciding what to include: where exactly are the boundaries of this new field? This problem has caused conference organizers enough headaches over the past decade or so, trying to find a common theme to embrace at least a dozen quite different research activities. Proceedings with curious titles like 'Ultrafast Phenomena VII', '3rd Fast Ignitor Workshop' and even 'Generation and Application of Short Wavelengths II' line the bookshelves of most researchers who have anything to do with high-power, femtosecond lasers. In spite of these schizophrenic tendencies, the global implications of future scientific endeavour using high-intensity lasers have been recently recognized by the OECD (Kato, 2001).

A science historian looking back 20 years from now may conclude that the field was at something of a crossroads at this point. Having grown out of an unlikely consortium of laser physicists, atomic physicists and hobbying plasma physicists from the laser fusion community, the research discipline of short pulse, high-intensity laser interactions with matter has matured to the point where several sub-fields can be clearly identified, such as 'high-harmonic generation', 'femtosecond x-ray diffractometry', or 'extreme-intensity physics'. The diversity of specialist input, together with a plasma-theory base augmented by contributions from established scientific fields such as astrophysics, nonlinear optics and computational physics, gives this topic an adventurous, pioneering feel, where speculative ideas compete for journal space with inexplicable experimental results.

For the graduate student just entering the fray, there is a bewildering variety of topics to choose from: atomic physics, laser physics, plasma physics and lately even particle physics; yet there are precious few defining criteria. One of the aims of this book then, is to supply just that: a cumulative definition of 'short pulse interactions' based on the physical phenomena which occur when a large number of laser-photons impinge on ordinary matter over a very short time.

In some respects it is easier to state what this book is *not* about, namely: quantum chemistry, spectroscopy, fusion or attosecond lasers; topics which crop up regularly enough, but as related fields or direct applications of the physics discussed in one of the earlier chapters. The recipe for its content instead comprises key-words such as: high (or ultra-high) intensity; femtosecond laser; relativistic; ultrafast; collective behavior; kinetic effects; extreme nonlinear, and so on. Most of the phenomena considered here involve laser field strengths way in excess of ionisation thresholds, so in Chapters 3–5, the core of the book, deals primarily with the interaction of light with free electrons or plasmas. By contrast, Chapter 2 begins with material on the borderline between its natural and ionized states, a regime which was the subject of many early studies with femtosecond lasers and which today constitutes an important scientific field in its own right. Together, these four chapters lay the foundations for the far-reaching practical applications reviewed in Chapter 7, which also takes a look at some of the more esoteric phenomena which are poised to form the basis of serious research programs in the future.

A large proportion of the illustrations in the book were created using numerical simulation codes developed by the author. Chapter 6 is therefore included to dispel some of the mystique surrounding this kind of theoretical calculation. Numerical schemes are given for these codes, together with a brief description of their use, and where to find them. Most of the programs are available from a dedicated web-site, the address of which can be found on page 215.

At the time of going to press, I am conscious of the fact that hundreds of new papers are appearing every month on topics directly related to those described herein. The rate of growth of this field — fuelled by seemingly unquenchable advances in laser technology — is such that in ten years' time each chapter will need a book of its own. Needless to say, a much more detailed account of atoms in intense fields can already be found, for example, in the work by Suter, *The Physics of Laser-Atom Interactions*; and a detailed treatment of nonlinear wave propagation in gases and underdense plasmas in *Laser Physics at Relativistic Intensities*, by Borovsky, Galkin, Shiryaev and Auguste. To counter these misgivings about impending out-datedness, I take heart from the fact that Kruer's book, *Laser Plasma Interactions* remains a must for people working in the femtosecond business, even though this book deals primarily with the long-pulse interactions typically found in inertial confinement fusion research. The present work should of course be seen as complementary to, and not competitive with Kruer's classic!

One could also argue that a textbook on this subject is premature: why not wait until the dust has settled and organize some collected monographs instead? Actually the material here has its origins in two review articles on femtosecond interactions penned by this author, but with a much more verbose approach in deriving and discussing important physical models. This will hopefully serve as an antidote to the increasingly concise style typical of the specialist journals (even for reviews), where the newcomer often comes to grief trying to figure out in which units the first equation is supposed to be written. The intention of this book, as the title suggests, is to introduce the reader — whether graduate student or non-specialist professional — to the wonders of high-intensity laser interactions with matter.

This mission could not have got underway without the positive intervention of Tony Bell when it was no more than a sketch on a sheet of A4; nor, for that matter, without the guiding hands of Malcolm Haines and Bucker Dangor, who set me on this trail in the Blackett Laboratory nearly 20 years ago. The valiant proof-reading efforts of former Jena colleagues Laszlo Veisz (who also checked many of the equations), Stefan Düsterer and Heiner Schwoerer (without whose help I could not have written a word on CPA) are gratefully acknowledged; as are my hosts at Jena University, Eckhart Förster and Roland Sauerbrey, who allowed me to inflict their students with large chunks of this material over the course of various lectures and computer practicals. Hearty thanks are also due to the following people who readily granted me permission to plunder their published and unpublished work: Elke Andersson, (Fig. 7.3); Thierry Auguste and Pascal Monot (Fig. 4.18); Marco Borghesi (Fig. 4.20); Christine Coverdale (Fig. 4.19); Eugene Clark (Fig. 5.20); Jacques Delettrez (Fig. 5.8); David Meyerhofer (Fig. 3.6); Patrick Mora (Fig. 4.16); Warren Mori (Fig. 4.17); Peter Norreys (Fig. 5.30); Heiner Schwoerer (Figs. 1.3, 1.4, 1.5); Donald Umstadter (Fig. 3.9) and Matt Zepf (Figs. 5.22, 5.30). I am also grateful to Lenore Betts at IC Press, whose enthusiasm and patience played a large part in rescuing this long-overdue project. Finally, a big thank-you is owed to my family, whose members have displayed extraordinary tolerance over the frequent disappearances of their spouse/father to the cellar during the past year.

Contents

Chapter 1

Introduction: Historical Background

The field of high-intensity laser interaction with matter, although barely two decades old, is already bursting with enough exotic phenomena to keep researchers busy for years to come. Since the invention of 'chirped pulse amplification' in 1985, progress in short pulse laser technology has been unrelenting, see Fig. 1.1. Pulse durations have come down from a picosecond to less than 5 femtoseconds (10^{-15} s); whereas focused intensities have skyrocketed six orders of magnitude. At present, several laboratories around the world are now promising intensities in excess of 10^{21} Wcm^{-2}.

In view of the impending escalation suggested by this diagram, a shift into the unchartered territory of highly relativistic laser particle physics, it is perhaps appropriate to take a brief look at the various disciplines from which this new field has grown. These contributory fields are numerous and diverse: they include laser physics, atomic physics, plasma physics, and lately even astrophysics and elementary particle physics. Many of the theoretical models described in later chapters can be traced back to one of these more classical areas. This does not mean that researchers in this business have been able to get their theories or ideas 'off the shelf' — on the contrary: the extreme conditions under which light and matter are forced to coexist during such interactions have posed a continual challenge to both theoreticians and experimentalists alike.

The key to understanding the underlying physics in these interactions is to realize that ordinary matter — whether solid, liquid or gas — will be rapidly ionized when subjected to high intensity irradiation. The electrons released are then immediately caught in the laser field, and oscillate with a characteristic energy (the right hand column in Fig. 1.1) which then dictates the subsequent interaction physics.

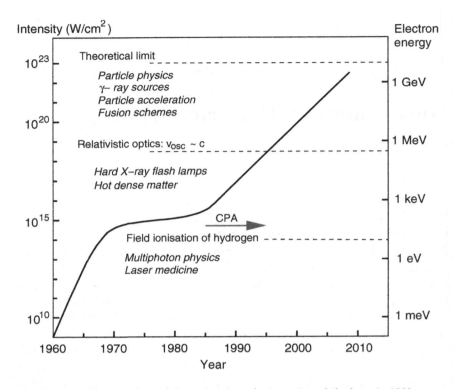

Fig. 1.1   Progress in peak intensity since the invention of the laser in 1960.

1.1   Multiphoton Physics

One of the most familiar examples of light inflicting a change in material properties is the photoelectric effect, predicted by Einstein a century ago (Einstein, 1905) and validated experimentally by Millikan a decade later (Millikan, 1916). This process, the ejection of an electron from an atom by a single photon, occurs when the photon energy $\hbar\omega$ matches the height of the atomic potential barrier, I_p, which the electron experiences in the vicinity of the ion, i.e. $\hbar\omega = I_p$. Even for the outer shells of most atoms, this energy still runs into many electron-volts, equivalent to photon wavelengths well into the ultraviolet (UV) range. For inner shells ($I_p \sim$ keV), one needs hard x-rays to induce photo-ionisation.

With standard lasers (operating wavelengths 0.25 μm – 13.4 μm), one cannot observe the photoelectric effect on normal material because $\hbar\omega \ll$

I_p. As lasers grew more powerful in the 1960s and '70s, however, (Fig. 1.1) it became possible to consider *multiphoton* ionisation, expressed by the condition

$$n\hbar\omega = I_p.$$

Thus, instead of one very energetic photon, an electron absorbs n photons of moderate energy (for example, laser photons with $\hbar\omega \approx$ eV) and is subsequently ejected. Some of the pioneering work on this was done at CEA-Saclay, France by the group of Mainfray and Manus (1991), including the discovery of *above-threshold ionisation* by Agostini *et al.* (1979). Another ground-breaking event during this period, prompted by the development of chirped pulse amplification, was the prediction of *high-harmonic generation* via recombination of multiphoton-ionized electrons by Shore and Knight (1987). This phenomenon was confirmed experimentally soon afterwards (McPherson *et al.*, 1987; Ferray *et al.*, 1988) by a intensive series of experimental investigations in a number of labs worldwide. Today it is possible to generate hundreds of harmonics down to nanometre wavelengths, a feat which has heralded a new era of highly compact, coherent XUV light sources.

1.2 Single-Electron Interaction with Intense Electromagnetic Fields

This subject actually goes much further back than most newcomers to the femtosecond business would probably guess; predating the invention of the laser at the beginning of the 1960s by some margin. One of the first analyses of the behavior of free electrons in the presence of intense radiation was made by Volkov (1935), who introduced the concept of a *dressed state* to describe the enhanced inertia experienced by an electron when oscillating in an electromagnetic field. Later, motivated by the first experiments with synchrotrons, Schwinger made a detailed analysis of the radiated power emitted by accelerated electrons (Schwinger, 1949), pointing out that radiation is preferentially emitted in the direction of motion at high energies.

These early ideas were refined and explored with more urgency when the invention of the laser brought the prospect of an experimental means to study relativistic photon-electron physics in the laboratory (Brown and Kibble, 1964; Vachaspati, 1962; Eberly and Sleeper, 1968; Sarachik and Schappert, 1970). These authors defined the figure of merit for 'laser-electron' interaction by a dimensionless parameter q (or q^2), given by:

$$q = \frac{eE_L}{m\omega c}, \tag{1.1}$$

where e, m and c are the electronic charge, electron mass and speed of light respectively; E_L is the laser electric field strength and ω the light frequency. Needless to say, these theoretical works all lamented the impossibility of achieving truly relativistic conditions ($q > 1$) with the optical lasers available at the time, but speculated that they might be might one day be reached with 'future' technology. Forty years later, this wishful thinking has become reality, and in Chapter 3 we will see how these visionary models have inspired new lines of experimental investigation.

Independently of the laser's arrival, astrophysicists were beginning to suggest mechanisms for cosmic ray generation in the vicinity of pulsars via the interaction of intense electromagnetic (EM) radiation with free electrons (Ostriker and Gunn, 1969; Gunn and Ostriker, 1971). The numbers involved here are of course vastly different from laser-plasma interactions: pulsar radiation has a frequency between 0.3 and 30 Hz, and magnetic field strengths near the star surface in the region of 10^{12} Gauss; causing gargantuan oscillation amplitudes of particles in the surrounding plasma. The interaction physics is very similar however, and is just one of many instances of *scalable laboratory astrophysics*, where one can emulate the conditions in astrophysical objects using high-power lasers.

1.3 Nonlinear Wave Propagation

The fact that plasmas can support large-amplitude, nonlinear waves has been known for almost as long as plasma physics established itself as a mainstream branch of science. Early seminal works by Akhiezer and Polovin (1956) and Dawson (1959) set the scene for numerous studies on the behavior of both large-amplitude Langmuir (electrostatic) waves and the propagation of high-intensity electromagnetic radiation in plasmas (Davidson, 1972). One attraction of this then rather obscure field was the tantalizing possibility of producing long-lived solitons in plasmas (Zakharov, 1972; Decoster, 1978; Shukla *et al.*, 1986).

The publication of a method for laser-acceleration of electrons in underdense plasmas by Tajima and Dawson (1979) sparked off a fresh wave of interest in wave propagation. This enthusiasm also encompasses members of the accelerator community, who are actively on the look-out for viable alternatives to conventional linear accelerators and synchrotrons as these devices approach their physical and economic limits. At the time of writing, the world record for laser-plasma acceleration of electrons lies at around 350 MeV, not breathtaking compared to the 50 GeV per electron/positron beam at SLAC, until one realizes the former was achieved with a few millimetres of plasma rather than several kilometres of supercooled accelerator structure. Although there is some way to go before laser-plasma acceleration can compete with existing facilities, scaling up to the Petawatt powers now available will soon make the tabletop GeV electron accelerator a reality.

1.4 Metal Optics

The story of laser interactions with solids also has a nineteenth century prologue. The simple observation that polished metals behave as almost perfect reflectors, whereas other materials either absorb or transmit light, could not be satisfactorily explained until Drude set out his 'Electron Theory' (Drude, 1900). Although solid state physics has advanced beyond recognition since then, the so-called *Drude model* of electron conduction still retains its appeal and usefulness in describing the main features of metal optics (see, for example: Ashcroft and Mermin, 1976, Chapter 1).

Drude's original idea — just three years after J. J. Thomson's discovery of the electron — was to suppose that the atoms in a metal somehow share a limited number of 'valence' electrons, forming a conduction band. These can wander as far as they like from their parent atoms, carrying current and heat through the material in the process. For an element with mass density ρ and atomic weight A, the free electron density is given by $n_e = N_A Z^* \rho / A$, where N_A is Avogadro's constant and Z^* is the number of valence electrons per atom.

The conductivity of a metal will depend on the rate at which these free electrons are slowed due to collisions with the ions. Mathematically this can be expressed by the relation: $\sigma_e = n_e e^2 \tau / m_e$, where τ is the collision or relaxation time. Even today, precise theoretical treatments to determine the relaxation time in solids remain a challenge, since the latter depends on details of the crystal structure, electronic configuration and so on. For this reason, nearly all the available data on metallic conductivity has been obtained through experimental measurements using Ohm's law: $j = \sigma_e E$. This technique — applying a DC voltage and finding the resulting current — can be recast in a form suitable for *optical* diagnostics via the AC conductivity:

$$\sigma(\omega) = \frac{\sigma_e}{1 - i\omega\tau}.$$

When combined with Maxwell's equations, this leads to a complex dielectric constant:

$$\varepsilon = 1 - \frac{\omega_p^2}{\omega(\omega + i\nu)},$$

where

$$\omega_p^2 = 4\pi n_e e^2/m_e$$

is the *plasma frequency* of the valence electrons, and $\nu \equiv \tau^{-1}$ is their relaxation rate or collision frequency. This expression basically predicts that a metal will only become transparent for radiation with wavelengths $\lambda < \lambda_p = 2\pi c/\omega_p$, which turns out to be in the UV (200–400 nm) range. Standard laser light (0.5–10 μm) will be reflected or absorbed, depending on the collision time.

Now, anyone with some basic plasma physics can see that the Drude model is almost identical to the standard theory of collisional laser absorption in plasmas, except that in the latter case, τ *can* be calculated with some accuracy. Suffice to say that metal optics provides an excellent starting point for studying reflectivity (and other transport properties) of short pulse laser-produced plasmas (Godwin, 1972). In Chapter 5 we will see how this leads to an almost seamless transition between the solid and plasma states of matter.

1.5   Long Pulse Laser-Plasma Interactions

A common misconception about femtosecond laser-plasma research is that it has little or nothing to do with the long pulse laser-plasma interaction issues relevant to inertial confinement fusion (ICF). It is certainly true that one cannot do ICF with a table-top laser (at least not yet, anyway!). The scientific (and often political) demarkation which prevails today masks a broad thematic overlap, and there are plenty of researchers who happily jump back and forth between the two fields on a daily basis. Whatever one's personal view of ICF as a future energy source, and its unavoidable relationship with nuclear weapons programs, it has to be said that the advances in the short pulse field would have been nowhere near as rapid without the considerable prior scientific and technical knowledge of laser science and interaction physics generated by the ICF programs over the last 30 years. Luckily, training in ICF physics is not necessary to work on short pulse interactions, but it is nonetheless useful to know where to find the original literature, on, say, hydrodynamics, parametric instabilities, energetic particle generation, and so on. A brief outline of ICF would therefore appear to be in order, even if to mainly draw contrasts between the new femtosecond phenomena and this *long pulse regime* in later chapters.

Laser fusion became official in 1972 (having previously been under military classification) with the publication of a classic but over-optimistic paper in Nature by Nuckolls *et al.*, (1972). In this work, the authors describe how a small micrometer-sized pellet filled with deuterium and tritium fuel can be compressed to enormous densities by irradiating it with laser beams focused symmetrically onto its surface (Brueckner and Jorna, 1974). By converting the laser energy into thermal plasma energy, the pellet shell material ablates radially outwards, thus pushing the fuel inwards via a rocket-like reaction. By virtue of the spherical symmetry, the fuel implodes, eventually reaching densities ρ of several hundred gcm^{-3} and temperatures T of 10 keV (10^7 degrees Kelvin), thus meeting the requirements for thermonuclear confinement encapsulated by the product:

$$nT\tau > 10^{15} \text{ keV s cm}^{-3}. \quad (1.2)$$

This condition — known as the *Lawson criterion* — basically states that the fuel must burn up and release its energy before the capsule blows apart, leading to a requirement for the *areal fuel density* ρR, where R is the final capsule radius.

The standard *hot-spot* scenario (Lindl, 1995) currently being pursued by the major laboratories in the USA, France and Japan, requires a central $\rho R = 0.3$, which, after allowing for all the inefficiencies and target design considerations, translates into a laser energy requirement of around 1 MJ to achieve gain. We will return to these constraints later in section 7.5, where some speculative new ideas to reduce this driver requirement with the help of an additional short pulse laser are described.

These more recent schemes aside, the whole process of target ablation, implosion and ignition is essentially determined by hydrodynamics: any plasma physics involved is almost invariably destructive, putting additional constraints on the laser and target design. These *coronal* processes (so-called by analogy with stellar objects) are illustrated in Fig. 1.2: as we

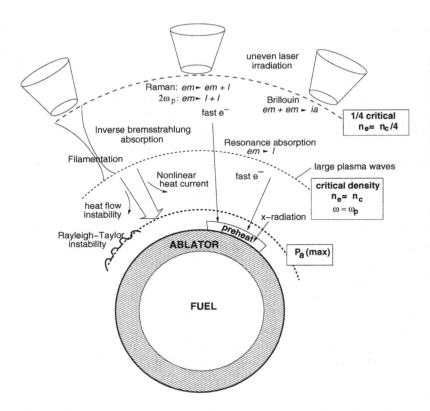

Fig. 1.2   Laser-plasma interactions in the corona of an imploding micro-balloon.

shall see in later chapters, nearly all of them turn up again in femtosecond interactions. Parametric instabilities such as Raman and Brillouin
scattering (Kruer, 1988) are generally bad for the implosion because they
generate fast electrons. Because of their long range, these electrons preheat
the target core, making the compression less efficient. Resonance absorption (which we will meet in Sec. 5.5.1) is also undesirable for this reason: in
fact, up to 50% of the laser energy can be wasted on superheated electrons
in this way. The classic signature of such 'anomalous absorption' processes
is a bi-Maxwellian electron distribution, see Fig. 5.17. In the case of resonance absorption, the hot electron component has a temperature T_H given
by (Forslund *et al.*, 1975a; Estabrook *et al.*, 1975):

$$T_H \simeq 14\,(I\lambda^2)^{1/3} \text{ keV}, \tag{1.3}$$

where I is the laser intensity in units of 10^{16} Wcm^{-2} and λ the laser wavelength in microns. The dependence on the product $I\lambda^2$ (originating in the quiver motion of the electrons in the laser field) means that long wavelength lasers (e.g.: CO_2) tend to produce hot electrons much more readily (than, say KrF) for a given intensity. This was a major factor in the strategic switch to shorter wavelengths ($\lambda < 0.5\mu$m) in ICF programs at the beginning of the 1980s. A major advantage of *indirectly* (X-ray) driven schemes is that hot electron generation is almost completely avoided (Lindl, 1995).

We have already seen that the rules of the game describing ultrafast, ultra-high-intensity (UHI) interactions have to be rewritten. For example, the short timescales make it necessary to discard steady-state or adiabatic models in favour of fully dynamic non-equilibrium formulations. On the other hand, some of the computational tools developed over 30 years ago to investigate hot electron generation in nanosecond laser-plasma context are used today to model femtosecond interactions at intensities of 10^{21} Wcm^{-2}, not least in the *fast ignitor* context (Sec. 7.5). In fact many of the codes used for UHI interactions can be traced back to ICF applications: atomic physics packages, fluid codes, particle-in-cell or Vlasov codes. These have usually been adapted to meet contemporary demands: for instance by using time-dependent ionization models to cope with the breakdown of local thermal equilibrium (LTE); by allowing for relativistic electron dynamics; or by introducing wave propagation into hydrodynamic models to handle the laser pressure and absorption more accurately. Details of these models are deferred to Chapter 6, but note that there are many areas — including many optical and diagnostic techniques — where the connection with ICF can turn out to be very fruitful.

1.6 Femtosecond Lasers

Hopefully, it will be clear by now that the intention of this book is not to supply assembly instructions for a Terawatt laser system, but rather to examine the physics which can be explored with the help of such a device. The material contained in the chapters which follow is thus very much aimed at those researchers either physically or mentally gathered around the target chamber (see Fig. 5.15), whose knowledge of the laser operation consists primarily of the four magic numbers: wavelength, energy, pulse duration and focal spot size. That said, the only way to appreciate where these numbers are conjured up from is to lift the lid off the box and take a peek at the optical wizardry inside.

Fig. 1.3   Front end of the Jena Ti:Sapphire femtosecond laser.

If we do this, we will find something resembling a huge BRIO-Builder^{TM} set with an awful lot of glass in it, see Fig. 1.3. Make no mistake, though: these devices represent the state-of-the-art in optical engineering.[1]. We can make more sense of this optical jungle with the help of the schematic diagram in Fig. 1.4, where we notice that the 'laser' actually comprises several autonomous units, namely: an *oscillator*, producing low-energy short pulses at regular intervals; a *stretcher*, converting the femtosecond pulse into a 50–200 ps *chirped* pulse; one or more amplifier stages to increase the pulse energy by a factor of 10^7–10^9; and finally a *compressor*, performing the exact optical inverse of the stretcher to deliver an amplified pulse of the same duration as the oscillator.

The key technique in this business is *chirped-pulse amplification* (CPA), which since its first application to lasers by Gerard Mourou and co-workers in 1985 (Strickland and Mourou, 1985; Maine *et al.*, 1988), is now exploited by nearly all existing high-power, femtosecond systems. At first sight, the procedure seems unwarranted: why bother with all the stretching and compressing when one could amplify the short pulse from the oscillator straight away? The answer is simply that most of the optical components will

---

[1]A tip for theoreticians being shown around a laser lab for the first time: resist the temptation to fiddle with the pieces on the table — it can take several days to put right again.

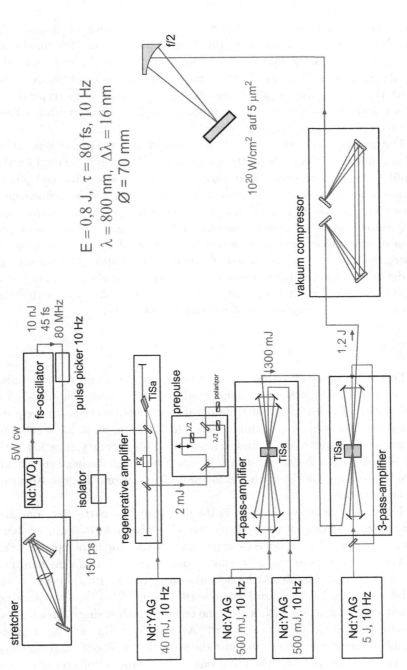

Fig. 1.4 Schematic diagram of the Jena TW laser.

be overheated and damaged if the fluence exceeds a level of around 0.16 $Jcm^{-2}\tau_{ps}^{1/2}$, where τ_{ps} is the pulse duration in picoseconds. For maximum efficiency, however, the fluence should be near the saturation level of the amplifying medium, which is 1 Jcm^{-2} for Ti:sapphire. Therefore, the pulse length throughout the amplifier chain has to be at least 40 ps (typical values are 100–200 ps) so that the components (e.g. mirror coatings) have time to cool by thermal conduction.

The stretcher-compressor combination separates the pulse generation and amplification stages, permitting standard techniques to be used for the amplifier chain, and furthermore leaving room for more advanced phase-compensation devices at either end. Before we consider these refinements, let us return to the 'front end', where the pulse is generated. Femtosecond laser sources are *mode-locked*: the output pulse is made up of a superposition of many electromagnetic waves (or laser modes), and is *transform limited*, so that $\tau_p \sim 1/\Delta\nu$, where $\Delta\nu$ is the bandwidth. Clearly a large bandwidth is essential to generate a short pulse. Consider, for example, a 10 fs Gaussian pulse, for which $\Delta\nu\tau = 0.44$. This gives $\Delta\nu = 4.4 \times 10^{13}$ Hz, which for a central wavelength of 800 nm, translates to:

$$\Delta\lambda = \Delta\nu\frac{\lambda^2}{c} = 94 \ nm.$$

This is an absolute requirement of the lasing medium in the oscillator, as well as the subsequent amplification stages. Most materials have bandwidths far below this, but titanium-doped sapphire crystals have a remarkable gain curve ranging from 700 nm to 1100 nm, with $\Delta\lambda = 230$ nm; the largest bandwidth of any known material. This property, together with its high saturation fluence of nearly 1 Jcm^{-2} and high damage threshold, make Ti:sapphire the ideal choice for most femtosecond laser systems in operation today.

How do we create such a pulse in the first place? In particular, a means of converting a continuous-wave (cw) pump laser beam into a train of short pulses is needed. This is done by exploiting another fortuitous property of the Ti:sapphire crystal: namely, that it also acts as a nonlinear focusing lens. This so-called nonlinear Kerr nonlinearity (Shen, 1984) occurs when the intensity inside the crystal exceeds 10^{11} Wcm^{-2}, so that by blocking the unfocused cw mode with a slit, one ends up with a single, short pulse bouncing to-and-fro in the oscillator. As it stands, this arrangement is unstable because dispersion will cause the pulse to spread out longitudinally (red wavelengths are faster than blue ones). To compensate, a set of prisms is inserted between the crystal and back mirror, which imposes an equal but opposite dispersion on the pulse (blue faster than red). The combination of cavity plus dispersion-correction constitutes a highly stable and reliable femtosecond laser, which these days can even be purchased 'off-the-shelf'. The true art of femtosecond laser physics comes when we try to amplify the pulse, as we shall see shortly.

The fs pulse from the oscillator typically contains just a few nJ of energy; yet we wish to increase its power to the TW level and beyond. As already mentioned, to avoid damaging optical elements, the pulse must first be stretched by a factor of 10^3–10^4 or so. This is usually done with a pair of diffraction gratings, which impose a positive chirp on the pulse, so that longer wavelengths emerge before shorter wavelengths. Early CPA systems (Maine *et al.*, 1988) actually used enormous km-length optical fibres to broaden the bandwidth via self-phase modulation. However, the stretch-factors obtained by this method are limited because there is no practical means of correcting the nonlinear, high-order phase distortion introduced by the fibre, ultimately restricting the final pulse length to values of just under 1 ps.

The stretched pulse can now be amplified — a process which is usually split into two or more stages, depending on the final beam energy required. The Jena system comprises three different modules: a *regenerative* preamplifier and two multipass *power* amplifiers. Most of the gain of the system ($\sim 10^7$) is obtained by the regenerative amplifier, which works in a very similar fashion to the laser resonator itself. The difference is that it seeded by the chirped pulse, which subsequently makes up to 20 round trips through a low gain medium, before being switched out by a combined Pockels cell/polarizer combination. Since the regenerative amplifier eventually becomes saturated at a few mJ, additional cavity-free techniques are then used to amplify up to and beyond the 1 J level. The multipass configuration in Fig. 1.4 is typical of modern tabletop-TW systems, producing gains of 10^2–10^3.

The amplified, long chirped pulse is then recompressed using a grating pair (or quadruplet in the Jena system shown), ideally reducing the pulse length back down to a value slightly above the one originally emitted by the oscillator. Some pulse lengthening is inevitable due to a combination of nonlinear dispersion effects and gain-narrowing. The latter arises because the amplifier medium preferentially enhances central wavelengths over peripheral ones, leading to a reduction in bandwidth and hence lengthening of pulse duration.

Due to the high-quality beam profile which can be produced by Ti:Sapphire amplifier systems, the pulses can be focused to an almost transform-limited spot containing most of the laser energy, see Fig. 1.5. This is the form of the laser pulse we want for interaction physics: the maximum possible photon density for a given wall-plug energy.

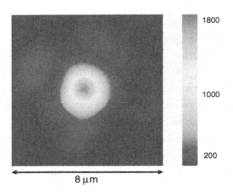

Fig. 1.5   Sharp end of the Jena Ti:sapphire femtosecond laser system: a focal spot of 3 μm diameter containing more than 50% of the pulse energy. The peak intensity reached here is 4×10^{19} Wcm^{-2}.

Naturally there are variations of this generic short pulse system depending on the intended application. What I have just described is a multi-TW system typical for a university laboratory. For a high-end user facility, offering powers above 100 TW, one still has to go to one of the dedicated laser laboratories listed in Table 1.1. On the other hand, there is also a growing number of smaller, *high-repetition-rate* (kHz) systems for high throughput applications such as x-ray sources. More recently, state-of-the-art *few-cycle* (< 10 fs) lasers have opened up a new field of attosecond physics, a theme which crops up again in Chapter 7.

As seen in the table, there are currently three or four Petawatt Nd:glass lasers coming on line over the next two years (VULCAN has already passed the PW mark). These will deliver focused intensities of about 10^{21} Wcm^{-2} — perhaps a bit higher after some tuning and by increasing the energy — but only up to a point. The damage threshold on the compression gratings or focusing optics is around 0.3 Jcm^{-2}, so a Petawatt facility with 400 J pulse energy needs gratings with least 40 cm in diameter. To be on the safe side, the VULCAN laser at RAL uses 1 m-diameter gratings, right at the

Table 1.1 Multi-Terawatt laser systems and laboratories worldwide.

Name	Laboratory	Country	Type	λ (nm)	Energy (J)	Pulse length (fs)	Power (TW)	Focal spot (μm)	Intensity (Wcm^{-2})
Petawatt[a]	LLNL	USA	Nd:glass	1053	700	500	1300	-	$> 10^{20}$
VULCAN[b]	RAL	UK	Nd:glass	1053	423	410	1030	10	1.06×10^{21}
PW laser[c]	ILE	Japan	Nd:glass	1054	420	470	1000	30	10^{20}
PHELIX[d]	GSI	Germany	Nd:glass	1064	500	500	1000	-	-
LULI 100TW	LULI	France	Ti:Sa	800	30	300	100	11	2×10^{19}
APR 100 TW	APR	Japan	Ti:Sa	800	2	20	100	(1)	(8×10^{21})
HERCULES	FOCUS	USA	Ti:Sa	800	1.2	27	45	(1)	(10^{22})
ALFA 2	FOCUS	USA	Ti:Sa	800	4.5	30	150		10^{19}
Salle Jaune	LOA	France	Ti:Sa	800	0.8	25	35		$> 10^{19}$
Lund TW	LLC	Sweden	Ti:Sa	800	1.0	30	30	10	$> 10^{19}$
MBI Ti:Sa	MBI	Germany	Ti:Sa	800	0.7	35	20		5×10^{19}
Jena TW	IOQ	Germany	Ti:Sa	800	1.0	80	12	3	10^{19}
ASTRA	RAL	UK	Ti:Sa	800	0.5	40	12		5×10^{19}
USP	LLNL	USA	Ti:Sa	800	1 (10)	100 (30)	10 (100)		5×10^{19}
UHI 10	CEA	France	Ti:Sa	800	0.7	65	10		5×10^{19}

[a] 1996–1999
[b] Petawatt performance achieved on October 5, 2004.
[c] Projected upgrade of PWM — PetaWatt Module.
[d] Commissioned for end 2005.

limit of present manufacturing capability.

So just as with the old (non-CPA) pulsed technology, the only way to increase the power without destroying the laser components is to reduce the pulse length further: a point where the smaller Ti:sapphire systems have a distinct advantage thanks to their broader bandwidth. Fortunately, an extension of CPA — optical parametric chirped pulse amplification, or OPCPA — has been proposed by Ross *et al.*, (1997) designed to achieve exactly that. This technique takes a conventional laser pump beam at any wavelength (Nd:glass, Ti:sapphire or KrF) and amplifies a large bandwidth of the chirped pulse, resulting in 10–20× shorter pulses after recompression than with the usual CPA scheme. Thus it is possible that the PW-class lasers may be upgradable to the 10 PW level via OPCPA without too much additional cost or reconstruction.

Chapter 2

Interaction with Single Atoms

Now that the Petawatt era has arrived, it may seem curious to devote a chapter on a class of laser-matter interaction which appears to be threatened with extinction. Admittedly, the primary subject of this book is the behavior of *ionized* material (plasma) under intense laser irradiation. In fact, for most of the phenomena we will meet in later chapters, the laser field strength dwarfs the interatomic fields by many orders of magnitude, and the question of *how* the target arrives in this state is usually of minor concern. Nevertheless, multiphoton physics was and still is one of the main driving forces behind the rapid development of short pulse laser technology. Add to this the obvious theoretical and experimental overlap between strong-field atomic physics and high-intensity laser-plasma interactions, then the introduction offered here becomes imperative.

The compromise struck in this chapter is to emphasize phenomena which can also play a significant role in laser-plasma interactions, such as ionization and propagation effects, where the target is in a borderline neutral/ionized state while the laser is incident. Readers looking for a more comprehensive or rigorous treatment of atoms in strong fields are referred to the many excellent reviews (Freeman and Bucksbaum, 1991; Burnett *et al.*, 1993; Protopapas *et al.*, 1997) already available.

The natural baseline often used to define 'high' laser intensity is the hydrogen atom. From the Bohr model, one can derive many of the quantities needed to get started in laser-atom interactions. First, we find that at the Bohr radius,

$$a_B = \frac{\hbar^2}{me^2} = 5.3 \times 10^{-9} \text{ cm},$$

17

the electric field strength is:

$$E_a = \frac{e}{a_B^2} \qquad \text{(cgs)}$$

$$= \frac{e}{4\pi\varepsilon_0 a_B^2} \qquad \text{(SI)} \qquad (2.1)$$

$$\simeq 5.1 \times 10^9 \ \text{Vm}^{-1}.$$

This immediately leads us to the *atomic intensity* — the intensity at which the laser field matches the binding strength of the electron to the atom:

$$I_a = \frac{cE_a^2}{8\pi} \qquad \text{(cgs)}$$

$$= \frac{\varepsilon_0 cE_a^2}{2} \qquad \text{(SI)} \qquad (2.2)$$

$$\simeq 3.51 \times 10^{16} \ \text{Wcm}^{-2}.$$

A laser intensity of $I_L > I_a$ will guarantee ionization for any target material, though in fact this can occur well below this threshold value via multiphoton effects, as seen shortly in Section 2.2.

2.1   Multiphoton Ionization

An electron will be ejected from an atom if it receives a large enough kick to propel it from its bound state to the free continuum outside. It can do this by absorbing a single high-frequency photon as in the photoelectric effect, or several lower frequency photons, as illustrated in Fig. 2.1a). The likelihood of the latter scenario depends strongly on the light intensity (or photon density), and according to perturbation theory, the n-photon ionization rate is given by:

$$\Gamma_n = \sigma_n I_L^n. \qquad (2.3)$$

The cross-section σ_n obviously decreases with n, but the I_L^n-dependence ensures that an nth-order ionization event will eventually occur provided that the intensity is high enough and sufficient atoms or ions are left. The laser made such intensities ($> 10^{10}$ Wcm^{-2}) readily available, and it did not take long for the first measurements of multiphoton ionization to be reported (Voronov and Delone, 1965; Agostini *et al.*, 1968).

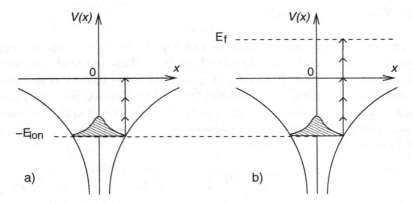

Fig. 2.1   a) Schematic picture of multiphoton ionization (MPI). An electron with binding energy E_{ion} simultaneously absorbs n photons with energy $\hbar\omega$ and is subsequently released from the atom with minimal kinetic energy. b) Above-threshold ionization (ATI): the electron absorbs *more* photons than necessary for ionization, leaving the atom with significant momentum.

This physically appealing picture went almost unchallenged until the late 1970s, when experimentally determined electron energy spectra started to reveal distinct peaks *beyond* the ionization energy E_{ion}, separated by the photon energy $\hbar\omega$ (Agostini *et al.*, 1979; Gontier *et al.*, 1980). This remarkable effect, *above-threshold ionization* (ATI), allows an electron to absorb more photons than strictly necessary to free it from the atom, see Fig. 2.1b). The final kinetic energy of the electron is then given by an extended version of Einstein's formula:

$$E_f = (n + s)\hbar\omega - E_{ion}, \qquad (2.4)$$

where n is the number of photons needed for multiphoton ionization; s is the excess absorbed. This process still conserves momentum because it takes place in the field of the parent ion.

The theoretical interpretation of ATI is controversial, especially when the electron oscillation energy becomes larger than the photon energy. In this limit, the energy spectrum starts to become dominated by peaks with energies larger than $n\hbar\omega$, strongly suggesting that ATI is a non-perturbative process (Kruit *et al.*, 1983), even at intensities in the region of 10^{13} Wcm^{-2}.

2.2   Tunneling Ionization

An important assumption in perturbative MPI is that the atomic binding potential remains undisturbed by the laser field. This will hold true until the intensity starts get within I_a, at which point the laser field becomes strong enough to distort the Coulomb field felt by the electron. This case was first considered some 40 years ago by Keldysh (1965) and Perelomov *et al.*, (1966), who introduced a parameter γ separating the multiphoton and tunneling regimes, given by:

$$\gamma = \omega_L \sqrt{\frac{2E_{\text{ion}}}{I_L}} \sim \sqrt{\frac{E_{\text{ion}}}{\Phi_{\text{pond}}}}. \qquad (2.5)$$

where

$$\Phi_{\text{pond}} = \frac{e^2 E_L^2}{4m\omega_L^2} \qquad (2.6)$$

is the so-called *ponderomotive potential* of the laser field, expressing the effective *quiver energy* acquired by an oscillating electron. As a rule of thumb, tunneling applies for strong fields and long wavelengths, i.e. for $\gamma < 1$; multiphoton ionization for $\gamma > 1$, although this picture becomes cloudier for many-electron atoms (Lambropoulos, 1985).

Before we come to the actual ionization rates, it proves instructive to first consider a simple classical picture of this phenomenon (Bethe and Saltpeter, 1977), in which the Coulomb potential is modified by a stationary, homogeneous electric field, see Fig. 2.2:

$$V(x) = -\frac{Ze^2}{x} - e\varepsilon x.$$

We see now that on the RHS of the atom, the Coulomb barrier has been suppressed, and for $x \gg x_{\text{max}}$ is *lower* than the binding energy of the electron. Quantum mechanically, the electron may tunnel through this barrier with some finite probability (given by Keldysh's formula, see Eq. (2.11)). If the barrier falls below E_{ion}, the electron will escape spontaneously: this is known as over-the-barrier (OBT) or barrier suppression (BS) ionization. More quantitatively, we differentiate V(x) to determine the position of the barrier,

$$x_{\text{max}} = (Ze/\varepsilon),$$

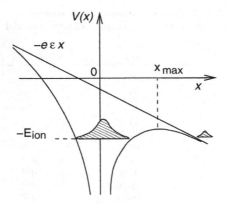

Fig. 2.2 a) Schematic picture of tunneling or barrier-suppression ionization by a strong external electric field.

and then set its height $V(x_{\max}) = E_{\text{ion}}$ to get the threshold field strength at which OTBI occurs, namely:

$$\varepsilon_c = \frac{E_{\text{ion}}^2}{4Ze^3}. \tag{2.7}$$

Equating this critical field to the peak electric field of the laser yields an effective *appearance intensity* for ions created with charge Z:

$$I_{\text{app}} = \frac{c}{8\pi}\varepsilon_c^2 = \frac{cE_{\text{ion}}^4}{128\pi Z^2 e^6}, \tag{2.8}$$

or, expressed in more convenient form:

$$I_{\text{app}} \simeq 4 \times 10^9 \left(\frac{E_{\text{ion}}}{\text{eV}}\right)^4 Z^{-2} \ \text{Wcm}^{-2}. \tag{2.9}$$

Note that E_{ion} is the ionization potential of the ion or atom with charge $(Z-1)$. The simplest example is of course hydrogen, for which $Z = 1$ and

$$E_{\text{ion}} = E_h = \frac{e^2}{2a_B} = 13.61 \ \text{eV}.$$

Making use of Eq. (2.1), the critical field for hydrogen is therefore:

$$\varepsilon_c = \frac{E_h^2}{4e^3} = \frac{e}{16a_B^2} = \frac{E_a}{16},$$

Table 2.1 Appearance intensities of selected ions according to the barrier-suppression ionization model.

Ion	E_{ion} (eV)	I_{app} (Wcm^{-2})
H^+	13.61	1.4×10^{14}
He^+	24.59	1.4×10^{15}
He^{2+}	54.42	8.8×10^{15}
C^+	11.2	6.4×10^{13}
C^{4+}	64.5	4.3×10^{15}
N^{5+}	97.9	1.5×10^{16}
O^{6+}	138.1	4.0×10^{16}
Ne^+	21.6	8.6×10^{14}
Ne^{2+}	40.96	2.8×10^{15}
Ne^{7+}	207.3	1.5×10^{17}
Ar^{8+}	143.5	2.6×10^{16}
Xe^+	12.13	8.6×10^{13}
Xe^{8+}	105.9	7.8×10^{15}

and the appearance intensity:

$$I_{app} = \frac{I_a}{256} \simeq 1.4 \times 10^{14} \ Wcm^{-2}. \tag{2.10}$$

This is why ionization becomes a serious issue in short pulse interactions long before the laser field reaches inter-atomic binding levels. In fact, this simple barrier-suppression model holds up remarkably well for other many-electron atoms, some thresholds for which are listed in Table 2.1. Experimental measurements with noble gases show particularly good agreement with these values (Augst *et al.*, 1989; Auguste *et al.*, 1992b), so that Eq. (2.9) serves as an excellent guide to predicting when a target will start to ionize.

An important caveat for the validity of the BSI model is that other ionization mechanisms, notably ATI and collisional (avalanche) ionization, are suppressed, so that the relevant atoms or ions are sufficiently numerous when the pulse reaches its maximum intensity. Short laser pulse duration is therefore a prerequisite for observing tunnel ionization (Lambropoulos, 1985). For high density (solid) targets, collisional ionization rates will usually take over from field ionization once enough electrons have been released (see Sec. 5.1).

As already mentioned, to compute the ionization rate itself, one has to determine the probability of the electron wave function 'leaking' through the tilted Coulomb barrier of Fig. 2.2 — a purely quantum mechanical exercise. According to Keldysh (1965), for hydrogen-like ions (atoms stripped down to the last *1s* electron) this is given by:

$$\alpha_i = 4\omega_a \left(\frac{E_i}{E_h}\right)^{\frac{5}{2}} \frac{E_a}{E_L(t)} \exp\left[-\frac{2}{3}\left(\frac{E_i}{E_h}\right)^{\frac{3}{2}} \frac{E_a}{E_L(t)}\right], \qquad (2.11)$$

where E_i and E_h, are the ionization potentials of the atom and hydrogen respectively, E_a is the atomic electric field defined by Eq. (2.1), E_L is the instantaneous laser field, and

$$\omega_a = \frac{me^4}{\hbar^3} = 4.16 \times 10^{16}\ \text{s}^{-1} \qquad (2.12)$$

is the atomic frequency. For more complex many-electron atoms or ions, this rate takes on a more complicated form, with coefficients depending on the configurational quantum numbers n, l and m (Ammosov *et al.*, 1986). In a remarkably detailed and carefully analyzed experiment measuring ion production from noble gases irradiated by a 1 ps laser, Auguste *et al.*, (1992b) were able to confirm the absolute ionization rates predicted by this tunneling theory over an intensity range spanning 5 orders of magnitude.

2.3 Residual Energy

An important implication of multiphoton ionization — whether via ATI or tunneling — is that the kinetic energy of electrons pulled out from atoms by the laser field is typically smaller than both the ionization potential and the quiver energy (Burnett and Corkum, 1987). This low *residual energy* arises because within an optical cycle, ionization occurs at the peak of the electric field where the electron quiver velocity is actually zero. On the other hand, any departure from linear polarization will increase the random energy acquired by 'newly born' electrons, and ultimately increase the final temperature of the plasma created. Hence, one observes significant differences in the ATI spectra of electrons ionized by linearly and circularly polarized lasers (Burnett and Corkum, 1987).

The residual energy issue is of central importance to novel X-ray laser schemes using short pulses (see Sec. 7.4), which rely on the rapid creation of a highly ionized, cold plasma which subsequently recombines and lases back to the ground state. If the plasma is too hot, the scheme will saturate and the X-ray efficiency will be small. A number of theoretical (Penetrante and Bardsley, 1991; Pert, 1995) and experimental (Offenberger *et al.*, 1993;

Dunne *et al.*, 1994; Blyth *et al.*, 1995) efforts have therefore focused on determining and minimising the plasma temperature resulting from a field-ionized gas target. In particular, *inverse-bremsstrahlung* (137) heating and collective heating due to parametric instabilities (see Section 4.4) could impair the effectiveness of such schemes. Consequently, short wavelengths ($\lambda = 1/4$ μm) and pulse lengths shorter than 100fs should improve the chances of success for this type of scheme.

2.4   Ionization-Induced Defocusing

We have seen from Table 2.1 that most gases can be multiply ionized at intensities above 10^{16} Wcm^{-2}. One might therefore think that a plasma should be already created by the leading foot of a TW laser pulse, allowing the main part to propagate through a fully-ionized, uniform length of plasma. This is exactly what was hoped for during early experiments studying the propagation of high-intensity pulses through 'gas-fill' targets — large chambers typically containing He or N$_2$ gas at high (atmospheric) pressure.

Unfortunately, the situation is complicated by a phenomenon known as *ionization-induced defocusing* (Auguste *et al.*, 1992a; Leemans *et al.*, 1992; Rae, 1993). At the foot of an intense pulse, where the field is close to the ionization threshold, the gas at the center of the beam will be ionized more, giving rise to a steep *radial* density gradient. This differential ionization rate also acts longitudinally in the time domain, in which case it leads to *spectral blue-shifting* of a portion of the transmitted light (Le Blanc *et al.*, 1993).

The refractive index of the plasma created after ionization is given by:

$$\eta(r,t) = \left(1 - \frac{n_e(r,t)}{n_c}\right)^{\frac{1}{2}}, \qquad (2.13)$$

where $n_e(r,t)$ is the local electron density and n_c the critical density for the laser, related to its frequency ω_L by: $\omega_L^2 = 4\pi e^2 n_c/m$. If more electrons are produced at the beam center, the refractive index will have a minimum there, forming a defocusing lens for the following portion of the beam. The result is that for high gas pressures, the laser beam is deflected well before it can reach its nominal focus (Auguste *et al.*, 1992a).

To get find out how much this effect changes the propagation of a laser pulse, we can apply of some simple geometrical optics. The trajectory of a light ray $x(t)$ in a refractive medium formally obeys the following second order differential equation (Born and Wolf, 1980):

$$\frac{d}{ds}\left(\eta(x)\frac{dx}{ds}\right) = \nabla\eta(x), \qquad (2.14)$$

where ds is an element of length along the ray. If the refractive index is slowly varying compared to a wavelength (i.e. $|\eta/\nabla\eta| \gg \lambda$), and the beam propagation is largely confined to one principal axis, then we can apply the so-called *paraxial approximation*. In terms of the wave vector $k = (k_\parallel, k_\perp)$, where $k_\parallel$ points along the beam axis (z, say), then paraxial propagation implies that $k_\perp \ll k_\parallel$. This turns out to be an excellent approximation for describing short pulse laser propagation in tenuous or transparent media of all kinds, including neutral gases, glass, partially ionized gases and fully ionized plasmas. In this case, a more useful form of Eq. (2.14) can be easily derived (see, for example Marchand *et al.*, 1987), setting $x = r + \hat{z}z$, and taking $ds \approx dz$:

$$\frac{dr}{dz} = \frac{k_\perp}{k(z)},$$

$$\frac{dk_\perp}{dz} = k_0\nabla_\perp\eta(r,z), \qquad (2.15)$$

whee $k_0 = \omega_0/c$ is now the vacuum wave vector of the laser and $k(z) = k_0\eta(r,z)$.

Defining the divergence of the beam from its propagation axis as $\theta = k_\perp/k_\parallel$, and noting that for a highly underdense plasma ($n_e/n_c \ll 1$), we can approximate the refractive index as

$$\eta(r) \simeq 1 - \frac{1}{2}\frac{n_e(r)}{n_c},$$

we have

$$\frac{d\theta}{dz} \simeq -\frac{1}{2}\frac{\partial}{\partial r}\left(\frac{n_e(r)}{n_c}\right).$$

For a laser spot size σ_L, the total deflection of the beam will scale as:

$$\theta_I \sim \frac{1}{\sigma_L}\int\frac{n_e(0)}{n_c}dz, \qquad (2.16)$$

meaning that rays are bent away from regions of higher electron density.

On the other hand, a Gaussian beam focused in vacuum is 'diffraction limited', and satisfies the relation (Siegman, 1986, — see also Sec. 4.6.1):

$$\theta_D = \frac{\sigma_L}{Z_R}, \qquad (2.17)$$

where $Z_R = 2\pi\sigma_L^2/\lambda$ is the Rayleigh length. If we compare θ_I to θ_D over the same propagation length Z_L, we find that ionization-induced refraction will dominate ($\theta_I(z_R) > \theta_D$) when

$$\frac{n_e}{n_c} > \frac{\lambda}{\pi Z_R}.$$

If the ionization process is still incomplete at this point, the density will effectively be *clamped* at value $O(\lambda/\pi Z_R)$, because no further focusing can occur.

More significant, however, is the fact that the maximum intensity reached can be well below the diffraction limited value. To get a more quantitative feel for this behavior, we turn to numerical modeling, and solve the coupled set of differential equations for the laser field and the plasma density. These comprise a paraxial wave equation, together with rate equations for the relevant ion species Z_i, whose ionization coefficients are determined by the tunneling model — Eq. (2.11). To avoid a lengthy digression into paraxial wave equations — essentially one step up in complexity from the ray equations Eq. (2.15) — the details of this model are deferred until Chapter 6, Sec. 6.3. Basically, it is equivalent to the technique first used by Rae (1993) to study this propagation regime.

In the following example, the numerical solution for a 1 μm, 80 fs laser pulse with nominal vacuum focal spot size of 4.5 μm and peak intensity of 10^{15} Wcm^{-2}. The pulse is initialized with a radial phase modulation corresponding to an $f/10$ lens, and enters a neutral H$_2$ gas at different pressures. In Fig. 2.3a) we see the earlier onset of defocusing as the pressure is increased, leading to the drastically reduced peak intensities shown in (b), confirming what we predicted above using heuristic arguments.

To get round this problem, experiments requiring high intensities are usually done either with a preformed plasma (Durfee III and Milchberg, 1993; Mackinnon *et al.*, 1995), or using a gas jet configuration in which the beam is focused in vacuum *before* it actually enters the gas (Auguste *et al.*, 1994). Interaction at the maximum intensity allowed by the focusing optics is then guaranteed, and one can more-or-less neglect the ionization physics.

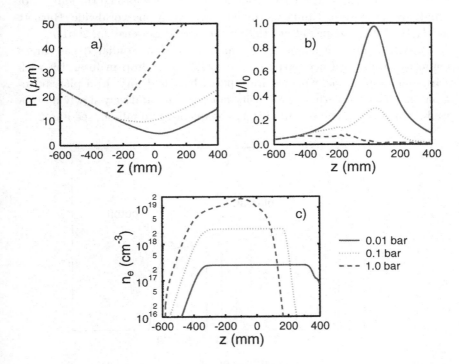

Fig. 2.3 Ionization-induced defocusing of a laser pulse in an extended hydrogen gas target at different fill pressures: a) beam width; b) peak intensity; c) electron density at the pulse center.

2.5 High Harmonic Generation in Gases

What distinguishes field-ionization processes from impact ionization by another particle is that the ejected electron is immediately placed at the mercy of the laser field. Not only will it start to oscillate with an amplitude of many Bohr radii, but as the field reverses, there is a good chance that it will be sent back close to its parent ion, where it can *recombine*, emitting a single, high-frequency photon. This process, known as optical or high harmonic generation (OHG/HHG), was predicted in the late 1980s by Shore and Knight (1987) and immediately confirmed in experiments by the Chicago and Saclay short pulse laser groups (McPherson *et al.*, 1987; Ferray *et al.*, 1988). Owing to its huge potential as a compact, readily accessible source of coherent x-rays, HHG from atoms has since formed the subject of hundreds of theoretical and experimental investigations worldwide (Burnett *et al.*, 1993; Protopapas *et al.*, 1997; Salieres and Lewenstein, 2001).

A typical high-harmonic spectrum from a laser-irradiated gas target comprises three distinct parts: i) an initial steep drop in intensity with harmonic number, as expected from perturbation theory; ii) a plateau region, in which the harmonic intensity falls off comparatively slowly up to a well-defined cutoff energy; and iii) a subsequent rapid fall-off, see Fig. 2.4.

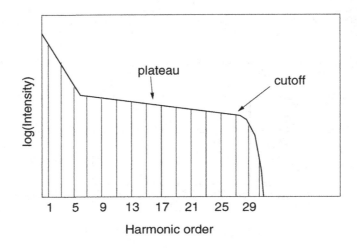

Fig. 2.4   Schematic diagram of high-harmonic spectrum generated from atoms.

The cutoff energy U_c — predicted by Krause *et al.*, (1992) using a numerical approach — is given by the now famous rule:

$$U_c = I_p + 3.17\, U_p, \tag{2.18}$$

where I_p and U_p are the ionization potential of the atom (equivalent to E_{ion} in Section 2.2) and the ponderomotive potential (given by Eq. 2.6) respectively.

Physically, this relation can be explained with the help of classical electrodynamics, as demonstrated by Corkum (1993). In this model, the electron motion is assumed to be completely determined by the laser field as soon as it is 'born' via multiphoton ionization (either ATI or tunneling). The subsequent equations of motion for a linearly polarized laser

$\boldsymbol{E} = \hat{x}E_0 \cos\omega t$ are then just:

$$v = v_{\rm os}\sin\omega t + v_i,$$
$$x = -\frac{v_{\rm os}}{\omega}\cos\omega t + v_i t + x_i,$$

where

$$v_{\rm os} \equiv \frac{eE_0}{m\omega} \qquad (2.19)$$

is the *electron quiver velocity*, and x_i, v_i are the electron's position and velocity just after ionization. We now suppose that this occurs at time $t = t_0$, and let $x(t_0) = v(t_0) = 0$: the electron is born with zero velocity close to the ion center. The orbit can then be rewritten thus:

$$v(\phi) = v_{\rm os}(\sin\phi - \sin\phi_0),$$
$$x(\phi) = \frac{v_{\rm os}}{\omega}\{\cos\phi_0 - \cos\phi + (\phi_0 - \phi)\sin\phi_0)\}, \qquad (2.20)$$

where $\phi = \omega t$ and $\phi_0 = \omega t_0$. What we are interested in here are orbits in which the electron returns to $x = 0$ (the ion center) at some later time t_1, along with the electron's kinetic energy $U_c = \frac{1}{2}mv^2$ at this point. The latter will depend on ϕ_0, the phase of the laser that the electron is born into, as illustrated by the phase portraits in Fig. 2.5. These have be obtained by plotting $v(\phi)$ *vs.* $x(\phi)$ according to the solution Eq. (2.20) for different values of ϕ_0.

Closer inspection of the recrossing energy $U_c(x = 0)$ reveals that this has a maximum value of $3.17U_p$ for $\phi_0 = 17^\circ$ and 197°, in agreement with Eq. (2.18). Of course the electron can acquire instantaneous energies up to $U_{max} = (2v_{os})^2 = 8U_p$, but only at distances too remote from the ion to permit recombination. Although this classical version of the recollision model does a remarkably good job in predicting the maximum harmonic number in many different experiments with short laser pulses (L'Huillier and Balcou, 1993), it cannot tell us anything about the harmonic intensities or the shape of the spectrum in general. To do this accurately, one has to treat the electron quantum mechanically with some form of the time-dependent Schrödinger equation (TDSE) — a much more challenging undertaking which has received considerable attention over the last 10 years (Krause *et al.*, 1992; Burnett *et al.*, 1993; Protopapas *et al.*, 1996; Walser *et al.*, 2000). Another important issue is many-atom, or phase-coherence effects resulting from the finite focusing geometry of a Gaussian

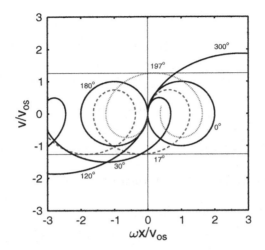

Fig. 2.5   Phase-space portrayal of recollision model. The orbits all start at the origin $(0,0)$ and proceed clockwise. The maximum velocity at the recrossing point $x = 0$ is indicated by the horizontal limits at $v_m/v_{os} = \pm\sqrt{3.17/2}$.

pump beam, which in turn can limit the efficiency of the harmonics (Balcou and L'Huillier, 1993; Tisch *et al.*, 1994).

The experimental pursuit of high harmonics from gas targets has concentrated on finding ways to extend the plateau region, thereby producing harmonics with ever smaller wavelengths. The obvious option of increasing the laser intensity — and thus U_p — only works up to a certain *saturation intensity*, beyond which most of the harmonic-producing atoms become ionized. Ions themselves (for example He^+) are also candidates for HHG by virtue of their higher ionization potential I_p, and have been investigated experimentally (Preston *et al.*, 1996). The most promising route to date, however, is to use extremely short (few-cycle) pulses, which appears to enable atoms to survive a bit longer at higher intensities, extending the range of observed harmonics well beyond the $I_p + 3U_p$ limit (Zhou *et al.*, 1996; Christov *et al.*, 1997; Schnürer *et al.*, 1998).

Chapter 3

Interaction with Single Electrons

3.1 Motion of an Electron in an Electromagnetic Plane Wave

The starting point for many early investigations on nonlinear laser-matter interaction was the orbit of a single electron in a strong electromagnetic plane-wave. This well-known problem can be solved exactly (Landau and Lifshitz, 1962; Gunn and Ostriker, 1971; Eberly and Sleeper, 1968), and is conveniently described in terms of an 'average rest frame', which drifts with a particular velocity with respect to the laboratory frame (the magnitude of which will be derived shortly). Although this ground has been well covered in the literature, and is frequently revisited (Bardsley *et al.*, 1989; Hartemann, 1998; San Roman *et al.*, 2000), it will prove worthwhile to review the main results in order to introduce some notation, and also to get a feel for the nonlinear electron dynamics which lies behind the more complex laser-plasma interactions which we will come to in the Chapters 4 and 5.

The motion of a particle in the presence of electromagnetic fields $\boldsymbol{E}$ and $\boldsymbol{B}$ wave is described by the Lorentz equation,

$$\frac{d\boldsymbol{p}}{dt} = -e(\boldsymbol{E} + \frac{1}{c}\boldsymbol{v} \times \boldsymbol{B}), \qquad (3.1)$$

together with an energy equation

$$\frac{d}{dt}\left(\gamma mc^2\right) = -e(\boldsymbol{v} \cdot \boldsymbol{E}), \qquad (3.2)$$

where $\boldsymbol{p} = \gamma m\boldsymbol{v}$, and $\gamma = (1 + p^2/m^2c^2)^{\frac{1}{2}}$ is the relativistic factor.

An elliptically polarized plane-wave $\boldsymbol{A}(\omega, \boldsymbol{k})$ traveling in the positive

x-direction can be represented by the wave vector

$$A = (0, \delta a_0 \cos\phi, (1 - \delta^2)^{\frac{1}{2}} a_0 \sin\phi), \qquad (3.3)$$

where $\phi = \omega t - kx$ is the phase of the wave; a_0 is the normalized amplitude (v_{os}/c), and δ is a *polarization* parameter such that $\delta = \{\pm 1, 0\}$ for a linearly polarized wave and $\pm 1/\sqrt{2}$ for a circular wave. To simplify the working, we now introduce the normalizations: $t \rightarrow \omega t, x \rightarrow kx, v \rightarrow v/c, p \rightarrow p/mc$, and $A \rightarrow eA/mc^2$. (This is equivalent to setting $\omega = k = c = e = m = 1$, as with the atomic units in the previous chapter.) Using the relations $E = -\partial A/\partial t$ and $B = \nabla \times A = (0, -\partial A_z/\partial x, \partial A_y/\partial x)$, the perpendicular component of Eq. (3.1) becomes:

$$\frac{dp_\perp}{dt} = \frac{\partial A}{\partial t} + v_x \frac{\partial A}{\partial x},$$

which after integrating gives:

$$p_\perp - A = p_{\perp 0}, \qquad (3.4)$$

where $p_{\perp 0}$ is a constant of motion representing the initial perpendicular momentum of the electron. This might be non-zero if the electron had just been ejected from an atom via multiphoton ionization, say. The longitudinal components of Eq. (3.1) and Eq. (3.2) yield a pair of equations which can be subtracted from each other thus:

$$\frac{dp_x}{dt} - \frac{d\gamma}{dt} = -v_y \left(\frac{\partial A_y}{\partial t} + \frac{\partial A_y}{\partial x} \right) - v_z \left(\frac{\partial A_z}{\partial t} + \frac{\partial A_z}{\partial x} \right).$$

Because the EM wave is a function of $t - x$ only, the terms on the RHS vanish identically, so we can immediately integrate the RHS to get:

$$\gamma - p_x = \alpha,$$

where α is a constant of motion still to be determined. Using the identity $\gamma^2 - p_x^2 - p_\perp^2 = 1$ and choosing $p_{\perp 0} = 0$, we can eliminate γ to get a relationship between the parallel and perpendicular momenta:

$$p_x = \frac{1 - \alpha^2 + p_\perp^2}{2\alpha}. \qquad (3.5)$$

Since Eq. (3.4) now tells us that $p_\perp$ is equal to the laser vector potential, Eq. (3.5) represents the general (Lorentz-covariant) solution for the motion of free electrons in an electromagnetic wave (Bardsley *et al.*, 1989). To proceed, we need to specify α and integrate Eq. (3.4) and Eq. (3.5). The latter is simplified by changing variables. Noting that

$$\frac{d\phi}{dt} = \frac{\partial \phi}{\partial t} + \frac{p_x}{\gamma}\frac{\partial \phi}{\partial x} = \frac{\alpha}{\gamma},$$

we have

$$\boldsymbol{p} = \gamma \frac{d\boldsymbol{r}}{dt} = \gamma \frac{d\phi}{dt}\frac{d\boldsymbol{r}}{d\phi} = \alpha \frac{d\boldsymbol{r}}{d\phi}. \tag{3.6}$$

3.1.1 *Laboratory frame*

In the laboratory frame, the electron is initially at rest before the EM wave arrives, so that at $t = 0$, $p_x = p_y = 0$ and $\gamma = 1$. From the conservation relation Eq. (3.5) it follows that $\alpha = 1$ in this case. This leads to the following expression for the momenta in the lab frame:

$$p_x = \frac{a_0^2}{4}\left[1 + (2\delta^2 - 1)\cos 2\phi\right],$$
$$p_y = \delta a_0 \cos\phi, \tag{3.7}$$
$$p_z = (1 - \delta^2)^{1/2} a_0 \sin\phi.$$

With the help of Eq. (3.6) we can integrate expressions Eqs. (3.7a–3.7c) to obtain the lab-frame orbits valid for arbitrary polarization δ:

$$x = \frac{1}{4}a_0^2\left[\phi + \frac{2\delta^2 - 1}{2}\sin 2\phi\right],$$
$$y = \delta a_0 \sin\phi, \tag{3.8}$$
$$z = -(1 - \delta^2)^{1/2} a_0 \cos\phi.$$

This solution, which exhibits a self-similarity in the variables $(x/a_0^2, y/a_0)$, is shown graphically in Fig. 3.1. We notice immediately that regardless of polarization, the longitudinal motion has a secular component which will grow in time or with propagation distance. In fact, in the presence of the EM wave, the electron immediately starts to *drift* with an average momentum $p_D \equiv \overline{p_x} = a_0^2/4$, corresponding to a velocity (see also Exercise 2 on p. 52):

$$\frac{v_D}{c} = \overline{v_x} = \frac{\overline{p_x}}{\overline{\gamma}} = \frac{a_0^2}{4 + a_0^2}, \tag{3.9}$$

where the overscore denotes averaging over the rapidly varying EM phase ϕ.

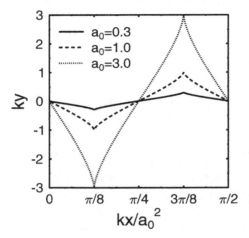

Fig. 3.1   Electron orbits in a large amplitude, linearly polarized electromagnetic plane wave. For a 1 μm laser wavelength, the pump strengths a_0 correspond roughly to intensities of 10^{17}, 10^{18} and 10^{19} Wcm^{-2} respectively.

For circularly polarized light ($\delta = \pm 1/\sqrt{2}$), the longitudinal oscillating component at 2ϕ vanishes identically, and the transverse motion is circular with radius a_0 and momentum $p_\perp = a_0/\sqrt{2}$. This combines with the linear drift in Eq. (3.9) to give a *helical* orbit with pitch angle

$$\theta_p = p_\perp/p_D = \sqrt{8}a_0^{-1}. \tag{3.10}$$

3.1.2   *Average rest frame*

The fact that the electron drifts in the lab-frame leads naturally to another possible choice of constant α. If we instead require that this drift velocity vanish for arbitrary pump strengths, then we set $\overline{p_x} = 0$ in Eq. (3.5) to get:

$$1 + \overline{A^2} - \alpha^2 = 0.$$

Averaging over a laser cycle to remove rapidly varying terms, and noting that $\overline{\cos^2 \phi} = 1/2$ gives:

$$\alpha = \left(1 + \frac{a_0^2}{2}\right)^{1/2} \equiv \gamma_0. \tag{3.11}$$

Plugging this back into Eq. (3.5) gives the momenta:

$$p_x = \left(2\delta^2 - 1\right) \frac{a_0^2}{4\gamma_0} \cos 2\phi, \tag{3.12}$$

$$p_y = \delta a_0 \cos \phi,$$

$$p_z = (1 - \delta^2)^{\frac{1}{2}} a_0 \sin \phi. \tag{3.13}$$

Noting that in this case, $\boldsymbol{p} = \gamma_0 d\boldsymbol{r}/d\phi$, we can integrate again to get the orbits:

$$x = \left(\delta^2 - \frac{1}{2}\right) q^2 \sin 2\phi,$$

$$\boldsymbol{r}_\perp = 2(\delta q \sin \phi, -(1 - \delta^2)^{\frac{1}{2}} q \cos \phi), \tag{3.14}$$

where $q = a_0/2\gamma_0$. Eliminating ϕ for linear polarization ($\delta = 1$), we obtain the famous *figure-of-eight* shown in Fig. 3.2.

$$16x^2 = y^2(4q^2 - y^2). \tag{3.15}$$

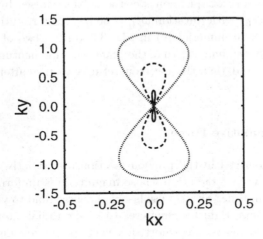

Fig. 3.2 Characteristic orbits of free electrons in a plane electromagnetic wave in the average rest frame. The trajectories correspond to the same light intensities as in Fig. 3.1

In the extreme intensity limit, $a_0 \gg 1$, we find $\gamma_0 \to a_0/\sqrt{2}$ and $q \to 1/\sqrt{2}$, so that the orbit becomes a 'fat-8' with proportions $\mid x \mid_{\max} = 1/4, \mid$

$y\,|_{\max}=\sqrt{2}$. Note that the laboratory frame orbits described by Eq. (3.8) can also be obtained by back-transformation of Eq. (3.14), see Exercise 2c. For circularly polarized light ($\delta = \pm 1/\sqrt{2}$), the longitudinal component vanishes identically, and the electron simply describes a circle with radius $a_0/\sqrt{2}\gamma_0$.

3.1.3   *Finite pulse duration*

The exact analytical solutions for plane waves are of course useful to have, but do not completely describe the details of electron motion caused by a real laser pulse. We can generalize these results to some extent by imposing a *temporal envelope* on the wave vector of the form:

$$\boldsymbol{A}(x,t) = a_0 f(t) \cos \phi, \qquad (3.16)$$

where $f(t)$ is considered to be slowly varying relative to the laser cycle: $df/dt \ll \omega f$. In this adiabatic approximation, we can directly substitute $a_0 f(t)$ for a_0 in the solutions given by Eqs. (3.7,3.8). Alternatively, the finite-difference form of Eq. (4.1) can be integrated directly, see Sec. 6.4. As an example, consider the motion caused by a 'sin^2' pulse: $f(t) = \sin(\pi t/2t_L)$ with duration $\omega t_L = 600/\pi$. The resulting electron orbit (computed here numerically) in Fig. 3.3 verifies two of the important points made above: namely, that the transverse momentum is conserved: $p_y(t) = A_y(t)$, and that the electron returns to rest after the laser has passed.

3.2   Ponderomotive Force

The solutions derived in the previous section are, strictly speaking, only valid for plane waves; radiation whose magnitude is uniform in space and slowly varying in time. Short pulse lasers of course tend to violate this condition on all fronts: tight focusing creates strong radial intensity gradients over a few wavelengths; ultrashort, few-cycle pulses are highly dispersive and demand a completely non-adiabatic treatment. For the time-being, the emphasis will be on the first of these cases: the cigar-shaped pulses typical of laser systems producing pulses with durations in excess of 50 fs.

There are few topics in laser-plasma interactions that have caused such persistent argument and controversy as the curiously named *ponderomotive force*. On a purely heuristic level, it is relatively straightforward to

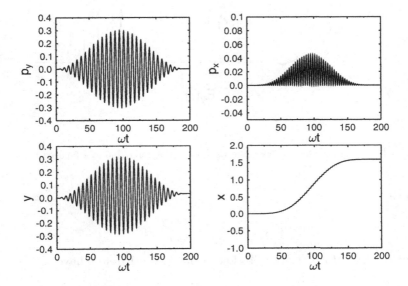

Fig. 3.3   Electron orbit generated by a finite-duration laser pulse.

define: namely as the gradient of the time-averaged 'oscillation potential' introduced in the previous chapter. Historically, problems have arisen when attempting to reconcile the single-particle definition with a more general plasma fluid picture (Hora, 1981). More recently, attention has turned to the relativistic version of this effect — an effort which has acquired poignancy now that high-intensity lasers have made direct tests of theory possible.

Rather than attempt a rigorous derivation here, I will stick to heuristic arguments in order to gain an intuitive feel for ponderomotive behavior. First consider the non-relativistic case of a single electron oscillating near the center of a focused laser beam, see Fig. 3.4. In the limit $v/c \ll 1$, the equation of motion (3.1) for the electron becomes:

$$\frac{\partial v_y}{\partial t} = -\frac{e}{m} E_y(\boldsymbol{r}). \qquad (3.17)$$

The EM wave is taken to be propagating in the +ve x-direction as before, but this time with a radial intensity dependence, which for the time-being we will assume to be in the y-direction only. Taylor expansion of the electric

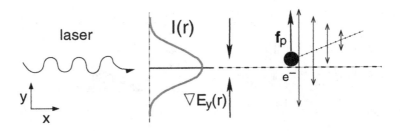

Fig. 3.4   Schematic view of the radial ponderomotive force due to a focused beam.

field then gives:

$$E_y(\boldsymbol{r}) \simeq E_0(y)\cos\phi + y\frac{\partial E_0(y)}{\partial y}\cos\phi + ...,$$

where $\phi = \omega t - kx$ as before. To lowest order, we therefore have

$$v_y^{(1)} = -v_{os}\sin\phi; \quad y^{(1)} = \frac{v_{os}}{\omega}\cos\phi,$$

where v_{os} is defined as usual by Eq. (2.19). Substituting back into Eq. (3.17) gives

$$\frac{\partial v_y^{(2)}}{\partial t} = -\frac{e^2}{m^2\omega^2}E_0\frac{\partial E_0(y)}{\partial y}\cos^2\phi.$$

Multiplying by m and taking the cycle-average yields the ponderomotive force on the electron:

$$f_p \equiv m\overline{\frac{\partial v_y^{(2)}}{\partial t}} = -\frac{e^2}{4m\omega^2}\frac{\partial E_0^2}{\partial y}. \tag{3.18}$$

We see immediately that this expression is just the (-ve) gradient of the ponderomotive potential Eq. (2.6) introduced in Chapter 2. Physically, the force will tend to push electrons away from regions of locally higher intensity. A single electron will therefore inevitably drift away from the center of focused laser beam, picking up a velocity $v \sim v_{os}$ in the process. This conversion of oscillatory to directed energy will be analyzed shortly, but first consider the fully relativistic version of Eq. (3.18). Rewriting our original Lorentz equation (3.1) in terms of the vector potential $\boldsymbol{A}$, we have:

$$\frac{\partial \boldsymbol{p}}{\partial t} + (\boldsymbol{v}.\nabla)\boldsymbol{p} = \frac{e}{c}\frac{\partial \boldsymbol{A}}{\partial t} - \frac{e}{c}\boldsymbol{v}\times\nabla\times\boldsymbol{A}. \tag{3.19}$$

To progress further, we separate the timescales of the electron motion into slow and fast components: $p = p^s + p^f$, and use the following identity:

$$v \times (\nabla \times p) = \frac{1}{m\gamma} p \times \nabla \times p$$

$$= \frac{1}{2m\gamma} \nabla \mid p \mid^2 - \frac{1}{m\gamma}(p.\nabla)p.$$

To lowest order, the fast (transverse) component of the electron momentum follows the vector potential: $p^f = eA/c$ as before. Averaging over a laser cycle, we arrive at the following expression for the relativistic ponderomotive force:

$$f_p = \frac{dp^s}{dt} = -mc^2 \nabla \overline{\gamma}, \qquad (3.20)$$

where $\gamma = (1 + p_s^2/m^2c^2 + a_0^2/2)^{1/2}$. This derivation is far from rigorous, but the result is actually equivalent to more sophisticated analyses using covariant (Startsev and McKinstrie, 1997) or Lagrangian (Bauer *et al.*, 1995) formulations.

3.3 Ejection from Focused Laser Beam

Solving the electron motion via Eq. (3.20) for a general laser intensity profile is only possible numerically, since the ponderomotive force is a nonlinear function of the electron's momentum and position. What we *can* do analytically is to determine the electron's final state, i.e. its momenta after having been ejected from the beam focus. We have already seen that in the low-intensity (non-relativistic) case, the electron will simply be accelerated out of the focal region at 90° to the laser axis. At higher intensities, the relativistic drift motion starts to kick in, causing the electron to be directed forwards as well as sideways, see Fig. 3.5. To determine the ejection angle θ, we do some simple relativistic kinematics. We first observe that the final kinetic energy of the electron is just given by:

$$\Delta U = (\gamma - 1)mc^2. \qquad (3.21)$$

This energy is extracted from the electromagnetic field via multiphoton momentum transfer. Since the parallel momentum is conserved, we must have:

$$p_{\parallel} = n\hbar k = \frac{n\hbar\omega}{c} = \frac{\Delta U}{c} = (\gamma - 1)mc. \qquad (3.22)$$

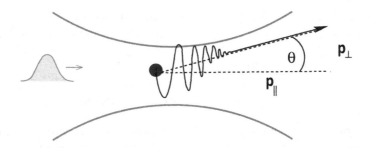

Fig. 3.5   Relativistic electron motion in extended laser focus.

Recalling the relationship from Eq. (3.5) between $p_\parallel$ and $p_\perp$ found earlier (with $\alpha = 1$), we have

$$p_\parallel = \frac{p_\perp^2}{2mc},$$

so the emission angle is given simply by:

$$\tan\theta = \frac{p_\perp}{p_\parallel} = \sqrt{\frac{2}{\gamma - 1}}, \tag{3.23}$$

or

$$\cos\theta = \sqrt{\frac{\gamma - 1}{\gamma + 1}}.$$

This surprisingly simple one-to-one relationship between exit angle and energy implies that the laser acts both as accelerator and spectrometer for free electrons placed near its focus. This property was elegantly exploited by Moore *et al.*, (1995), who determined the energies of electrons released during multiphoton ionization of neon and krypton gas targets by a 1 ps glass laser focused to 10^{18} Wcm^{-2}. As we saw in the previous chapter, barrier suppression creates electrons at well-defined appearance intensities corresponding to successive ionization states. Energy spectra taken within a narrow angular spread should therefore pick up electrons originating from predominantly lower or higher ionization states depending on the viewing angle θ, see Fig. 3.6. As the figure shows, the agreement between the peak angles for given 'release' energies and the theoretical curve is extremely good (see also: Meyerhofer, 1997).

This verification of single electron dynamics was extended to the MeV regime in a slightly more controversial manner by Malka *et al.*, (1997). In

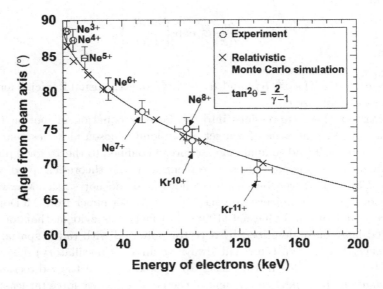

Fig. 3.6 Energies of electrons ejected from a laser focus by the relativistic pondero-motive force. *Courtesy:* D. Meyerhofer, Laboratory for Laser Energetics; Fig. 3 from Meyerhofer (1997). ©IEEE 1997.

their case, the electrons were created by the interaction of a second (ns) laser with a plastic target. These electrons, with energies of a few keV, were then allowed to drift to the center of a target chamber into which the main 500 fs pulse with $I\lambda^2 = 10^{19}$ Wcm$^{-2}\mu$m^2 was focused. Spectra were then measured at two different angles $\theta = 39^o$ and 46° and compared to a generalized form of Eq. (3.23) (Hartemann *et al.*, 1995):

$$\tan\theta = \frac{\sqrt{2(\frac{\gamma}{\gamma_0} - 1)/(1 + \beta_0)}}{\gamma - \gamma_0(1 - \beta_0)}, \tag{3.24}$$

where $\beta_0 = v_0/c$, the initial drift velocity of the electron before the laser pulse arrives, and $\gamma_0 = (1 - \beta_0^2)^{1/2}$, see also Exercise 4 at the end of the chapter.

To analyze their results, the authors deployed a single-particle model to compute the electron trajectories in the spatially and temporally varying electromagnetic fields of a Gaussian focus, which neglecting phase factors, look like:

$$E_y = \frac{\sigma_0}{\sigma(x)} E_0 \exp\left\{\frac{-y^2}{\sigma(x)^2}\right\},$$
$$B_z = E_y,$$

where $\sigma(x) = \sigma_0(1+x^2/R_L^2)$ and $R_L = \pi\sigma_0^2/\lambda_L$ is the usual Rayleigh length for a focal spot size σ_0.

Inserting these expressions into the Lorentz equation of motion (3.1) and integrating the orbits for a set of randomly chosen particles near the beam center will lead to final angular spread confined to the xy focal plane. Although this procedure gives agreement with the theoretical prediction Eq. (3.23), it is flawed because the above fields do not satisfy Maxwell's equations! (See Comments provoked by the Malka paper from McDonald (1998) and Mora and Quesnel (1998)). In fact, it is evident that for any focused laser pulse described by a vector potential with radial spatial dependence $A_y = A_0(r)$, there will always be an axial (oscillatory) magnetic field $B_x = \frac{\partial A_y}{\partial z}$. This in turn leads to a force $v_y B_x$ in the z-direction *of the same order* as the y-component of the ponderomotive force (at least for $\sigma_0 \gg \lambda_L$ — see Exercise 3). The net result is that a symmetric, tightly focused laser will tend to eject *rings* of electrons in the forward direction.

3.4   Vacuum Acceleration Schemes

The idea of using lasers to accelerate particles is a compeling one, primarily because of the extremely high electric field strengths $O(10^{12})$ Vm^{-1} which these 'devices' now have to offer. As the simple example in Section 3.1 showed, however, conversion of electromagnetic energy into directed kinetic energy is not that easy. The annoying habit which electrons have of surrendering their energy back to the wave was recognized well before lasers were invented. What is now commonly known as the Lawson-Woodward (LW) theorem (Woodward, 1947; Lawson, 1979) states that an isolated, relativistic electron cannot gain energy by interacting with an EM field. This rather bleak prediction fortunately comes with a number of provisos attached:

(1) the laser field is in vacuum, with no interfering walls or boundaries,
(2) the electron is highly relativistic along the acceleration path,
(3) no static electric or magnetic fields are present,
(4) the interaction region is infinite,

(5) ponderomotive forces are neglected.

It follows that in order to accelerate electrons, one or more of the above conditions must be violated: for example, by focusing the laser to a finite spot size as in Sec. 3.3, which introduces both a radial ponderomotive force *and* limits the interaction region. The introduction of a plasma violates nearly all of the LW assumptions, and a whole host of possibilities open up, which are deferred to Chapters 4 and 7. For now, we concentrate on pure *vacuum* schemes which rely solely on the interaction between a single electron and an electromagnetic wave in various ingenious configurations. A thorough treatment of this subject can be found in the review by Esarey (1995).

Tightly focused, stationary beam

In Sec. 3.3, we saw how a finite focal spot size will cause an electron to drift away from the beam axis, converting its quiver energy into forward-directed kinetic energy in the process. The energy gain from this essentially adiabatic process is roughly the ponderomotive energy: $\Delta U \sim mc^2 a_0^2/4$. This reflects the fact that the electron simply slides down the ponderomotive potential without picking up any longitudinal component: $v_x.E_x = 0$.

A series of studies by Y.K. Ho's group at the Fudan University in Shanghai has shown that this limitation can be overcome, albeit at extremely high intensities ($a_0 > 100$, or $I\lambda^2 > 10^{21}$ Wcm$^{-2}\mu$m^2). Wang *et al.*, (2001) show that this process — dubbed 'electron capture' by the authors — works best when the electron is injected *obliquely* into the laser focuses with an incident momentum in the multi-MeV/c range (Kong *et al.*, 2000). Physically, what happens during capture is that the electrons move with the laser phase, thus sampling the field for much longer: the ponderomotive picture (which assumes cycle-averaging) is then no longer applicable. A more recent study by Salamin and Keitel (2002) points out the importance of including higher order terms in the electromagnetic fields in order to correctly describe a tightly (few μm) focused Gaussian beam.

Tailored laser focus

An attractive variation of the above scheme is to confine the electrons to the laser axis by tailoring the radial beam profile so that it creates a ponderomotive potential well. This can be done by clever optics, superimposing higher-order light modes in such a way as to produce an intensity minimum at $r = 0$ (Stupakov and Zolotorev, 2001). Numerical studies of this arrangement predict that extremely high-brightness particle beams might be produced in this fashion.

Sub-cycle acceleration

The most direct way of getting round the LW theorem is to make the laser pulse so short that it comprises only half a cycle, and the electric field does not change sign. This idea may seem fanciful, but technically it is not so far off. Few-cycle pulses (~ 5 fs @ 0.8 μm) are now routinely available, and at increasing intensities, see Chapter 7. Quite a number of schemes have been proposed on sub- or half-cycle pulses. One of the more recent ideas considers crossed beams to create an axial, DC accelerating field capable of producing TeV energy gains for GeV injection energies at 10^{24} Wcm^{-2} (Salamin and Keitel, 2000b).

Other schemes

The number and variety of 'alternative' acceleration schemes is too great to do justice to here, but it is worth mentioning a few of them. The vacuum beat-wave scheme (Esarey *et al.*, 1995; Salamin and Keitel, 2000a) deploys two copropagating laser beams of differing frequencies, beating together to produce a longitudinal axial field via the $v \times B$ force. Highly charged ions can be exploited to overcome the LW theorem by 'holding on' to tightly bound electrons during the laser build up, allowing electrons ionized at the peak amplitude to enjoy phase-matched acceleration during the latter half of the pulse (Hu and Starace, 2002). Finally, static magnetic fields provide a simple way of maintaining the phase-matching condition between electron and fast electrostatic or electromagnetic waves (Katsouleas and Dawson, 1983; Chernikov *et al.*, 1992).

3.5 Relativistic Thomson Scattering

It is a well established fact that accelerated charges act as radiation sources, a property that is exploited to generate x-rays from electrons circulating in a synchrotron, for example. In the present context, it is more appropriate to talk about *scattering* of the incident laser fields. Moreover, the relativistic motion resulting from high laser intensities leads to a high harmonic content in the re-emitted radiation. Thus, free electrons provide an ideal medium for nonlinear scattering and frequency upshift of high intensity laser pulses, a theme which will recur throughout the chapters which follow. Using the electron orbits $r(t), v(t)$ derived earlier in Sec. 3.1, we can in principle compute the harmonic content of the emitted radiation by substituting these solutions into the standard formulae for light emission by an accelerated charge, discussed extensively in Chapter 14 of Jackson (1975) and which are briefly summarized for convenience here.

The starting point for these formulae are the Liénard-Wiechert potentials, satisfying the wave equations for the scalar and vector radiation potentials caused by a charge in relativistic motion:

$$\phi(r,t) = \left[\frac{e}{(1-\boldsymbol{\beta}.\boldsymbol{n})R}\right]_{\text{ret}}, \qquad (3.25)$$

$$A(r,t) = \left[\frac{e\boldsymbol{\beta}}{(1-\boldsymbol{\beta}.\boldsymbol{n})R}\right]_{\text{ret}}, \qquad (3.26)$$

where $\boldsymbol{\beta} = v/c$, $\boldsymbol{n}$ is a unit vector from the charge in the direction of observation, and the notation $[\ ]_{\text{ret}}$ indicates that the argument is to be evaluated at the *retarded* time $t' = t - R(t')/c$, see Fig. 3.7. From Eqs. (3.25

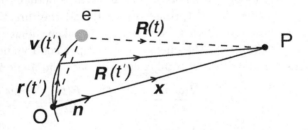

Fig. 3.7   Geometry for calculating radiation fields from a relativistic, accelerated charge.

and 3.26), taking special care with the derivatives of retarded quantities, one can obtain the general radiation field seen at point P due to the charge at $x(t)$:

$$\boldsymbol{E}(\boldsymbol{x},t) = e\left[\frac{\boldsymbol{n}-\boldsymbol{\beta}}{\gamma^2\kappa R^2}\right]_{\text{ret}} + \frac{e}{c}\left[\frac{\boldsymbol{n}\times\{(\boldsymbol{n}-\boldsymbol{\beta})\times\dot{\boldsymbol{\beta}}\}}{\kappa^3 R}\right]_{\text{ret}}, \qquad (3.27)$$

where $\kappa = dt'/dt = 1 - \boldsymbol{\beta}.\boldsymbol{n}$ and $\dot{\boldsymbol{\beta}} = d\boldsymbol{\beta}/dt$ is the acceleration. Using this field, we can compute the energy flux or Poynting vector at the observation point P:

$$S = \frac{c}{4\pi} \boldsymbol{E} \times \boldsymbol{B} = \frac{c}{4\pi} \mid E \mid^2 \boldsymbol{n},$$

so that the radiated power per unit solid angle can be written:

$$\frac{dP(t)}{d\Omega} = R^2(\boldsymbol{S}.\boldsymbol{n}) = \frac{c}{4\pi} R^2 \mid E \mid^2 . \qquad (3.28)$$

To analyze this further, we assume that the field can be expressed as a Fourier integral:

$$\boldsymbol{E}(t) = \int_{-\infty}^{\infty} e^{-i\omega t} \boldsymbol{E}(\omega) d\omega.$$

Substituting this into Eq. (3.27), and applying Parseval's theorem for power spectra to Eq. (3.28) eventually leads to an *intensity* distribution:

$$\frac{d^2 I}{d\omega d\Omega} = \frac{e^2 \omega^2}{4\pi^2 c} \left| \int_{-\infty}^{\infty} \boldsymbol{n} \times (\boldsymbol{n} \times \boldsymbol{\beta}) e^{i\omega(t - \boldsymbol{n}.\boldsymbol{r}/\boldsymbol{c})} dt \right|^2 . \qquad (3.29)$$

In arriving at this expression the observation point P is assumed to be far away from the charge motion, so that $R(t') \simeq x - \boldsymbol{n}.\boldsymbol{r}(t')$; the additional phase factor $\exp(i\omega x)$ can be ignored. The integral is actually taken over retarded time t' rather than t (primes omitted) and the direction vector $\boldsymbol{n}$ is assumed to be constant in time: hence it is taken along $\overline{OP}$ rather than $\boldsymbol{R}(t')$. Eq. (3.29) gives the energy radiated per solid angle per unit frequency. This is related to the total energy W per solid angle Ω by:

$$\frac{dW}{d\Omega} = \int_{-\infty}^{\infty} \frac{d^2 I(\omega, \boldsymbol{n})}{d\omega d\Omega} d\omega = \int_{-\infty}^{\infty} \frac{dP(t)}{d\Omega} dt. \qquad (3.30)$$

One special case of Eq. (3.29) is periodic motion, for which the radiated power P per solid angle Ω can be decomposed into multiples m of the fundamental frequency ω_0, whereby:

$$\frac{dP_m}{d\Omega} = \frac{\omega_0^2}{2\pi} \frac{d^2 I_m}{d\omega d\Omega}$$

$$= \frac{e^2 \omega_0^4 m^2}{(2\pi c)^3} \left| \int_0^{2\pi/\omega_0} \boldsymbol{n} \times (\boldsymbol{n} \times \boldsymbol{v}) e^{im\omega_0(t - \boldsymbol{n}.\boldsymbol{r}/\boldsymbol{c})} dt \right|^2 . \qquad (3.31)$$

We can apply this formula directly to the periodic particle orbits Eq. (3.15) in the average rest frame discussed in Sec. 3.1.2. In the laboratory frame however, the motion is not purely harmonic (as evident from Fig.3.2b), so we have to either go back to Eq. (3.29), or Lorentz-transform the solution obtained from Eq. (3.31) back from the average rest frame. The latter procedure is quite subtle, and was famously demonstrated by Sarachik and Schappert (1970). The resulting solution for the harmonic power content is intractable for arbitrary polarization and laser amplitude — but see Hartemann (1998). For circularly polarized light, however, a useful form can be found, giving a power spectrum in the average rest frame R:

$$\frac{dP_R^m}{d\Omega_R} = \frac{2m^2 A(\omega_R^2)}{\gamma_0^2} \left[\frac{\cot^2\theta_R}{2q^2} J_m^2(\sqrt{2}q\,m\sin\theta) + J'^2_m(\sqrt{2}q\,m\sin\theta_R) \right],$$
(3.32)

where

$$A(\omega_R^2) = \frac{e^2\omega_R^2 a_0^2}{8\pi c},$$

J_m is the usual Bessel function and J'_m its derivative; θ_R is the angle between the $\mathbf{n}$ and the laser wave-vector in the average rest frame (see Fig. 3.8) and $q = a_0/\gamma_0$, $\omega_R = \omega_0/\gamma_0$ corresponding to the orbits on page 34. This formula corresponds to synchrotron radiation from a circulating electron with velocity $\sqrt{2}qc$ at a radius of $\sqrt{2}q/k_R$. This result can be Lorentz-

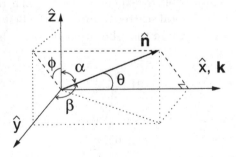

Fig. 3.8 Coordinate system for determining angular dependence of nonlinear Thomson scattering. See Sarachik & Schappert (1970).

transformed back to the lab frame to get the *observed* power spectrum. After proper scaling of angles, time intervals and frequencies, one finally obtains (Sarachik and Schappert, 1970):

$$\frac{dP_L^m}{d\Omega_L} = D(a_0, \theta_L)\frac{dP_R^m}{d\Omega_R},$$
(3.33)

where

$$D(a_0, \theta_L) = \frac{\gamma_0^4}{\left(1 + \frac{a_0^2}{2} \sin^2 \frac{\theta_L}{2}\right)^4}.$$

The spectrum also no longer consists of integer harmonics of ω_0, but at shifted frequencies given by:

$$\omega_L^m = \frac{m\omega_0}{\left(1 + \frac{a_0^2}{2} \sin^2 \frac{\theta_L}{2}\right)}. \tag{3.34}$$

In the extreme intensity limit $(a_0 \gg 1)$, the radiation is predominantly forward and confined to an angle $\theta_L = \sqrt{8}/a_0$ from the axis of propagation, with a harmonic cutoff at $M = m_{\max} = 3(a_0^2/2)^{3/2}$. This *headlamping* effect is typical of relativistic electrons, and in this case can also be physically understood from the forward drift motion seen earlier. In fact, for circular light, $\theta_L = \theta_p$, the pitch angle of the helical electron orbit given by Eq. (3.10).

For linearly polarized light, the angular distribution is more complex, but follows the qualitative behavior seen with circular light, shifting from sideways to forward due to the increase in drift velocity $v_D \to c$. A more quantitative analysis is possible by expanding the angular radiation pattern Eq. (3.31) for $a_0 < 1$. The total scattered power in the first three harmonics — integrated over solid angle in the laboratory frame — has the following leading terms:

$$P_1 \simeq W_0 \frac{8\pi}{3} a_0^2,$$

$$P_2 \simeq W_0 \frac{14\pi}{5} a_0^4,$$

$$P_3 \simeq W_0 \frac{621\pi}{224} a_0^6, \tag{3.35}$$

where $W_0 = e^2 \omega_0^2 c / 8\pi$ is a characteristic scattered power per electron for a given laser frequency ω_0. The expression for P_1 is readily identified as the classical Thomson scattering result: dividing by the Poynting vector for the incoming light, $S = cE_0^2/8\pi$, yields the Thomson cross-section:

$$\sigma_T = \frac{P_1}{S} = \frac{8\pi}{3} \frac{W_0 a_0^2}{S} = \frac{8\pi}{3} r_0^2,$$

where $r_0 = e^2/mc^2$ is the classical electron radius.

A more thorough discussion of the Sarachik and Schappert work, including an explicit analysis of the angular distribution and some corrections to the higher order coefficients can be found in the article by Castillo-Herrera and Johnston (1993). Other work revisiting this problem by Bardsley *et al.*, (1989) and Mohideen *et al.*, (1992) has reexamined the single electron dynamics, taking into account finite pulse shapes, ionization and space-charge effects. The first two effects mainly give rise to irregularities in the amplitude and phase of the electron orbits which can lead to 'residual heating'; the third effect is potentially the most damaging with regards to harmonic generation in a plasma. As we will see later in Chapter 4, the collective restoring force due to charge separation not only suppresses the drift motion, but produces density nonlinearities which tend to cancel the harmonic-producing relativistic effects.

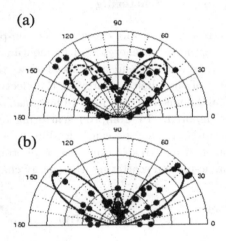

Fig. 3.9 Angular measurements (points) of a) 2nd harmonic light and b) 3rd harmonic light generated by relativistic electrons in a high intensity laser focus. The solid/dashed curves represent theoretically computed radiation patterns. *Courtesy:* D. Umstadter, University of Michigan; Figs. 2 and 3 from Chen *et al.*, (1998b). ©Nature 1998.

Despite the enormous increases in laser intensity seen over the last two decades, experimental evidence for harmonic generation by free electrons has been relatively late in coming. Early observations of 2nd harmonic photons were made by Englert and Rinehart, (1983), but these were at rather low intensities by today's standards. Truly relativistic scattering has been demonstrated in a landmark paper by Chen, Maksimchuk and Umstadter (1998b), in which angularly resolved measurements of the 2nd and 3rd laser harmonics were shown to qualitatively match the theoretical predictions, thus providing the first unambiguous signature of relativistic figure-of-8 electron motion in a laser field, see Fig. 3.9.

One of the most effective ways of exploiting nonlinear Thomson scattering is to use relativistic electron beams in place of a plasma (Leemans *et al.*, 1997). In this case, the harmonic content is modified by the relative doppler shift of the electron beam (which scales as γ^2), an effect which can be optimized by scattering a short laser pulse off a counterstreaming e-beam (Ride *et al.*, 1995; Hartemann, 1998; He *et al.*, 2003).

3.6   Nonlinear Compton Scattering

Traditionally in electrodynamics, one classifies electron-photon scattering processes according to the recoil energy and momentum acquired by the electron. From the quantum mechanical point of view, this is simply determined by the ratio of the photon energy to the electron rest mass, or $\hbar\omega/mc^2$. For classical Thomson scattering, we have $\hbar\omega/mc^2 \ll 1$, so that recoil effects can be ignored. In Compton scattering one assumes that $\hbar\omega/mc^2 \sim 1$ (or upwards of a few percent), leading to a gain in momentum for the electron, and a corresponding loss of photon energy change, see Fig. 3.10. For single-photon scattering, the energy change ΔE is easily

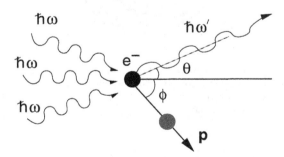

Fig. 3.10   Geometry for multiphoton-electron scattering event in which electron is initially at rest.

found by momentum and energy conservation:

$$\Delta E = \frac{\frac{\hbar\omega}{mc^2}(1 - \cos\theta)}{1 + \frac{\hbar\omega}{mc^2}(1 - \cos\theta)} \hbar\omega, \tag{3.36}$$

which corresponds to the usual wavelength shift of the scattered photon,

$$\Delta\lambda = \lambda' - \lambda = \lambda_c(1 - \cos\theta),$$

where $\lambda_c = h/mc = 0.0243\text{Å}$ is the Compton wavelength.

This situation becomes more complex when the laser intensity — and therefore photon density — becomes high enough to cause *multiphoton* scattering. We have already seen in the previous section that electrons can gain relativistic energies in the 'Thomson regime' if $n\hbar\omega/mc^2 \sim 1$. Moreover, if the electron is *already* relativistic, the photon frequency gets upshifted by γ in the electron rest frame, so that the relevant parameter is $\gamma n\hbar\omega/mc^2$ and the net energy shift takes on a somewhat more complicated form (Meyerhofer, 1997). For a head-on collision ($\theta = 180^o$), the maximum backscattered photon energy is:

$$\hbar\omega' = \frac{4n\hbar\omega\gamma^2}{1 + \frac{4n\hbar\omega\gamma^2}{mc^2} + a_0^2}.$$

The factor a_0^2 arises from the mass shift or 'dressing' experienced by the electron due to the electromagnetic field. Clearly in this regime, the delineation between nonlinear Thomson and Compton scattering becomes somewhat blurred.

The first experiments to observe multiphoton Compton scattering were performed at the Stanford Linear Accelerator Center (SLAC), where a 47 GeV ($\gamma \simeq 92000$) was 'collided' with a TW, 1 ps, 1 μm laser pulse (Bula *et al.*, 1996). Nd:glass laser photons have an energy of around 1 eV, giving $\gamma\hbar\omega/mc^2 \sim 1$, so that at an intensity of 10^{18} Wcm^{-2}, there is a significant probability of multiple scattering, which scales as $P_n \sim a_0^{2n}$. Sure enough, scattered electrons corresponding to $n = 2, 3$ and 4-photon events were detected in quantities broadly in line with interaction rates predicted by QED theory.

Besides exploring nonlinear QED physics, the photon-e-beam collider configuration (laser/synchrotron) can also be used to generate hard, short pulse x-rays (Leemans *et al.*, 1997), a topic which is examined in Chapter 7. This idea is physically and mathematically very similar to the synchrotron/FEL combination now being pursued at a number of centers

worldwide, reflecting the strong overlap which short pulse laser technology
has created between the plasma physics and accelerator communities.

Exercises

(1) The intensity of an electromagnetic (EM) wave with frequency ω is related to its field strength (in cgs units), by $I = cE_0^2/8\pi$.

a) Find the relationship between the normalized pump strength, $a_0 \equiv eE_0/m\omega c$ and the product $I\lambda_L^2$, with I in Wcm^{-2} and λ_L in microns.

b) Determine the transverse oscillation amplitude $x_{\text{osc}} = v_{\text{osc}}/\omega$ of an
electron in a beam focused to an intensity of 10^{18} Wcm^{-2} for: i) a
Ti:Sapphire laser, ii) a KrF* laser, iii) a free-electron laser (FEL) with
wavelength 1 Å.

(2) The electron orbit in a linearly polarized plane wave $\boldsymbol{A} = a_0 \cos(\omega t - kx)\hat{y}$ is given by:

$$kx = \frac{a_0^2}{4}(\phi + \frac{\sin 2\phi}{2}),$$
$$ky = -a_0 \sin\phi,$$

where $\phi = \omega t - kx$ and a_0 is the normalized field strength as defined
in Problem 1.

a) Eliminate ϕ from the first of these equations to demonstrate that
when the motion is averaged over a laser cycle, the electron drifts with
a velocity

$$v_D = \frac{a_0^2 c}{4 + a_0^2}.$$

b) What are the doppler-shifted frequency and wave vector ω', k' in a
reference frame R moving with velocity v_D?

c) Using the result of (b) together with the Lorentz transformations
$x' = \gamma_D(x + v_D t), t' = \gamma_D(t + v_D/c^2 x)$, find the electron orbit in R.
(Hint: recall that the phase ϕ is Lorentz-invariant). Show that this can

be written in implicit form:

$$16\tilde{x}^2 = \tilde{y}^2(4q^2 - \tilde{y}^2),$$

where $q = a_0/2\gamma_0$; $\tilde{x} \equiv k'x'$; $\tilde{y} \equiv k'y'$. Sketch the orbit for i) $a_0 = 0.1$, ii) $a_0 = 1.0$ and iii) $a_0 = 10$. What is the maximum ratio of the transverse:longitudinal oscillations (y'_{max}/x'_{max})?

(3) The electron in Problem 2 now finds itself in a weakly *inhomogeneous* laser field $E_y = E_0(r)\sin\phi$; $B_z = E_0(r)/c\sin\phi$, such that the field gradient is small compared to a laser wavelength: $\nabla E_0 \ll E_0/\lambda_L$. The plane wave solution comprises oscillations in x and y about a 'guiding center' moving with velocity v_D.

a) By expanding the fields about the guiding center $(x = v_D t, y = 0)$, show that averaged over a laser period, the electron experiences a *ponderomotive force*:

$$\boldsymbol{f}_p = -\frac{e^2}{4\omega^2 m}\nabla E_0^2.$$

(Hint: calculate each component f_{px}, f_{py}, f_{pz} separately).

b) When applied to a plasma fluid, it is usually more appropriate to speak of a *volume* force:

$$\boldsymbol{F}_p = n_e \boldsymbol{f}_p = -\frac{\omega_p^2}{16\pi\omega^2}\nabla E_0^2.$$

Determine the laser intensity at which this ponderomotive force becomes comparable to the thermal plasma pressure $P_e = n_e k_B T_e$.

(4) By including an arbitrary drift v_0 parallel to the laser axis in the energy and momenta conservation relations Eq. (3.21) and Eq. (3.22), show that the ejection angle of an electron is given by Eq. 3.24.

Chapter 4

Laser Propagation in Underdense Plasmas

The physics considered in this chapter is not exclusive to short laser pulses: many of the nonlinear wave phenomena which are confronted here actually predate femtosecond lasers by a good decade or two. Nevertheless, the vast increases in intensity achieved by technological advances — around six orders of magnitude in the last ten years — have ensured that many previously somewhat esoteric ideas, like relativistic self-guiding and plasma-based GeV particle accelerators, are now firmly on the experimental agenda.

Just as with femtosecond laser-solid interactions, which we will meet in the next chapter, much of the 'classical' long pulse theory has to be re-derived because the assumptions about amplitude or timescale are invalid. We will come to many specific examples of this later, but to get a feel for the nature of the processes which occur here, it is perhaps helpful to first look at a few general consequences of high intensity on wave propagation through plasmas. Consider the effect of high field amplitude in a pulse of finite width. Electrons in the plasma see an oscillating electric field $E_0 \sin \omega t$ and according to Eq. (3.7) will acquire a quiver velocity

$$v_0(r, t) = \frac{eE_0(r)}{m\omega_0} \cos \omega t.$$

A laser pulse with a Gaussian radial profile will jiggle the electrons harder at the center than in its wings, resulting in a relativistic 'drag' on the phase fronts in the middle of the beam. In other words, the transverse oscillating current $J \sim n_e v_0(r, t)$, which governs the refractive properties of the medium, has a spatial dependence. This is the physical basis of a number of nonlinear propagation effects, such as self-focusing, filamentation and self-modulation, which all have analogs in other nonlinear media (Shen, 1984). In the present context, the onset of these effects is characterized by the quiver velocity approaching the speed of light:[1] $v_{os} \sim c$, which as for single electrons, implies intensities typically exceeding 10^{18} Wcm^{-2}.

Another parameter which determines the nature of the interaction is of course the pulse duration. In the literature, one frequently comes across 'finite-pulse-length-' or 'finite-rise-time-' effects, which simply mean that one considers phenomena on a timescale shorter than the pulse duration τ_p. Here we have to be a little careful what we mean by 'slow' and 'fast' because the responsiveness of the medium depends on the plasma density through ω_p. For example, a 1 ps pulse passing through a plasma at 10^{17} cm^{-3} can still be considered 'long' because the plasma period is comparatively short so that the product $\omega_p \tau_p \simeq 18$. In such a case we can sometimes neglect electron inertia so that the medium response can be treated adiabatically. We will come to such a case later in this chapter, when we consider self-focusing of light in plasmas.

Since laser pulses propagate at nearly the speed of light in underdense plasmas, we also have to distinguish between processes which follow the pulse (copropagate) and those which are left behind in its wake. The photon bullet analogy of a rigid light packet ploughing through a jelly-like plasma is sometimes helpful here, though we shall see later that there are many effects which can cause the pulse to disintegrate.

Most of this chapter will be devoted to this new kind of electromagnetic propagation in plasmas, where the linear dispersion relations no longer hold, and we are confronted with phenomena such as wakefields, nonlinear frequency shifts, beam-breakup, particle trapping and wavebreaking. Before we come to the theoretical models of *short* pulses, however, it will be instructive to recall some of the prior work on nonlinear wave propagation. In fact, the equations derived over 40 years ago by Akhiezer and Polovin (1956) actually contain many features of the more recent models developed specifically for femtosecond laser pulses.

4.1   Nonlinear Plane Waves in a Cold Plasma

Our starting point is the Lorentz equation of motion (4.1) for the electrons in a cold, unmagnetized plasma, plus the Maxwell equations. The ions are assumed to be singly charged ($Z = 1$) and are treated as a homogeneous, neutralizing background: $Zn_i = n_0$. The neglect of thermal motion is often

---

[1]Strictly speaking, the quiver *momentum* approaches or exceeds $m_e c$: the velocity has, of course, to stay below c

justified for underdense plasmas because the temperature remains small compared to the typical oscillation energy. From Chapter 2, we recall that the field ionization process typically creates plasmas with temperatures of a few eV — well below the multi-keV oscillation energies which the electrons acquire from the laser field. Our starting equations are then as follows:

$$\frac{\partial \mathbf{p}}{\partial t} + (\mathbf{v} \cdot \boldsymbol{\nabla})\mathbf{p} = -e(\mathbf{E} + \frac{1}{c}\mathbf{v} \times \mathbf{B}), \tag{4.1}$$

$$\boldsymbol{\nabla} \cdot \mathbf{E} = 4\pi e(n_0 - n_e), \tag{4.2}$$

$$\boldsymbol{\nabla} \times \mathbf{E} = -\frac{1}{c}\frac{\partial \mathbf{B}}{\partial t}, \tag{4.3}$$

$$\boldsymbol{\nabla} \times \mathbf{B} = -\frac{4\pi}{c}en_e\mathbf{v} + \frac{1}{c}\frac{\partial \mathbf{E}}{\partial t}, \tag{4.4}$$

$$\boldsymbol{\nabla} \cdot \mathbf{B} = 0, \tag{4.5}$$

where $\mathbf{p} = \gamma m\mathbf{v}$ and $\gamma = (1+p^2/m^2c^2)^{1/2}$. Just as with any other propagation problem involving electromagnetic plane waves, we look for solutions of the form $f(\omega t - \mathbf{k}.\mathbf{r})$, or $f(\tau)$, where $\tau = t - \mathbf{i} \cdot \mathbf{r}/v_p$, and $v_p = \omega/k$ is the phase velocity of the wave. The temporal and spatial derivatives can then be rewritten thus:

$$\frac{\partial}{\partial t} = \frac{\partial}{\partial \tau}$$

$$\boldsymbol{\nabla} \cdot = -\frac{\mathbf{i}}{v_p}\frac{\partial}{\partial \tau} \cdot$$

$$\boldsymbol{\nabla} \times = -\frac{\mathbf{i}}{v_p}\frac{\partial}{\partial \tau} \times ,$$

where $\mathbf{i} = \mathbf{k}/|\mathbf{k}|$ is a unit vector in the direction of wave propagation. Our set of Maxwell-fluid equations (4.1)–(4.5) then become:

$$\left(\frac{\mathbf{i} \cdot \mathbf{v}}{v_p} - 1\right)\frac{d\mathbf{p}}{d\tau} = e(\mathbf{E} + \frac{1}{c}\mathbf{v} \times \mathbf{B}), \tag{4.6}$$

$$-\mathbf{i} \cdot \frac{d\mathbf{E}}{d\tau} = 4\pi e v_p(n_0 - n_e), \tag{4.7}$$

$$\mathbf{B} = \frac{c}{v_p}\mathbf{i} \times \mathbf{E} + \mathbf{B}_0, \tag{4.8}$$

$$-\mathbf{i} \times \frac{d\mathbf{B}}{d\tau} = -\frac{4\pi}{c}e v_p n_e\mathbf{v} + \frac{v_p}{c}\frac{d\mathbf{E}}{d\tau}, \tag{4.9}$$

$$\mathbf{i} \cdot \frac{d\mathbf{B}}{d\tau} = 0 . \tag{4.10}$$

The term B_0 represents an external magnetic field, which is not considered here, so it is set to zero. Thus, from Eq. (4.8) and Eq. (4.10) we can write $i \cdot B = E \cdot B = 0$, meaning that the B-field is perpendicular to both the wave vector and E-field. Note that the partial derivatives have now been replaced by total derivatives in the variable τ. Taking the dot product of Eq. (4.9) with the direction vector i, we may then eliminate E using Eq. (4.7) to give an equation for the density:

$$n_e = \frac{v_p n_0}{v_p - i \cdot v} \; . \tag{4.11}$$

Similarly, the cross-product of the direction vector with Eq. (4.6) gives

$$\left(\frac{i \cdot v}{v_p} - 1\right) i \times \frac{dp}{d\tau} = ei \times E + \frac{e}{c} i \times (v \times B)$$

$$= e\frac{v_p}{c} B + \frac{e}{c}\left[(i \cdot B)v - (i \cdot v)B\right]$$

$$= e\frac{v_p}{c} B \left(1 - \frac{i \cdot v}{v_p}\right) \; .$$

Thus we arrive at an explicit equation for B, namely:

$$B = -\frac{c}{ev_p} i \times \frac{dp}{d\tau} \; . \tag{4.12}$$

Likewise, taking $i \times (4.9)$ and making use of Eq. (4.8), we obtain an equation for $dB/d\tau$:

$$\frac{dB}{d\tau} = \frac{4\pi e n_e \beta_p}{\beta_p^2 - 1} i \times v, \tag{4.13}$$

where $\beta_p = v_p/c$.

We can now eliminate B from the previous two equations by subtracting Eq. (4.13) from $d/d\tau(4.12)$, leaving a transverse wave equation

$$i \times \frac{d^2p}{d\tau^2} + \frac{4\pi e^2 n_e \beta_p^2}{\beta_p^2 - 1} i \times v = 0 \; . \tag{4.14}$$

The longitudinal component of the fluid motion is found by differentiating $i\cdot(4.6)$ and this time using Eq. (4.12) to eliminate B and Eq. (4.11) to eliminate n_e, so that:

$$\frac{d}{d\tau}\left[\left(\frac{i \cdot v}{v_p} - 1\right) i \cdot \frac{dp}{d\tau}\right] = \frac{4\pi e^2 v_p n_0 i \cdot v}{v_p - i \cdot v} - \frac{1}{v_p}\frac{d}{d\tau}\left[v \cdot \frac{dp}{d\tau} - (i \cdot v)(i \cdot \frac{dp}{d\tau})\right] . \tag{4.15}$$

To render equations (4.14) and (4.15) into something more tractable, we specify the wave vector $\boldsymbol{k}$ to be in the x-direction. Thus, we have $\boldsymbol{i} = \hat{x}$, $\boldsymbol{i} \cdot \boldsymbol{p} = p_x$ and $\boldsymbol{i} \times \boldsymbol{p} = (0, -p_z, p_y)$. With these simplifications, and defining $\boldsymbol{u} = \boldsymbol{v}/c$, Eq. (4.11) becomes:

$$n_e = \frac{\beta_p n_0}{\beta_p - u_x}. \tag{4.16}$$

From this expression we can immediately deduce a salient feature of nonlinear plasma waves: namely, that the density becomes very large in regions where the fluid velocity approaches the phase velocity.

Taking the y and z components of Eq. (4.14) and making use of Eq. (4.16) and the usual definition $\omega_p^2 = 4\pi e^2 n_0/m$, gives us the coupled transverse wave equations:

$$\frac{d^2 p_z}{d\tau^2} + \frac{\omega_p^2 \beta_p^2}{\beta_p^2 - 1} \frac{\beta_p u_z}{\beta_p - u_x} = 0, \tag{4.17}$$

$$\frac{d^2 p_y}{d\tau^2} + \frac{\omega_p^2 \beta_p^2}{\beta_p^2 - 1} \frac{\beta_p u_y}{\beta_p - u_x} = 0, \tag{4.18}$$

where now p_y and p_z have been normalized to mc, so that $\boldsymbol{p} = \gamma \boldsymbol{u}$. The longitudinal wave equation follows from Eq. (4.15), which on applying the same choice of wave vector, simplifies to:

$$\frac{d}{d\tau}\left[(u_x - \beta_p)\frac{dp_x}{d\tau} + u_y\frac{dp_y}{d\tau} + u_z\frac{dp_z}{d\tau}\right] = \frac{\omega_p^2 \beta_p^2 u_x}{\beta_p - u_x}. \tag{4.19}$$

Apart from the cyclic rotation of the axes ($x \to y; y \to z; z \to x$), equations (4.18) and (4.19) are identical with Akhiezer and Polovin (1956, Eq. (15) in). They represent a closed set of equations for nonlinear plasma waves of arbitrary amplitude and fixed phase velocity v_p. Once solved for $\boldsymbol{p}$, the corresponding electric and magnetic fields (normalized here to $m\omega_p c/e$) can be obtained straightforwardly from:

$$E_x = -\frac{1}{\beta_p}\frac{d}{d(\omega_p\tau)}\left(\beta_p p_x - (1 + p^2)^{\frac{1}{2}}\right),$$

$$E_y = -\frac{dp_y}{d(\omega_p\tau)}, \tag{4.20}$$

$$E_z = -\frac{dp_z}{d(\omega_p\tau)},$$

$$B_x = 0,$$

$$B_y = \frac{1}{\beta_p}\frac{dp_z}{d(\omega_p\tau)},$$

$$B_z = -\frac{1}{\beta_p}\frac{dp_y}{d(\omega_p\tau)}. \qquad (4.21)$$

The components of B in Eq. (4.20) and the transverse component of E are obtained from Eq. (4.12) and Faraday's law Eq. (4.3). The longitudinal field is found by taking $i\cdot$(4.6) and using the energy equation $d\gamma/dt = -v\cdot E$ to eliminate E_x and E_y from the resulting equation for $dp/d\tau$. The details are left to the reader. A rather simple expression for the potential can be found by setting $E_x = \beta_p^{-1}d\phi/d\tau$, integrating, and observing that $\phi = p_x = 0$ and $\gamma = 1$ at $\tau = -\infty$, giving

$$\phi = \gamma - \beta_p p_x - 1. \qquad (4.22)$$

The full set of fluid equations can in general not be solved analytically. Various limiting cases can be found in the original work by Akhiezer and Polovin (1956). For a thorough account of the types of solution admitted, see the review by Decoster (1978). The analysis here is restricted to one example of particular relevance to short pulse propagation, namely pure longitudinal plasma oscillations. This solution is due to Noble (1984), who analyzed these cold plasma equations with novel particle acceleration concepts in mind. To proceed, we set $p_y = p_z = 0$ in Eq. (4.19), which after dropping the x-subscript, simplifies to

$$\frac{d}{d\tau}\left[(u-\beta_p)\frac{dp}{d\tau}\right] = \frac{\omega_p^2\beta_p^2 u}{\beta_p - u}.$$

Writing $p = \gamma u = u/\sqrt{1-u^2}$ and rearranging gives a 2nd order differential equation for the longitudinal velocity alone:

$$\frac{d^2}{d\tau^2}\left[\gamma(1-\beta_p u)\right] = \frac{\omega_p^2\beta_p^2 u}{\beta_p - u}. \qquad (4.23)$$

The intermediate steps in arriving at Eq. (4.23) are left as an exercise for the reader, see p. 125.

This equation can be integrated once to give:

$$\frac{1}{2}Y^2 = \beta_p^2\omega_p^2(\gamma_m - \gamma),$$

where

$$Y = \frac{d}{d\tau}\left[\gamma(1 - \beta_p u)\right], \ \gamma_m = (1 - u_m^2)^{-1/2},$$

and $u_m = (v/c)_{max}$ is the maximum oscillation velocity of the wave. The waveform can thus be determined from the solution of:

$$\frac{d}{d\tau}\left[\gamma(1 - \beta_p u)\right] = \pm\sqrt{2}\omega_p\beta_p(\gamma_m - \gamma)^{1/2}. \tag{4.24}$$

Once u is found, the density and electric field can immediately be determined using Eq. (4.16) and Eq. (4.20a) respectively:

$$n_e(\tau) = \frac{\beta_p n_0}{\beta_p - u(\tau)}, \tag{4.25}$$

$$E(\tau) = \frac{Y}{\beta_p} = \pm\sqrt{2}(\gamma_m - \gamma(\tau))^{1/2}. \tag{4.26}$$

An exact analytical solution of Eqs. (4.24–4.26) can be obtained in the limiting case of $\beta_p = 1$, corresponding to a highly underdense plasma. For details, the reader is referred to Noble (1984). The numerical solution of these equations for arbitrary phase velocity gives us more immediate insight into the nonlinear nature of these waves however, and three examples of these are shown in Fig. 4.1 for various values of β_p and u_m. The first case ($u_m = 0.1$) shows the classical linear Langmuir wave with electric field and density $90°$ out of phase. The solution for $\beta_p \simeq 1$ and $u_m = 0.9$, see Fig. 4.1b) is typical of nonlinear plasma waves: a sawtooth electric field and spiked density, accompanied by a *lengthening* of the oscillation period by a factor γ. Physically, this is due to the enhanced inertia of electrons as their velocity becomes relativistic. A spiky waveform is also possible for smaller amplitudes if the oscillation velocity is close to the phase velocity, see Fig. 4.1c). In a thermal plasma, this spikiness will be prone to particle trapping and wavebreaking (see the next section) eventually leading to damping of the wave.

4.2 Wavebreaking

One question which naturally arises with longitudinal plasma oscillations is: how large can they actually get? This issue is obviously crucial for applications such as particle acceleration because one would like the electric field to be as high as possible in order to make the most out of a finite

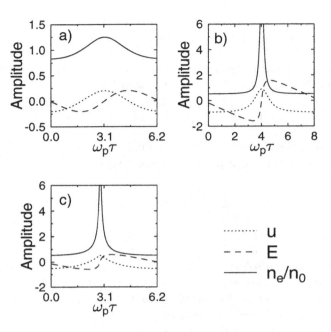

Fig. 4.1 Nonlinear plasma waves in an underdense plasma: a) $\beta_p = 1, u_m = 0.2$, b) $\beta_p = 1, u_m = 0.9$ elect and c) $\beta_p = 0.6, u_m = 0.55$.

length of plasma, see Sec. 7.1. Practical considerations aside, this problem has kept theoreticians busy for nearly half a century. Akhiezer & Polovin provided the lead with their solutions for cold plasma oscillations on p. 61. To find the maximum electric field from Eq. (4.26), we first note that the fluid velocity cannot exceed the phase velocity; if it does, then some of the electron charge sheets may 'cross' each other, and the wave will lose its coherence — it *breaks*. In fact, there are several definitions of wave-breaking depending on the type of oscillation involved, but for want of anything better at the moment, we just set $u_m = u_p$, or equivalently, $\gamma_m = \gamma_p$. An extremum of the electric field occurs for $\gamma = 1$, corresponding to the point in the oscillation when the electrons are momentarily stationary. Thus we have in physical units:

$$E_{\max} = \frac{mc\omega_p}{e}\sqrt{2}(\gamma_p - 1)^{1/2}. \qquad (4.27)$$

For non-relativistic phase velocities, we have $\gamma_p - 1 \simeq \beta_p^2/2$, so that

$$E_{\text{max}} = \frac{m\omega_p v_p}{e}, \qquad (4.28)$$

which is the well-known 'cold wave-breaking limit' derived by Dawson and Oberman (1962) using a more physically motivated Lagrangian sheet model. In this picture, wavebreaking can be thought of as the crossing of neighbouring charge sheets, accompanied by a density singularity.

This picture changes in a *warm* plasma because thermal effects act to reduce the maximum attainable wave amplitude. There are two reasons for this: first, plasma pressure resists the tendency for the density to explode; second, thermal electrons moving in the direction of the wave may be trapped at a lower wave amplitude than cold particles would be. These corrections were found by Coffey (1971) using a so-called 'waterbag' model for the electron distribution function. His result for the wavebreaking limit for a non-relativistic plasma wave in a warm plasma is:

$$\frac{eE_{\text{max}}}{m\omega_p v_p} = \left(1 - \frac{\mu}{3} - \frac{8}{3}\mu^{1/4} + 2\mu^{1/2}\right)^{1/2}, \qquad (4.29)$$

where $\mu = 3k_B T_e / m v_p^2$.

Extending this result to the relativistic case is not just a matter of making thermal corrections to Eqs. (4.24–4.26); a relativistically correct equation of state is needed before one can write down the appropriate fluid equation of motion (or Euler equation). This was first done self-consistently by Katsouleas and Mori (1988), who generalized Coffey's waterbag model to include relativistic fluid momenta. From this they found the equivalent equation to Eq. (4.23) for the fluid velocity:

$$\frac{d^2}{d\tau^2} F(u) = \frac{\omega_p^2 \beta_p^2 u}{\beta_p - u}, \qquad (4.30)$$

where

$$F(u) = \frac{1 - \beta_p u}{(1 - u^2)^{1/2}} \left[1 + \mu\beta_p^2 \frac{1 - u^2}{(\beta_p^2 - u^2)^{1/2}}\right].$$

Eq. (4.30) can be integrated once as for the cold plasma case, finally giving a maximum electric field in the limit $\beta_p \simeq 1 \pm \sqrt{2}\omega_p \beta_p (\gamma_m - \gamma)^{1/2}$ of:

$$\frac{eE_{\text{max}}}{m\omega_p c} = \mu^{-1/4}(\ln 2\gamma_p^{1/2}\mu^{1/4})^{1/2}, \qquad (4.31)$$

valid for

$$\gamma_p \gg \frac{1}{2\mu^{1/2}} \log 2\mu^{1/4}\gamma_p^{1/2}.$$

This result, together with the non-relativistic limit, is depicted in Fig. 4.2.

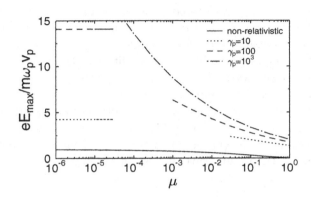

Fig. 4.2 Wavebreaking amplitude of longitudinal plasma oscillations for different phase velocities. The relativistic curves are depicted up to the limit of validity of Eq. (4.31) — the full numerical solutions join up with the cold wavebreaking limits (shown on the right for $\gamma_p = 10$ and $\gamma_p = 100$).

4.3   Wakefield Generation

So far, we have considered the *types* of large amplitude wave that can propagate in a cold plasma without worrying about how they are created. However, we know that plasma waves can be generated by all sorts of means: particle beams, instabilities, turbulence and mode conversion, to name but a few. The last of these is one of the main techniques for coupling electromagnetic wave energy into a plasma. A low intensity laser pulse will normally pass through a homogeneous, underdense plasma without disturbing it. We can demonstrate this by linearizing Eqs. (4.18–4.20); that is, letting $p_{x,y,z} \ll 1$, and assuming that $u_x \ll \beta_p$, so that we can neglect all terms $O(u^2)$ and higher. The geometry is again chosen so that the laser (EM wave) propagates along the x-axis: $E_L = (0, E_y, 0), B_L = (0, 0, B_z), p_y = A_y$. Effectively, this procedure leads to a decoupling of the longitudinal and transverse wave components, so that we recover the usual wave equation

$$\frac{d^2 u_y}{d\tau^2} + \frac{\omega_p^2 \beta_p^2 u_y}{\beta_p^2 - 1} = 0, \tag{4.32}$$

which has the solution

$$u_y = u_o e^{-i\omega\tau},$$

provided that

$$-\omega^2 + \frac{\omega_p^2 \beta_p^2}{\beta_p^2 - 1} = 0.$$

Recalling that $\beta_p = v_p/c = \omega/kc$, the above relation can be rearranged to read:

$$\omega^2 = \omega_p^2 + c^2 k^2, \tag{4.33}$$

which is of course the usual dispersion relation for electromagnetic waves in a cold, collisionless plasma (Kruer, 1988).

Linearizing the longitudinal momentum equation yields

$$-\beta_p \frac{d^2 p_x}{d\tau^2} - \omega_p^2 \beta_p u_x = 0,$$

or, since $\gamma \simeq 1$,

$$\frac{d^2 u_x}{d\tau^2} + \omega_p^2 u_x = 0. \tag{4.34}$$

Again, writing $u_x = u_{x0} e^{-i\omega\tau}$, this gives the familiar dispersion relation for longitudinal waves in the limit $T_e \to 0$.

$$\omega^2 = \omega_p^2. \tag{4.35}$$

4.3.1 *Small pump strengths*

Given that the laser propagation is described by the transverse wave equation (4.18), the question naturally arises whether and how we can drive plasma waves with laser pulses. The key point is that Eqs. (4.18) and (4.19) are coupled through the nonlinear terms (second term on the LHS of either equation). Thus, as a first order improvement on the linearized equations, we retain a nonlinear *pump* term in Eq. (4.19) to give:

$$\frac{d^2 u_x}{d\tau^2} + \omega_p^2 u_x = \frac{1}{\beta_p} \frac{d}{d\tau}\left(u_y \frac{dp_y}{d\tau}\right)$$

$$\simeq \frac{1}{\beta_p} \frac{d^2}{d\tau^2}\left(\frac{u_y^2}{2}\right), \tag{4.36}$$

where we have set $p_y = \gamma u_y \simeq u_y$ if $u_{x,y} \ll 1$. From Eq. (4.16), we can write down an expression for the perturbed density:

$$\begin{aligned} n \equiv n_e - n_0 &= \frac{\beta_p n_0 - n_0(\beta_p - u_x)}{\beta_p - u_x} \\ &= \frac{n_0 u_x}{\beta_p - u_x} \\ &\simeq \frac{n_0 u_x}{\beta_p}. \end{aligned} \tag{4.37}$$

Substituting Eq. (4.37) in Eq. (4.36), we find

$$\frac{d^2 n}{d\tau^2} + \omega_p^2 n = \frac{n_0}{\beta_p^2} \frac{d^2}{d\tau^2} \frac{u_y^2}{2}.$$

Now recalling that the original change of variables involved a transformation:

$$\frac{\partial}{\partial t} = \frac{\partial}{\partial \tau}, \qquad \frac{\partial}{\partial x} = -\frac{1}{c\beta_p} \frac{\partial}{\partial \tau},$$

we can identify the derivatives on the left and right side of the equation with time and space derivatives respectively, so that in terms of the original Eulerian variables (t, x), we have:

$$\frac{\partial^2 n}{\partial t^2} + \omega_p^2 n = \frac{n_0}{2} \frac{\partial^2}{\partial x^2} \frac{v_y^2}{c^2}. \tag{4.38}$$

This equation describes driven plasma waves in the weak field limit, and is the starting point for the analysis of both parametric instabilities (e.g. Raman scattering, which we will come to in Sec. 4.4) and *wakefield excitation*. In the latter case, one considers the electromagnetic wave to be confined within a small 'packet', so that the driving term in Eq. (4.38) is impulse-like and induces a wake oscillation in the plasma as the wave passes through. The wakefield excitation process described by Eq. (4.38) was first analyzed in detail by Gorbunov and Kirsanov (1987), and Sprangle *et al.*, (1988). The idea of using laser to generate high-phase-velocity plasma actually came a decade earlier from Tajima and Dawson (1979), who proposed such a scheme as a way of accelerating particles to GeV energies, see Secs. 4.3.4 and 7.1.

4.3.2 *The quasistatic approximation*

At this point it is convenient to introduce another coordinate transformation; this time to a frame moving with the *group velocity* of the laser pulse $v_g \simeq c$. Thus we choose variables (ξ, τ) such that $\xi = x - ct, \tau = t$. Our partial derivatives then become:

$$\frac{\partial}{\partial x} = \frac{\partial}{\partial \xi}; \qquad \frac{\partial}{\partial t} = \frac{\partial}{\partial \tau} - c\frac{\partial}{\partial \xi} \simeq -c\frac{\partial}{\partial \xi} \qquad (4.39)$$

so that

$$\frac{\partial^2}{\partial x^2} = \frac{\partial^2}{\partial \xi^2}; \qquad \frac{\partial}{\partial t^2} = c^2\frac{\partial^2}{\partial \xi^2}.$$

The time τ is considered to be slowly varying during the transit time of the pulse, so that we can effectively set $\partial/\partial \tau = 0$ in this 'co-moving' frame. Note that this is an Eulerian, not Lorentz transformation. The latter can be done too, and in fact leads to a mathematically more symmetric system of coupled equations for the electromagnetic and electrostatic waves (McKinstrie and DuBois, 1988); however, one is left with the burden of reinterpreting the results in the laboratory frame.

Physically, the so-called *quasi-static approximation* (QSA) represented by Eq. (4.39) can be thought of as a kind of 'speedboat' approximation: a rigid photon bullet passes through initally undisturbed plasma, leaving a wake of plasma oscillations behind it. Looked at in terms of the distance from the leading edge of the pulse, the wake *retains its form*; only changes in pulse shape can alter the properties of the wake in the co-moving frame. Gorbunov and Kirsanov liken this to an EM wave packet 'emitting' plasmons in a manner reminiscent of Čerenkov radiation from relativistic particles passing through nonlinear media.

In this new coordinate system, letting $a = v_y/c$, Eq. (4.38) appears thus:

$$\left(\frac{\partial^2}{\partial \xi^2} + k_p^2\right) n = \frac{n_0}{2}\frac{\partial^2}{\partial \xi^2}a^2. \qquad (4.40)$$

Making use of Poisson's equation,

$$\frac{\partial^2 \phi}{\partial \xi^2} = -\frac{\partial E}{\partial \xi} = 4\pi e n,$$

we can also write down equations for the electric field and potential (normalized to $m\omega_p c/e$ and mc/e respectively) of the wakefield:

$$\left(\frac{\partial^2}{\partial \xi^2} + k_p^2\right) E = k_p^2 \frac{\partial}{\partial \xi} \Phi_L, \tag{4.41}$$

$$\left(\frac{\partial^2}{\partial \xi^2} + k_p^2\right) \phi = -k_p^2 \Phi_L, \tag{4.42}$$

where $\Phi_L = -\frac{1}{2} < a^2 >$ is the normalized ponderomotive potential of the laser pulse, which we take to be averaged over the laser period $2\pi/\omega_0$, so that in terms of the slowly varying pulse envelope $\Phi_L(\xi) = -\frac{1}{4} a(\xi)^2$.

Of these three equations for the plasma wake, Eq. (4.41) is probably the most intuitively accessible. The driving term in this case is the ponderomotive force, which as we already saw in Sec. 3.2, acts to push electrons away from regions of high intensity. As the pulse enters a region of fresh plasma therefore, it first pushes electrons forward. Half-way through, however, the ponderomotive force reverses its sign, and the electrons receive another kick in the opposite direction, see Fig. 4.3. Clearly, the amplitude of the longitudinal oscillation will be enhanced if the pulse length (transit time of the ponderomotive force) is roughly matched to the plasma period ω_p^{-1}.

Eq. (4.42) is a driven Helmholtz equation which can be solved by Green's function methods (see Mathews and Walker, 1970, Chapter 9). It has the formal solution:

$$\phi(\xi) = -\frac{k_p}{4} \int_\xi^\infty d\xi' \mid a(\xi') \mid^2 \sin[k_p(\xi - \xi')]. \tag{4.43}$$

Similar expressions can be found for the electric field and number density, see Exercise 2.

In order to get a feeling for the characteristics of this solution, let us consider a specific example, namely a 'sin^2'-pulse with amplitude given by:

$$a^2(\xi) = \begin{cases} a_0^2 \sin^2(\frac{\pi \xi}{\xi_L}), & 0 \le \xi \le \xi_L \\ 0, & \xi < 0, \xi > \xi_L \end{cases}$$

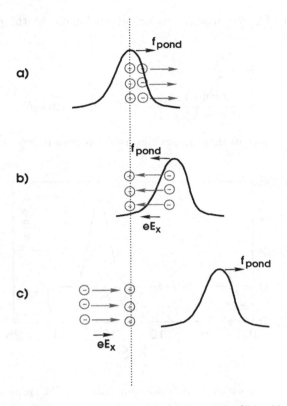

Fig. 4.3 Wakefield generation by double ponderomotive kick of laser pulse.

Behind the pulse ($\xi < 0$), we have, therefore:

$$
\phi(\xi) = -\frac{k_p a_0^2}{4} \int_0^{\xi_L} d\xi' \sin k_p(\xi - \xi') \sin^2\left(\frac{\pi\xi}{\xi_L}\right)
$$

$$
= -\frac{k_p a_0^2}{8} \left[\frac{\cos k_p(\xi - \xi')}{k_p} + \frac{\xi_L/2}{2\pi - k_p\xi_L} \cos\left\{ k_p\xi - k_p\xi'(1 - \frac{2\pi}{k_p\xi_L}) \right\} \right.
$$

$$
\left. - \frac{\xi_L/2}{2\pi + k_p\xi_L} \cos\left\{ k_p\xi - k_p\xi'(1 + \frac{2\pi}{k_p\xi_L}) \right\} \right]_0^{\xi_L}
$$

$$
= \frac{2\pi^2 \Phi_L}{(4\pi^2 - k_p^2\xi_L^2)} \left[\cos k_p(\xi - \xi_L) - \cos k_p\xi \right]. \tag{4.44}
$$

Within the pulse ($0 < \xi < \xi_L$), the solution looks slightly different (Esarey *et al.*, 1989; Gorbunov and Kirsanov, 1987), but we leave this as an exercise for the reader. The longitudinal wakefield left behind by the pulse is then simply:

$$E_z = -\frac{\partial \phi}{\partial \xi}$$

$$= \frac{2\pi^2 \Phi_L k_p}{(4\pi^2 - k_p^2 \xi_L^2)} \left[\sin k_p(\xi - \xi_L) - \sin k_p \xi \right]. \qquad (4.45)$$

Numerical examples of these quantities are displayed in Fig. 4.4.

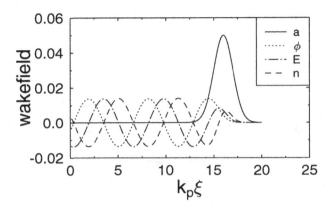

Fig. 4.4   Solutions of the small-amplitude wakefield equation.

By inspection, we see that Eq. (4.45) has a maximum for $k_p \xi_L = 2\pi$, or equivalently, for pulse lengths $\xi_L = \lambda_p$. (For square pulses this maximum occurs for $\xi_L = \lambda_p/2$, see Gorbunov and Kirsanov (1987) and Ex. 2). In this limit, we can use l'Hospital's rule to obtain:

$$E_z^{\max}(\xi) = \frac{\pi^2 \Phi_L}{\lambda_p} \cos k_p \xi, \qquad (4.46)$$

which basically scales with the laser intensity, or a_0^2. Likewise, the wake potential has a maximum value $\phi_{\max} = -\pi \Phi_L/2 \sin k_p \xi$. This result can be generalized to 3D wakes provided that the laser pulse is not too narrow, i.e. $\sigma \gg c/\omega_p$. In other words, we may consider a Gaussian beam profile $a = a_0 \exp(-r^2/2\sigma^2)$, so that $\Phi_L = -\frac{1}{4}a_0^2 \exp(-r^2/\sigma^2)$. In this case, the longitudinal field has the same form as above, but diminishes monotonically with increasing distance from the beam axis. On the other hand, a finite spot size leads additionally to a *radial* wakefield

$$E_r^{\max}(\xi, r) = -\frac{\partial \phi}{\partial r}$$

$$= -\frac{\Phi_L \pi r}{\sigma^2} \sin k_p \xi. \qquad (4.47)$$

Note that the radial field is 90° out of phase with the longitudinal component — a general feature of wakefields which has interesting consequences for particle acceleration by plasma waves — see Sec. 7.1.

There are many other aspects of wakefield generation which are beyond the scope of this introduction — the original works by Sprangle *et al.*, (1990b) and Gorbunov and Kirsanov (1987) provide a good starting point for further reading. Two important points which should be mentioned (which may both be seen from Eq. (4.44)) are: first, the wake far behind the pulse depends only weakly on the *shape* of the laser pulse; second, for very short pulses ($\xi_L \ll \pi/k_p$), or very long pulses ($\xi_L \gg \pi/k_p$), only a very small wake is produced, see Fig. 4.5.

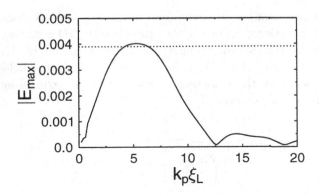

Fig. 4.5   Resonance curve for wakefield excitation for small pump strengths.

4.3.3   *Arbitrary pump strengths*

In this section we generalize the preceding theory by considering plasma waves generated by laser pulses of arbitrary intensity. From Chapter 3, we know that relativistic effects become important for electron quiver momenta $p_y/mc \simeq 1$. The standard way of improving on the linearized equation

Eq. (4.36) is to expand the nonlinear terms in powers of the normalized pump strength a^2. This is usually adequate for laser intensities between, say, 10^{15} and 10^{17} Wcm^{-2}, and we will come across an explicit example of this later when we look at the related phenomenon of *beat-wave excitation* in Sec. 4.3.4. For intensities beyond 10^{18} Wcm^{-2}, or equivalently $p_y/mc \gg 1$, we need a model which is fully nonlinear, and which is valid for arbitrary plasma densities (i.e. where $v_g \neq c$).

One of the first such models was developed by Sprangle *et al.*, (1990b), who formulated a fully nonlinear ordinary differential equation for the wake potential in the limit $v_g = c$. Many similar 1D formulations were independently presented by Bulanov *et al.*, (1989); Tsytovich *et al.*, (1989) and Berezhiani and Murusidze, (1990), and later generalized to arbitrary transformation velocity, i.e. v_g or v_p instead of c (Esarey *et al.*, 1997; Mori *et al.*, 1993; Dalla and Lontano, 1993; Kingham and Bell, 1997). This most general case will be considered here, which can subsequently be simplified to recover most of the results in the literature. Of particular interest here is the strongly relativistic regime, for which we must finally resort to numerical solution.

The starting point is once again the Lorentz equation (4.1) together with the Maxwell equations (4.2)–(4.5), but this time the 1D geometry dictated by the laser propagation direction is introduced immediately. Thus, if the electromagnetic wave $\boldsymbol{A} = \hat{y}A_y(x - v_g t)$ comprises only the fields E_y and B_z, we can solve for the transverse momentum p_y by expressing these in terms of the vector potential A_y:

$$\frac{dp_y}{dt} = -e\left(E_y - \frac{v_x}{c}B_z\right)$$
$$= \frac{e}{c}\left(\frac{\partial A_y}{\partial t} + v_x\frac{\partial A_y}{\partial x}\right) = \frac{e}{c}\frac{dA_y}{dt}.$$

If there is no initial drift in the y-direction, then

$$p_y = \frac{e}{c}A_y. \tag{4.48}$$

Likewise, the longitudinal component of Eq. (4.1) gives us:

$$\frac{dp_x}{dt} = -e\left(E_x + \frac{v_y}{c}B_z\right),$$

which, after setting $E_x = -\partial\phi/\partial x$, $B_z = \partial A/\partial x$, and using the identity

(4.48) leads to

$$\frac{d}{dt}(\gamma u) = c\frac{\partial \phi}{\partial x} - \frac{c}{2\gamma}\frac{\partial a^2}{\partial x}, \tag{4.49}$$

where $u = v_x/c$ as before, and the potentials ϕ and a have both been normalized to mc^2/e; thus $a \equiv eA_y/mc^2$, etc. The relativistic factor $\gamma = (1 - \boldsymbol{u}^2)^{-1/2}$ can conveniently be separated into longitudinal and transverse components, thus:

$$\begin{aligned} \gamma &= \gamma_\perp \gamma_\parallel \\ &= \frac{(1 + a^2)^{1/2}}{(1 - u^2)^{1/2}}. \end{aligned} \tag{4.50}$$

As with most problems involving fluids, one can write down a continuity equation for the mass, or in this case, charge density:

$$\frac{\partial n_e}{\partial t} + \nabla \cdot (n_e \boldsymbol{v}) = 0. \tag{4.51}$$

In one dimension this becomes

$$\frac{\partial n_e}{\partial t} + c\frac{\partial}{\partial x}(n_e v_x) = 0. \tag{4.52}$$

To obtain an equation for the electromagnetic modes we substitute $\boldsymbol{E} = -\nabla\phi - c^{-1}\partial\boldsymbol{A}/\partial t$ and $\boldsymbol{B} = \nabla \times \boldsymbol{A}$ into Ampere's law (4.4) and then make use of the Coulomb gauge $\nabla \cdot \boldsymbol{A} = 0$, to get:

$$\frac{1}{c^2}\frac{\partial^2 A_y}{\partial t^2} - \nabla^2 A_y = \frac{4\pi}{c}J_y = -4\pi e n_e \frac{v_y}{c}.$$

In general the wave equation above would contain a term involving the scalar potential $c^{-1}\nabla\partial\phi/\partial t$, which can usually be neglected anyway, because it is small compared to the transverse current. In this 1-D case it vanishes exactly. The choice of the Coulomb gauge also implies that $A_x = 0$, which we implicitly assumed when deriving Eq. (4.49). Normalizing as before, with

$$\frac{v_y}{c} = \frac{eA_y}{mc^2\gamma} = \frac{a}{\gamma},$$

and $n = n_e/n_0$, we reduce the wave equation to the following:

$$\frac{\partial^2 a}{\partial t^2} - c^2\nabla^2 a = -\omega_p^2 \frac{na}{\gamma}. \tag{4.53}$$

Eqs. (4.49)–(4.53), together with Poisson's equation

$$\frac{\partial^2 \phi}{\partial x^2} = k_p^2(n-1),\tag{4.54}$$

constitute a closed set for the coupled electromagnetic and plasma waves. To proceed further, the quasistatic approximation is again applied in a frame moving with the pulse. Unlike in Sec. 4.3.1, however, this time we use an arbitrary group velocity v_g, so that $\tau = t; \xi = x - v_g t$ and for the derivatives we have:

$$\frac{\partial}{\partial x} = \frac{\partial}{\partial \xi}$$

$$\frac{\partial}{\partial t} = \frac{\partial}{\partial \tau} - v_g \frac{\partial}{\partial \xi}.$$

Before applying this transformation to the momentum equation (4.49), we first note that from the identity (4.50),

$$\frac{\partial a^2}{\partial \xi} = \frac{\partial}{\partial \xi}\left[\gamma^2(1-u^2)-1\right]$$

$$= 2\gamma\frac{\partial \gamma}{\partial \xi} - 2u^2\gamma\frac{\partial \gamma}{\partial \xi} - 2\gamma^2 u\frac{\partial u}{\partial \xi}.$$

Thus,

$$\frac{d}{dt}(\gamma u) = \left(\frac{\partial}{\partial \tau} - v_g\frac{\partial}{\partial \xi} + cu\frac{\partial}{\partial \xi}\right)\gamma u$$

$$= c\frac{\partial \phi}{\partial \xi} - \frac{c}{2\gamma}\frac{\partial a^2}{\partial \xi}.$$

Or, substituting the above expression for $\partial a^2/\partial \xi$ and letting $\beta_g = v_g/c$,

$$\frac{1}{c}\frac{\partial}{\partial \tau}(\gamma u) = \frac{\partial \phi}{\partial \xi} - \frac{\partial \gamma}{\partial \xi} + u^2\frac{\partial \gamma}{\partial \xi} + \gamma u\frac{\partial u}{\partial \xi}$$

$$+ \beta_g u\frac{\partial \gamma}{\partial \xi} + \beta_g\gamma\frac{\partial u}{\partial \xi} - u^2\frac{\partial \gamma}{\partial \xi} - \gamma u\frac{\partial u}{\partial \xi}$$

$$= \frac{\partial}{\partial \xi}\left[\phi - \gamma(1-\beta_g u)\right].\tag{4.55}$$

Similarly, the continuity equation in these new coordinates becomes:

$$\frac{1}{c}\frac{\partial n}{\partial \tau} = \frac{\partial}{\partial \xi}\left[n(\beta_g - u)\right].\tag{4.56}$$

The electromagnetic wave equation is not needed explicitly here (and in any case takes on a somewhat untidy appearance in these coordinates), but we note that the evolution timescale τ of the pulse envelope is typically the Rayleigh diffraction time (see p. 98),

$$t_R = R_L/c = \frac{k_0 \sigma^2}{c},$$

which is much longer than a laser period. This effectively allows us to neglect $\partial/\partial\tau$ relative to $\partial/\partial\xi \sim ik_0$, given that the vector potential has the form $a = a(\xi, r, \tau) \exp[ik_0\xi]$ in the co-moving frame. We proceed by applying this quasistatic approximation to the fluid equations, just as we did previously for the small-amplitude wakefield (p. 67). This consists of setting $\partial/\partial\tau = 0$ in Eq. (4.55) and Eq. (4.56), and then integrating to yield the following conservation relations:

$$n(\beta_g - u) = const.$$

Or, requiring that $n(\xi = +\infty) = 1$,

$$n = \frac{\beta_g}{\beta_g - u}. \tag{4.57}$$

Likewise,

$$\phi - \gamma(1 - \beta_g u) + 1 = 0, \tag{4.58}$$

since $\phi = u = 0; \gamma = 1$ in the absence of a plasma wave (i.e. when $u = 0$ at $\xi = \infty$).

The set of partial differential equations for the fluid variables u, n and ϕ have thus been reduced to one ordinary differential equation for ϕ, plus three algebraic expressions (4.50), (4.57) and (4.58) relating u, n, ϕ and the pump strength a. Remember that the laser is regarded as being fixed on the fluid timescale, so we can determine the wakefield quantities independently of the laser evolution. In fact, it turns out that we can express the fluid quantities entirely in terms of the pump amplitude a. To do this, we take the square of Eq. (4.58) to obtain:

$$\begin{aligned}
(1 + \phi)^2 &= \gamma^2(1 - \beta_g u)^2 \\
&= \gamma^2[2(1 - \beta_g u) + \beta_g^2 u^2 - 1] \\
&= \gamma^2 \left[\frac{2(1 + \phi)}{\gamma} + \beta_g^2 - 1 - \beta_g^2 \frac{(1 + a^2)}{\gamma^2} \right],
\end{aligned}$$

where we have used $1 - \beta_g u = (1 + \phi)/\gamma$ and $u^2 = 1 - (1 + a^2)/\gamma^2$ to eliminate u. Expanding the bracket on the RHS and dividing both sides by $\gamma_g^2(1 + \phi)^2$, we get after a little rearranging:

$$1 - \frac{2\gamma}{\gamma_g^2(1 + \phi)} + \frac{\gamma^2}{\gamma_g^4(1 + \phi)^2} = \beta_g^2\left[1 - \frac{1 + a^2}{\gamma_g^2(1 + \phi)^2}\right].$$

The LHS is a perfect square, which we can easily solve to arrive at an explicit expression for γ:

$$\gamma = \gamma_g^2(1 + \phi)\left[1 - \beta_g\left(1 - \frac{1 + a^2}{\gamma_g^2(1 + \phi)^2}\right)^{1/2}\right]. \tag{4.59}$$

From Eq. (4.58), we can now solve for u in terms of a and ϕ:

$$u = \frac{1}{\beta_g}\left[1 - \frac{(1 + \phi)}{\gamma}\right]$$

$$= \frac{1}{\beta_g}\left[1 - \frac{1}{\gamma_g^2(1 - \beta_g\psi)}\right],$$

where

$$\psi = \left(1 - \frac{1 + a^2}{\gamma_g^2(1 + \phi)^2}\right)^{1/2}.$$

After some algebra, this reduces to:

$$u = \frac{\beta_g - \psi}{1 - \beta_g\psi}. \tag{4.60}$$

The point of this rather tedious manipulation becomes apparent when we use Eq. (4.60) to eliminate u from Eq. (4.57) to find the density:

$$n = \gamma_g^2\beta_g\left(\frac{1}{\psi} - \beta_g\right). \tag{4.61}$$

Then from Poisson's equation in the co-moving coordinates,

$$\frac{\partial^2\phi}{\partial\xi^2} = k_p^2(n - 1)$$

$$= k_p^2\gamma_g^2\left(\frac{\beta_g}{\psi} - 1\right)$$

$$= k_p^2\gamma_g^2\left\{\frac{\beta_g(1 + \phi)}{[(1 + \phi)^2 - \gamma_g^{-2}(1 + a^2)]^{1/2}} - 1\right\}. \tag{4.62}$$

Eq. (4.62) is nonlinear ordinary differential equation for the wake potential $\phi(\xi)$ — exact within the QSA — which we can integrate numerically for a given pulse amplitude $a(\xi)$ at a given time τ. An exact analytical solution for a square pump in the limit $\beta_g \to 1$ (see also Exercise 4 on p. 125) was derived by Berezhiani and Murusidze (1990), from which the scaling of the wake-variable maxima can be deduced:

$$\phi_{max} \sim \gamma_\perp^2 - 1; \quad E_{max} \sim \frac{\gamma_\perp^2 - 1}{\gamma_\perp}; \quad p_{max} \sim (\gamma u)_{max} = \frac{\gamma_\perp^4 - 1}{2\gamma_\perp^2}, \qquad (4.63)$$

where $\gamma_\perp = (1 + a^2)^{1/2}$ as on p. 73.

In general, once we have $\phi(\xi)$, we can immediately obtain u and n from Eq. (4.60) and Eq. (4.61) respectively, the results of which are displayed in Fig. 4.6. Note the correspondance between the QSA and the plane wave

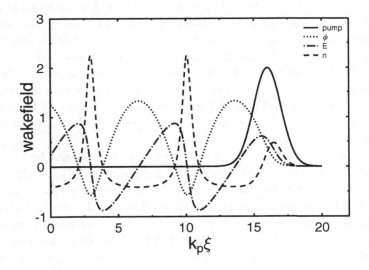

Fig. 4.6   Laser-wakefield excitation in the extreme nonlinear regime.

solutions of Akhiezer & Polovin derived in Sec. 4.1, see Fig. 4.1. Comparison of Eq. (4.22) with Eq. (4.58) and of Eq. (4.25) with Eq. (4.57) reveals that the plane wave solutions are *identical* to the quasi-static expressions if we let $\beta_p \to \beta_g$ and $\tau \to -\xi$. This symmetry makes sense: the plane wave ansatz explicitly excludes spatial derivatives ($\partial/\partial\xi = 0$) so that all variables are a function of the retarded time variable $\tau = t - x/v_p$ only. The QSA on the other hand, excludes *time* derivatives in the wake following the pulse. These two pictures are equivalent: the (transverse) laser pump can easily be introduced into the longitudinal equation (4.23) as a slowly varying envelope $p_\perp = a(\tau)$ without violating the initial plane wave assumption.

The 'new' nonlinear features for short pulse wakefield generation are therefore to some extent already contained within the original works of Akhiezer and Polovin (1956) and the subsequent analysis of Noble (1984). On the other hand, the QSA version is more readily accessible in terms of the physics, and can easily be generalized to include the evolution of the laser pulse in 2 or 3 dimensions (Esarey *et al.*, 1997; Bulanov *et al.*, 1995).

4.3.4 Beat waves

So far in this chapter I have concentrated on the nature of large amplitude plasma waves, and how they may be generated in underdense plasmas. With the widespread availability of high-power, short pulse lasers, this subject is now of primary importance in the context of plasma-based particle acceleration. Although we defer the details of so-called bench-top particle accelerators to Chapter 7, it seems a convenient point to briefly mention the forerunner of the LWFA here — namely, the beat-wave accelerator scheme (BWA). The BWA was proposed by Tajima & Dawson as an alternative means of generating high-phase-velocity plasma waves in the absence of femtosecond laser technology.

At that time, lasers had pulse lengths exceeding several hundred picoseconds, so Tajima and Dawson (1979) suggested using two pulses with slightly different wavelengths such that the beat frequency $\delta\omega = \omega_0 - \omega_1$ is matched to the plasma frequency. This effectively results in a train of shorter pulses which is able to drive up a *beat-wave* over many plasma periods, generating a wake almost identical to that created by a single pulse.

Mathematically, beat-wave excitation can be analyzed as follows. Consider the linearized equation for the perturbed plasma density Eq. (4.38), which setting

$$\tilde{n} = \frac{n}{n_0} = -\frac{c}{\omega_p}\frac{\partial\varepsilon}{\partial x} \quad \text{and} \quad \frac{v_y}{c} = a,$$

where $\varepsilon = eE_x/m\omega_p c$, can be rewritten:

$$\left(\frac{\partial^2}{\partial t^2} + \omega_p^2\right)\varepsilon = -\frac{\omega_p c}{2}\frac{\partial a^2}{\partial x}. \tag{4.64}$$

Recall that the RHS is just the ponderomotive force of the laser, which for single frequency illumination is only effective at driving the plasma wave if $\omega_p \tau_L \sim 1$. Now consider what happens when we mix two laser pulses at frequencies ω_0 and ω_1. The *combined* vector potential is then

$$a^2 = \frac{1}{4}\left[a_0 e^{i(\omega_0 t - k_0 x)} + a_1 e^{i(\omega_1 t - k_1 x)} + c.c.\right]^2$$

$$= \frac{a_0 a_1^*}{2} e^{i[(\omega_0 - \omega_1)t - (k_0 - k_1)x]} + \frac{a_0 a_1}{2} e^{i[(\omega_0 + \omega_1)t - (k_0 + k_1)x]} + c.c.$$

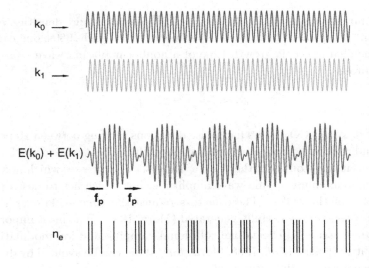

Fig. 4.7 Plasma wave generation by beating of two electromagnetic waves with frequency difference $\omega_0 - \omega_1 = \omega_p$.

The net force evidently comprises both fast $(\omega_0 + \omega_1)$ and slow $(\omega_0 - \omega_1)$ terms, which we can equate to the resonant mode on the LHS of Eq. (4.64) by letting

$$\varepsilon = \epsilon(t) e^{i(\omega_p t - k_p x)},$$

and using the phase matching condition $\omega_p = \omega_0 - \omega_1; k_p = k_0 - k_1$. This beat-wave drive process is illustrated schematically in Fig. 4.7. Assuming the plasma wave amplitude ϵ is slowly varying ($\partial\epsilon/\partial t \ll \omega_p\epsilon$, we arrive at a so-called *envelope* equation for the plasma wave:

$$\frac{\partial \epsilon}{\partial t} = \frac{\omega_p a_0 a_1^*}{4}. \qquad (4.65)$$

In other words, for constant pump amplitudes, the plasma wave amplitude initially grows linearly in time:

$$\epsilon(t) = \frac{a_0 a_1^*}{4}\omega_p t. \qquad (4.66)$$

For pump strengths $a_0 = 0.03$ ($I = 10^{15}$ Wcm^{-2}) and densities $n_e = 10^{17}$ cm^{-3}, typical of beat-wave experiments in the mid-1980s, one can see that the characteristic growth time of a nonlinear plasma wave ($\epsilon \sim 1$) is of the order

$$\tau_{\mathrm{pw}} \sim \frac{4}{a_0^2 \omega_p} \simeq 250 \text{ ps.}$$

Nd:glass and CO_2 lasers had pulse durations ranging between 10 ps and 2 ns, and thus were reasonably compatible with the beat-wave growth time. However, it turns out that there are a whole set of processes which not only limit the maximum plasma wave amplitude, but which act to destroy the coherence of the wake. These factors reduce the time scale over which plasma waves can be usefully generated (Mora, 1988). The most important of these — assuming effects like collisional damping and ion modulational instabilities (Pesme *et al.*, 1988; Mora *et al.*, 1988) can be avoided by driving the plasma wave sufficiently hard — are:

i) linear detuning, caused by a mismatch between beat frequency $\Delta\omega = \omega_0 - \omega_1$ and plasma frequency, such that $\delta \equiv \Delta\omega - \omega_p \neq 0$;

ii) relativistic detuning, caused by the decrease in local plasma frequency due to the relativistic mass increase of strongly oscillating electrons.

Both of these effects can in fact be included in Eq. (4.65) for the plasma wave amplitude using a quasilinear approach (Tang *et al.*, 1985; McKinstrie and Forslund, 1986; Karttunen and Salomaa, 1986; Gibbon, 1990), leading to the following envelope equation:

$$\frac{\partial \epsilon}{\partial t} - \frac{3i}{16} \mid \epsilon \mid^2 \epsilon - i\delta\epsilon = \kappa, \qquad (4.67)$$

where $\delta = (\Delta\omega - \omega_p)/\Delta\omega$ and $\kappa = a_0 a_1^*/4$, and we have normalized the time variable to ω_p. The technique used to obtain this *envelope equation* will be shown later when we consider two-dimensional waves. We can proceed to solve it by writing the field as a product of amplitude and phase $\epsilon = r\exp[i\theta]$, see Appendix A for a discussion of the equivalence between the various representations for the slowly-varying envelope approximation.

Equating real and imaginary parts, we get two coupled equations for the amplitude and phase respectively:

$$\dot{r} = \kappa\cos\theta$$
$$r\dot{\theta} = -\kappa\sin\theta + \delta r + \frac{3r^3}{16}. \qquad (4.68)$$

The the last term of the equation for $\dot{\theta}$ can be identified as a *nonlinear frequency shift*, giving an effective wave frequency

$$\omega_p' = \omega_p\left(1 - \frac{3r^2}{16}\right).$$

These equations can be integrated (see Exercise 6) and expressed in potential form (McKinstrie and Forslund, 1986):

$$\left(\frac{\partial r}{\partial t}\right)^2 + r^2\left(\frac{3r^2}{64} + \frac{\delta}{2}\right)^2 = \kappa^2.$$

The plasma wave amplitude will reach its maximum — or saturate — when $\dot{r} = 0$, leaving a cubic relation

$$\frac{3r^3}{16} + 2\delta r - 4\kappa = 0. \qquad (4.69)$$

For perfect frequency matching ($\delta = 0$), the maximum amplitude predicted by Eq. (4.69) is:

$$|\varepsilon|_{\text{sat}} = \left(\frac{64\kappa}{3}\right)^{1/3} = \left(\frac{16a_o a_1}{3}\right)^{1/3}, \qquad (4.70)$$

which is the standard result for plasma wave saturation by relativistic detuning, originally derived by Rosenbluth and Liu (1972). After reaching this amplitude, the plasma wave is driven back down again, resulting in an oscillating solution with period $\tau_r \simeq 17(a_0 a_1)^{-2/3}$ (Karttunen and Salomaa, 1989), see Fig. 4.8.

Notice that the relativistic saturation amplitude can actually be exceeded by compensating the nonlinear frequency shift with a beat frequency

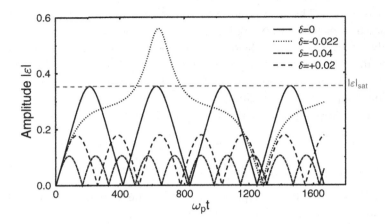

Fig. 4.8   Evolution of beat-wave amplitude for different frequency mismatches.

slightly below the plasma frequency ($\delta < 0$). In fact, from Eq. (4.69) one can show (Tang *et al.*, 1985) that there is an optimum mismatch for

$$\delta = -\frac{1}{2} \left(\frac{9a_0 a_1}{8} \right)^{2/3} ,$$

with a corresponding saturation amplitude

$$|\varepsilon|_{\max} = 4 \left(\frac{a_0 a_1}{3} \right)^{1/3} .$$

In practice, these parameters cannot be chosen so precisely, especially since the linear detuning parameter is sensitive to the background plasma density, which is hard to maintain within the few percent range demanded by the above formulae. On the other hand, modern short pulse lasers offer an opportunity to reexamine the beat-wave acceleration scheme with substantially higher pump strengths than those which were available in the mid-1980s. Indeed, the BWA may even have potential advantages over wakefield schemes in terms of optical guiding and phase-velocity control (Esarey *et al.*, 1988; Gibbon and Bell, 1988).

4.4   Instabilities

Murphy's Law of Plasma Physics states that if a plasma can go unstable, then it will do so in the most damaging manner possible. Researchers working in the field of magnetic confinement fusion will readily testify to this observation, many of whom have spent a lifetime coming up with ingenious ways of persuading plasmas of various geometrical shapes to stay around for longer. Of course the timescales we have to deal with here are femtoseconds, not microseconds or above, so one might guess that many instabilities will simply not have time to grow during typical interaction times. For instance, a pulse length of 100 fs is generally short enough to avoid significant ion motion, which virtually rules out hydrodynamic instabilities as well as certain parametric instabilities involving ion sound waves, such as stimulated Brillouin scattering (SBS).

Short pulses are a two-edged sword, however: higher intensities not only bring higher growth rates, but can also give rise to entirely new types of instability. In the context of this chapter, the most dangerous instabilities are those which directly afflict the laser pulse, ultimately leading to beam breakup. These include stimulated Raman scattering (SRS) and two-plasmon decay (TPD), both extremely well-known and studied within the ICF context (see, for example, Drake *et al.*, 1974; Forslund *et al.*, 1975b, and references therein) as well as the relativistic modulational (RMI) and filamentational (RFI) instabilities (Max *et al.*, 1974). For the one-dimensional parametric instabilities, the analysis basically follows the standard technique of Fourier-analysing the linearized fluid-Maxwell equations (Kruer, 1988) with one important modification: the laser pump strength is assumed to be relativistic ($a_0 > 1$). Physically this implies that the unstable EM and fluid modes experience a relativistic drag due to the transverse electron quiver velocity being close to the speed of light.

Our starting point is once again the set of cold, relativistic fluid equations derived previously in Sec. 4.3.3. These are the one-dimensional relativistic wave equation for the vector potential (4.53), combined with the one-dimensional continuity and momentum equations (4.52,4.49) for the electron fluid, completed with Poisson's equation (4.2). Normalizing all dependent variables as before, with $n = n_e/n_0, \boldsymbol{a} = e\boldsymbol{A}/mc^2, \phi \rightarrow e\phi/mc^2$

and $p = \gamma u$, we quote these equations again here for convenience:

$$\left(\frac{\partial^2}{\partial t^2} - c^2 \frac{\partial^2}{\partial x^2} + \omega_p^2 \frac{n}{\gamma} \right) \boldsymbol{a} = 0,$$

$$\frac{\partial p}{\partial t} + cu \frac{\partial p}{\partial x} = \frac{\partial \phi}{\partial x} - \frac{1}{2\gamma} \frac{\partial}{\partial x} \boldsymbol{a}^2,$$

$$\frac{\partial^2 \phi}{\partial x^2} = k_p^2 (n - 1), \tag{4.71}$$

$$\frac{\partial n}{\partial t} + c \frac{\partial}{\partial x} \left(\frac{np}{\gamma} \right) = 0,$$

$$\gamma = (1 + p^2 + \boldsymbol{a}^2)^{1/2}.$$

The laser amplitude is represented here as a vector rather than a scalar in order to retain an arbitrary polarization; the reasons for which will become clear later. In contrast to the wakefield analysis of Sec. 4.3.3, where a fully nonlinear solution for a driven plasma wave was sought, we look here for modes which grow exponentially from noise. This means that all 'decay' modes start off very small in amplitude compared to the pump, so that we can write:

$$u = \tilde{u},$$

$$n = 1 + \tilde{n},$$

$$\boldsymbol{a} = \boldsymbol{a}_0 + \tilde{\boldsymbol{a}},$$

where the tildas denote perturbations of the fluid and field quantities. Note that we take the plasma fluid to be initially at rest ($u_0 = 0$). Substituting these into the set of equations (4.71) and linearizing — i.e. discarding terms like $\tilde{p}^2, \tilde{n}\tilde{p}$, etc. — gives to zeroth order:

$$\left(\frac{\partial^2}{\partial t^2} - c^2 \frac{\partial^2}{\partial x^2} + \frac{\omega_p^2}{\gamma_0} \right) \boldsymbol{a}_0 = 0, \tag{4.72}$$

where $\gamma_0 = (1 + \boldsymbol{a}_0^2)^{1/2}$. Fourier analyzing (i.e. substituting $\partial^2/\partial t^2 \rightarrow -\omega_0^2$ and $\partial^2/\partial x^2 \rightarrow -k_0^2$ for a plane-wave pump), we recover the usual dispersion relation for a relativistic light wave (Kaw and Dawson, 1970):

$$\omega_0^2 = c^2 k_0^2 + \omega_p'^2, \tag{4.73}$$

where $\omega_p'^2 = \omega_p^2/\gamma_0$ is the effective plasma frequency for the EM wave.

To next (or first) order, we have for the perturbed EM mode:

$$\left(\frac{\partial^2}{\partial t^2} - c^2\frac{\partial^2}{\partial x^2} + \omega_p'^2\right)\tilde{a} = -\omega_p'^2\tilde{n}a_0 + \frac{\omega_p'^2}{\gamma_0^2}(a_0\cdot\tilde{a})a_0. \tag{4.74}$$

In arriving at Eq. (4.74) we have expanded the factor $\gamma = (1 + p^2 + a^2)^{1/2}$ by assuming that $\tilde{p}$ and $\tilde{a}$ are both small compared to a_0, so that to first order,

$$\gamma^{-1} \simeq \gamma_0^{-1} - \frac{a_0 \cdot \tilde{a}}{\gamma_0^3}.$$

For the fluid equations, a similar procedure is carried out, the ponderomotive force $\frac{\partial}{\partial x}(a^2/2)$ now comprising a *beat* term between pump and scattered EM wave. (Contrast this with the wakefield excitation in Sec. 4.3.1, where the ponderomotive term was just $\frac{\partial}{\partial x}(a_0^2/2)$.) Thus, the first-order momentum equation is:

$$\frac{1}{c}\frac{\partial u}{\partial t} = \frac{1}{\gamma_0}\frac{\partial\tilde{\phi}}{\partial x} - \frac{1}{\gamma_0^2}\frac{\partial}{\partial x}(a_0\cdot\tilde{a}),$$

and after subsitution into the time-derivative of the continuity equation, we finally arrive at:

$$\left(\frac{\partial^2}{\partial t^2} + \omega_p'^2\right)\tilde{n} = \frac{c^2}{\gamma_0^2}\frac{\partial^2}{\partial x^2}(a_0\cdot\tilde{a}). \tag{4.75}$$

Eqs. (4.74) and (4.75) describe the nonlinear forward Raman scattering (FRS) and Raman backscattering (RBS) instabilities, and can be thought of as a generalization of their counterparts in Kruer (1988) for a relativistic pump. The latter equations are fully recovered in the non-relativistic limit ($\gamma_0 \simeq 1$). To find the unstable modes, we follow the usual procedure of separating the fluid quantities into the product of a slowly varying envelope and rapidly varying phase (Drake *et al.*, 1974) and set:

$$\tilde{n} = \frac{n_1}{2}e^{-i\psi} + c.c.,$$

$$\tilde{a} = \frac{a_+}{2}e^{i\psi_+} + \frac{a_-}{2}e^{i\psi_-} + c.c.,$$

$$a_0 = \frac{a_0}{2}e^{i\psi_0} + c.c.,$$

where $\psi = kx - \omega t$; $\psi_0 = k_0 x - \omega_0 t$ and $\psi_\pm = (k \pm k_0)x - (\omega \pm \omega_0)t$. These phase matching conditions are depicted graphically in Fig. 4.9. Thus for FRS, we have a *four-wave* interaction between the pump k_0, the plasma wave k and the Stokes $k_- = k - k_0$ and anti-Stokes $k_+ = k + k_0$ decay modes, respectively. Substituting the above expression for the density into

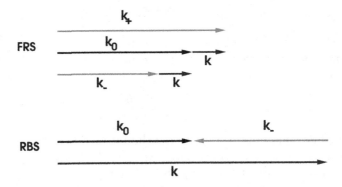

FRS

RBS

Fig. 4.9   Phase matching conditions for forward and backward Raman scattering.

Eq. (4.75), we pick out low frequency resonant terms with phase ψ by noting that

$$\psi = \psi_+ - \psi_0 = \psi_- + \psi_0,$$

giving rise to terms $a_+ a_0^*$ and $a_- a_0$ on the RHS respectively. The time- and space-derivatives again reduce to functions of ω and k, through the substitutions: $\partial/\partial t \rightarrow -i\omega$; $\partial/\partial x \rightarrow ik$, giving:

$$(-\omega^2 + \omega_p'^2)\frac{n_1}{2} = -\frac{c^2 k^2}{4\gamma_0^2}(\boldsymbol{a}_0 \cdot \boldsymbol{a}_- + \boldsymbol{a}_0^* \cdot \boldsymbol{a}_+),$$

which, denoting $D_e = -\omega^2 + \omega_p'^2$, yields:

$$n_1(\omega, k) = -\frac{c^2 k^2}{2\gamma_0^2 D_e}(\boldsymbol{a}_0 \cdot \boldsymbol{a}_- + \boldsymbol{a}_0^* \cdot \boldsymbol{a}_+). \qquad (4.76)$$

This expression for the perturbed density can now be substituted back into the RHS of Eq. (4.74) to determine the nonlinear currents (cubic in pump amplitude a_0) which drive the scattered EM modes at $\omega_\pm = \omega \pm \omega_0$ respectively:

$$\boldsymbol{J}_{\mathrm{NL}}\ (\omega_\pm, k_\pm) = -\omega_p'^2 \left(\frac{n_1}{2}e^{-i\psi} + c.c.\right)\left(\frac{\boldsymbol{a}_0}{2}e^{i\psi_0} + c.c.\right)$$

$$+ \frac{\omega_p'^2}{\gamma_0^2}\left(\frac{\boldsymbol{a}_0}{2}e^{i\psi_0} + c.c.\right)\left(\frac{\boldsymbol{a}_+}{2}e^{i\psi_+} + \frac{\boldsymbol{a}_-}{2}e^{i\psi_-} + c.c.\right)\left(\frac{\boldsymbol{a}_0}{2}e^{i\psi_0} + c.c.\right).$$

Note that the second cubic term above has been deliberately expressed as a vector dot-product multiplied by the pump amplitude; this is because the result depends on the laser polarization. In fact, the relativistic factor γ_0 is also polarization-dependent, so before proceeding we need to determine how this quantity — which depends on a_0^2 — is affected by the slowly-varying envelope approximation. Consider a more general expression for the pump wave vector

$$a_0 = \frac{a_0}{2}\sigma e^{i\psi_0} + c.c.,$$

where σ is a polarization vector such that

$$\sigma = \begin{cases} e_y & \text{represents linearly polarized light,} \\ \frac{1}{\sqrt{2}}(e_y + ie_z) & \text{represents circular light.} \end{cases} \tag{4.77}$$

Thus the square of the pump amplitude is given by:

$$a_0^2 = a_0 \cdot a_0 = \left(\frac{1}{2}a_0\sigma e^{i\psi_0} + \frac{1}{2}a_0^*\sigma^* e^{-i\psi_0}\right)^2,$$

$$= \frac{1}{4}a_0^2\sigma^2 e^{2i\psi_0} + \frac{1}{2}a_0 a_0^*\sigma \cdot \sigma^* + \frac{1}{4}a_0^{*2}\sigma^{*2}e^{-2i\psi_0}$$

$$= \begin{cases} \frac{1}{4}a_0^2 e^{2i\psi_0} + \frac{1}{2} \mid a_0 \mid^2 + \frac{1}{4}a_0^{*2}e^{-2i\psi_0}, & \text{(linear)} \\ \frac{1}{2} \mid a_0 \mid^2 . & \text{(circular)} \end{cases}$$

To get the above result, the identities $\sigma^2 = 0; \sigma \cdot \sigma^* = 1$ have been used for circularly polarized light, causing the harmonic terms to vanish. In the linear case, we can simplify the result by taking a time-average over the laser cycle, i.e. $< a^2 >= \frac{1}{2}|a_0|^2$. This is permissible when one is interested in driving terms evolving on a slow time-scale (as for example with the ponderomotive force). In this case, however, the γ-factor is multiplied with further quickly varying quantities, so to neglect the $e^{2i\psi_0}$ terms would actually lead to error. On the other hand, their inclusion leads to somewhat unwealdy expressions for the nonlinear currents, so in the following only the circular polarized case is considered, which has the advantage that the expansion for γ_0 is *exact* for arbitrary pump amplitude. Thus for circular light we can write

$$\gamma_0 = (1 + a_0^2)^{1/2} = \left(1 + \frac{\mid a_0 \mid^2}{2}\right)^{1/2}.$$

Evaluating the RHS of the nonlinear current J_{NL} and again applying the phase-matching conditions $\psi_\pm = \psi \pm \psi_0$, yields for the Stokes and anti-Stokes modes respectively (also circularly polarized):

$$J(\omega - \omega_0) = \frac{\omega_p'^2}{8\gamma_0^2}\left[\left(\frac{c^2 k^2}{D_e} + 1\right)(|a_0|^2 a_- + a_0^{*2} a_+)\right]$$

$$J(\omega + \omega_0) = \frac{\omega_p'^2}{8\gamma_0^2}\left[\left(\frac{c^2 k^2}{D_e} + 1\right)(|a_0|^2 a_+ + a_0^2 a_-)\right].$$

Equating these expressions to the LHS of Eq. (4.74) gives:

$$D_- a_- = \frac{\omega_p'^2}{4\gamma_0^2}\left(\frac{c^2 k^2}{D_e} + 1\right)(|a_0|^2 a_- + a_0^{*2} a_+) \qquad (4.78)$$

$$D_+ a_+ = \frac{\omega_p'^2}{4\gamma_0^2}\left(\frac{c^2 k^2}{D_e} + 1\right)(|a_0|^2 a_+ + a_0^2 a_-), \qquad (4.79)$$

where

$$\begin{aligned} D_\pm &= -\omega_\pm^2 + c^2 k_\pm^2 + \omega_p'^2 \\ &= -\omega^2 + c^2 k^2 \pm 2(k_0 k c^2 - \omega_0 \omega). \end{aligned} \qquad (4.80)$$

Finally, the scattered wave amplitudes $a_\pm$ can be eliminated from Eq. (4.78) and Eq. (4.79) to yield the dispersion relation in conventional form (Bers, 1983):

$$\frac{\omega_p^2 |a_0|^2}{4\gamma_0^3}\left(\frac{c^2 k^2}{D_e} + 1\right)\left(\frac{1}{D_+} + \frac{1}{D_-}\right) = 1. \qquad (4.81)$$

This result — strictly valid for constant pump strengths — was obtained independently by Sakharov and Kirsanov (1994) and Guérin *et al.*, (1995). A weakly nonlinear version of Eq. (4.81) valid for $a_0 \ll 1$ can be found in the extensive treatment by McKinstrie and Bingham (1992), who also consider the linearly polarized case. (Note that the definition of pump strength a_0 differs by a factor of 2 in some works). More sophisticated analyses generalizing the above analysis to arbitrary densities, scattering geometry and pump strengths have been made by Quesnel *et al.*, (1997), Barr *et al.*, (1999) and most recently by Barr and Hill (2003). In these works it is shown how all electronic parametric instabilities (SRS, RMI, TPD and RFI) can be unified under a single dispersion relation. For the sake of simplicity, the outcome of these analyses will be illustrated by considering Raman scattering in rarefied plasma.

4.4.1 *Forward Raman scattering*

Eq. (4.81) can be solved in various limits. For forward scattering (FRS), we take $\omega = \omega_p' + i\Gamma$, where Γ is the growth rate to be determined, such that $\Gamma \ll \omega_p' \ll \omega_0$. Thus, from Eq. (4.80), we have $D_\pm \simeq -2i\Gamma(\omega_p' \pm \omega_0)$, $D_e \simeq -2i\omega_p'\Gamma$. Substituting these expressions into Eq. (4.81) and setting $c^2 k_{max}^2 = \omega_p'^2$, after a little algebra we obtain:

$$\Gamma_{\text{FRS}} = \frac{\omega_p^2 a_0}{\sqrt{8}\omega_0(1 + a_0^2/2)}.\tag{4.82}$$

In the non-relativistic limit, $a_0 \ll 1$, this reduces to the standard result (Drake *et al.*, 1974; Forslund *et al.*, 1975b; Kruer, 1988):

$$\Gamma = \frac{\omega_p^2}{\sqrt{8}\omega_0}a_0.$$

In the other extreme, $a_0 \gg 1$, the growth rate becomes:

$$\Gamma_{\text{rel}} \simeq \frac{\omega_p^2}{\sqrt{2}\omega_0 a_0} \sim a_0^{-1},$$

and therefore falls off with increasing laser intensity. The maximum growth rate is easily found from Eq. (4.82) to be $\Gamma_{\text{max}} = \omega_p^2/4\omega_0$. A detailed discussion of the various branches of this instability and its merger with RMI at high intensities can be found in the works by Guérin *et al.*, (1995) and Quesnel *et al.*, (1997). For finite pulse lengths, the RFS growth depends both on the propagation distance L and the position relative to the foot of the pulse (Mori *et al.*, 1994), giving an asymptotic growth of

$$G \sim \exp\left(\sqrt{\frac{2P}{P_c}\frac{L}{Z_R}\omega_p\tau_L\frac{\omega_p}{\omega_o}}\right),\tag{4.83}$$

where P, τ_L, Z_R and P_c are, respectively: the laser power, pulse duration, Rayleigh length and critical power for self-focusing (see Sec. 4.6.2).

4.4.2 *Backward Raman scattering*

For backward scattering (RBS), the instability is inevitably strongly coupled in this regime, since the pump strength $a_0 \gg (\omega_p/\omega_0)^{1/2}$, and we can assume that the growth rate $\Gamma \gg \omega_p'$. Neglecting the upshifted wave

(ω_+, k_+) in Eq. (4.81) as non-resonant, we have:

$$D_- = \frac{\omega_p^2 a_0^2}{4\gamma_0^3} \left(\frac{c^2 k^2}{D_e} + 1 \right), \qquad (4.84)$$

this time with

$$D_- = \omega_p'^2 - (\omega - \omega_0)^2 + c^2 (k - k_0)^2.$$

Maximum growth occurs for

$$k = k_0 + \frac{\omega_0}{c} \left(1 - \frac{2\omega_p}{\omega_0} \right)^{1/2}$$

$$\simeq 2k_0, \qquad \text{for } \omega_p/\omega_0 \ll 1.$$

In this limit, we have

$$\frac{\Gamma_{\mathrm{RBS}}}{\omega_p} = \frac{\sqrt{3}}{2} \left(\frac{\omega_0}{2\omega_p} \right)^{1/3} \frac{a_0^{2/3}}{(1 + a_0^2/2)^{1/2}}, \qquad (4.85)$$

which is maximum for $a_0 = 2$, after which it falls off as $a_0^{-1/3}$ for $a_0 \gg 1$. In the non-relativistic limit, we also recover the standard result for the strongly coupled regime (Forslund *et al.*, 1975b; Darrow *et al.*, 1992):

$$\Gamma_{\mathrm{SC}} = \frac{\sqrt{3}}{2} \left(\frac{\omega_0 \omega_p^2 a_0^2}{2} \right)^{1/3}. \qquad (4.86)$$

4.5   Harmonic Generation in Underdense Plasmas

The key to efficient harmonic generation is to bring photons and electrons together at high number densities for at least a few optical cycles, so that the nonlinearities inherent in relativistic electron motion can be efficiently exploited. Given the practical difficulties of achieving these conditions with isolated electron 'clouds' or beams discussed earlier in Chapter 3, and the early saturation of HHG in gases, it is natural to ask whether a fully ionized plasma can be better exploited as a medium for frequency up-conversion. In this case, harmonic generation can be analyzed in terms of the collective *fluid* response of the plasma electrons in a uniform ion background, In other words, we require the plasma *current*, $\boldsymbol{J} = -en_e\boldsymbol{v}$, as opposed to individual electron contributions to the radiated power as in Eq. (3.29). For an axisymmetric pulse propagating through a uniform plasma, it turns out that the scattering contributions cancel except in the forward direction, where they add coherently. The calculation of harmonic emission reduces to determining the current source and using the wave equation to obtain the scattered vector potential.

In the context of short pulse laser-plasma interaction, such an analysis was first carried out by (Sprangle *et al.*, 1990a), who showed that the 3rd harmonic power scales as n_e^2 rather than n_e, as we might expect by summing over a set of coherent scatterers. In fact the actual reason for this is that the longitudinal electron motion in Eq. (3.13) sets up a space charge oscillating at twice the laser frequency. The resulting density perturbation beats with the laser field to form a nonlinear current source which nearly cancels the relativistic nonlinearity. A tutorial account of this effect is given by Mori (1994), upon which the following derivation is based. Starting with our old friend, the electromagnetic wave equation,

$$\nabla^2 \boldsymbol{A} - \frac{1}{c^2}\frac{\partial^2 \boldsymbol{A}}{\partial t^2} = -\frac{4\pi}{c}\boldsymbol{J} = \frac{4\pi e}{m_e c}\frac{n_e \boldsymbol{p}}{\gamma}, \qquad (4.87)$$

we see that the harmonics are driven by the nonlinear current on the right-hand-side. To evaluate this source term we need first the x-component of the Lorentz equation of motion, Eq. (4.1), which we can rewrite as:

$$\frac{\partial v_x}{\partial t} = -\frac{eE}{\gamma_0 m_e} - \frac{c^2}{2\gamma_0^2}\frac{\partial A^2}{\partial x} + O(v_x^2), \qquad (4.88)$$

where $A \to eA/m_e c^2$ as before, and we have retained only non-oscillatory terms in the relativistic factor, i.e. we expand about a relativistically strong pump, just as in the instability analysis of Sec. 4.4. Combining Eq. (4.88) with Poisson's equation

$$\frac{\partial E}{\partial x} = 4\pi e(n_0 - n_e)$$

and the continuity equation

$$\frac{\partial n_e}{\partial t} + \frac{\partial (n_e v_x)}{\partial x} = 0,$$

we write $n_e = n_0 + n$, normalize to n_0 and neglect terms $O(n^2)$ to obtain

$$\frac{\partial^2 n}{\partial t^2} + \frac{\omega_p^2}{\gamma_0}n = \frac{c^2}{2\gamma_0^2}\frac{\partial A^2}{\partial x}. \qquad (4.89)$$

Expanding to the same order, the (normalized) transverse current can be expressed as

$$J_y = \frac{\omega_p^2}{\gamma_0} \left[A_y + n A_y - \frac{1}{2\gamma_0^2} A_y^3 \right]. \tag{4.90}$$

The 2nd and 3rd terms represent the density bunching and relativistic non-linearities respectively; both are $O(A_y^3)$. To proceed further, we expand all quantities as a harmonic series:

$$f(\mathbf{r}, t) = f_1 \cos \phi + f_2 \cos 2\phi + \dots$$

and let $A_y = a_0 \cos \phi$, where $\phi = w_0 t - k_0 x$ as before. (I have added a subscript 0 to the laser frequency and wave vector this time, in order to distinguish harmonic frequencies from the fundamental mode). One can then show that Eq. (4.89) reduces to

$$n(2\omega_0) = \frac{c^2 k_0^2}{(4\omega_0^2 - \omega_p^2/\gamma_0)} \frac{a_0^2}{\gamma_0^2} \cos 2\phi,$$

so that in the limit $\omega_p/\omega_0 \ll 1$, the 3rd harmonic current becomes:

$$J(3\omega_0) \simeq -\frac{3}{32} \frac{\omega_p^4}{\omega_0^2} \frac{a_0^3}{\gamma_0^4} \cos 3\phi. \tag{4.91}$$

In other words, the relativistic and density-bunching nonlinearities for the 3rd harmonic source actually cancel to $O(\omega_p^2/\omega_0^2)$. To obtain the power efficiency, we equate Eq. (4.91) with the matching $\cos 3\phi$ term on the LHS of the expanded wave equation Eq. (4.87), taking care to eliminate terms which would lead to secular growth (Montgomery and Tidman, 1964). This procedure eventually yields a steady-state amplitude

$$a_3 = \frac{3}{256} \left(\frac{\omega_p}{\omega_0} \right)^2 \frac{a_0^3}{\gamma_0^3}. \tag{4.92}$$

Using the fact that the electric field strength of the harmonic, $E_n \propto \omega_n a_n$, we finally obtain an overall power scaling of (Rax and Fisch, 1992; Mori et al., 1993; Esarey et al., 1993)

$$\frac{P_3}{P_0} = \left(\frac{9}{256} \right)^2 \left(\frac{\omega_p}{\omega_0} \right)^4 \frac{a_0^4}{(1 + a_0^2/2)^3}. \tag{4.93}$$

This analysis can be carried further iteratively to get the powers of the next higher order harmonics (Esarey *et al.*, 1993):

$$\frac{P_5}{P_0} = 2.4 \times 10^{-5} \left(\frac{\omega_p}{\omega_0}\right)^8 \frac{a_0^8}{\gamma_0^{12}}$$

$$\frac{P_7}{P_0} = 2.3 \times 10^{-7} \left(\frac{\omega_p}{\omega_0}\right)^{12} \frac{a_0^{12}}{\gamma_0^{18}}. \tag{4.94}$$

The 3rd harmonic power has a maximum for $a_0 = 2$, after which it falls off slowly. Typical efficiencies are rather low: for example, a 1 TW Ti:sapphire laser with intensity 10^{18} Wcm^{-2} ($a_0 = 0.68$) focused into a plasma with density 10^{19} cm^{-3} will yield just 5 kW of 3rd harmonic radiation. If this result by itself is not disappointing enough, it turns out that including propagation effects reduces the efficiency even further. In fact there are two pieces of physics (initially neglected by Sprangle *et al.*, 1990a) which are responsible for this.

First, the fact that in a plasma, the phase velocities of the harmonics differ from the velocity of the fundamental. This is easily seen from the dispersion relation for the mth harmonic:

$$\omega_m^2 = c^2 k_m^2 + \frac{\omega_p^2}{\gamma_0}.$$

Thus

$$\frac{v_{\phi m}}{c} = \frac{\omega_m}{ck_m} = \frac{m\omega}{(m^2\omega^2 - \omega_p^2/\gamma_0)^{\frac{1}{2}}},$$

and the harmonic will be π out of phase with the pump after a length:

$$l_m = ct_m = \frac{\lambda_p^2}{\lambda} \frac{m\gamma_0}{(m^2 - 1)}. \tag{4.95}$$

For the Ti:sapphire example just considered above, with $\lambda_p = 2\pi c/\omega_p = 11$ μm and $\lambda_0 = 0.8$ μm, the dephasing length of the 3rd harmonic would be $56\mu m$, which is much smaller than a typical Rayleigh focusing length. The second problem is that harmonics generated by a Gaussian beam focused in finite-length medium will not be easily detected in the far-field due to phase cancellation between the source terms during focusing and defocusing (Ward and New, 1969).

Dephasing between pump and plasma harmonics manifests itself as oscillations in the magnitude of the harmonics around the values given in Eq. (4.92)–Eq. (4.94) (Rax and Fisch, 1992; Mori *et al.*, 1993). This could possibly be avoided by imposing a density modulation with period $2l_3$ via an ion acoustic wave or a multilayered medium (Rax and Fisch, 1993). The idea would be to suppress harmonic generation in the lower density regions, using them instead to allow the 3rd harmonic to coast back into phase with the pump, thereby restoring linear growth.

As far as radiation sources go, one of the key points of interest is the asymptotic scaling of the higher order harmonics. This was also considered by Mori *et al.*, (1993), who used a fully nonlinear analysis to show that near the critical density and for large a_0, the mth harmonic should scale as:

$$\frac{P_m}{P_0} \sim m^{-4}. \tag{4.96}$$

They also presented the first simulations of harmonic generation in an underdense plasma using a particle-in-cell (PIC) code specially adapted to follow a 'window' of plasma occupied by the laser pulse. While the simulation results largely agree with theory for the 3rd harmonic, the higher harmonics were orders of magnitude larger than predicted, and tend to form a plateau reminiscent of harmonic generation in gases (L'Huillier and Balcou, 1993). An example of such a cyclic-grid PIC simulation (performed by the present author) with parameters close to those used by Mori *et al.*, (1993) is shown in Fig. 4.10. The 3rd harmonic $E_3/E_0 = (\omega_3 a_3)/(\omega_0 a_0) = 5.6 \times 10^{-5}$ oscillates around the theoretical value given by Eq. (4.92). The oscillation amplitude in Fig. 4.10b differs from the ideal case (zero to twice the steady-state value) because of the finite pulse-shape.

Further simulations by Mori *et al.* showed that the harmonic content could be enhanced by introducing a density ramp, a result which can be partially understood in the light of the density-ripple analysis of Rax and Fisch (1993). Another possibility for enhancement was suggested by Zeng *et al.*, (1996), who found that harmonic amplitudes could be increased by at least an order of magnitude for laser pulse lengths much shorter than a plasma wavelength.

In a realistic focusing geometry, further phase differences are introduced between the pump and the harmonics. This effect has been known for some time in the context of harmonic generation via 3rd order nonlinearities in gases, where it can be shown that *no* harmonic signal emerges from an infinite medium (Ward and New, 1969). Esarey *et al.*, (1993) calculated this effect for 3rd harmonic generation in plasmas, and found that for a semi-infinite medium, the harmonic signal is 1/15 smaller than the 1D

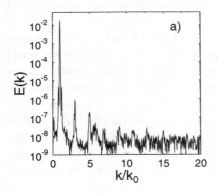
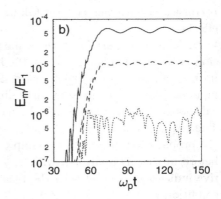

Fig. 4.10 a) Simulated harmonic spectrum generated by a short pulse with amplitude $a_0 = 0.2$ propagating through an underdense plasma with $\omega_0/\omega_p = 5$. b) Time-evolution of 3rd (solid), 5th (dashed) and 7th (dotted) harmonics.

result. In practice, this can be arranged by focusing the pulse onto either the leading or trailing edge of the medium (e.g. gas jet or fill).

Another related mechanism is stimulated backscattered harmonic (SBH) generation (Sprangle and Esarey, 1991), which is basically an extension of the nonlinear Raman backscatter (RBS) instability, see Sec. 4.4.2. Here, the plasma wave acts like a mirror grating, which at high pump strengths superimposes the harmonic content in the nonlinear density perturbation onto the reflected wave. Unlike the forward harmonics, the SBH radiation grows from the foot of the laser pulse towards the rear, up to a point where the plasma wave becomes saturated by particle trapping. In practice, therefore, the effectiveness of this scheme will be very sensitive to the background plasma temperature. For this reason, it may be easier to produce and detect SBH using electron beams instead (Sprangle and Esarey, 1991).

Experimentally there are still precious few data to compare these results with, largely because of the difficulty in excluding the harmonics generated by bound electrons (Sec. 2.5). Even at focused intensities of 10^{18} Wcm^{-2} multiphoton effects can occur near the ionization threshold in the foot, tail and wings of the pulse, which is typically 10^{14}–10^{16} Wcm^{-2} depending on the gas used. In an experiment designed to cover the transition between atomic and ionized media, Liu *et al.*, (1993) determined an upper bound for the efficiency of third harmonic generation in a plasma jet of $P_3/P_0 \sim 10^{-13}$ for an intensity of 10^{16} Wcm^{-2} and plasma density of 10^{17} cm^{-3}, which is consistent with theory after taking into account phase-matching and 3D effects.

A more recent study by the same group observed a *phase matched* version of this mechanism, in which the harmonics are preferentially scattered into a narrow ring in the forward direction (Chen *et al.*, 2000). Phase matching can be arranged in three-dimensional geometry by using tight focal spot containing higher-order Laguerre-Gaussian modes (Siegman, 1986). The requirement $\Delta\phi = k_3 z' - 3k_0 z = 0$ can then be satisfied by allowing the propagation directions of pump (z) and harmonic (z') to *differ* (unlike before in the 1D analysis), with $z = z' \cos\theta$. From the respective refractive indicies associated with the focal modes, one can write the matching condition as (Chen *et al.*, 2000):

$$\cos\theta = \frac{\left(1 - \frac{\lambda_0^2}{\gamma_0 n^2 \lambda_p^2} - \frac{\lambda_0^2}{\pi^2 n^2 r_3^2}\right)^{1/2}}{\left(1 - \frac{\lambda_0^2}{\gamma_0 \lambda_p^2} - \frac{\lambda_0^2}{\pi^2 r_0^2}\right)^{1/2}}, \tag{4.97}$$

where $n = 3$ is the harmonic mode number, r_0 is the spot size, and $r_3 \simeq \Delta\theta Z_R/4$ is the transverse width of the third-harmonic ring. In their experiment, Chen *et al.* found an absolute conversion efficiency for the 3rd harmonic, emitted at a cone angle of 5.6°, of 2×10^{-5} at $I = 1.7 \times 10^{17}$ Wcm^{-2}; 2 orders of magnitude higher than that predicted by Eq. (4.93) for forward-directed light.

4.6 Propagation of Finite-Width Laser Pulses

So far in this chapter the propagation effects of 'real' laser pulses have largely been ignored; the focusing optics, diffraction and refraction due to both density gradients and self-nonlinearity. Instead, the pulse was assumed to act like a rigid photon bullet in order to concentrate on the plasma-wave dynamics. In an *ionizing* plasma, we have seen how transverse variations in the background plasma density can lead to premature defocusing, see Sec. 2.4. In a uniform, fully ionized plasma, one of the first genuinely two-dimensional effects which crops up as we increase the laser intensity is *relativistic self-focusing* (Litvak, 1970; Max *et al.*, 1974; Sprangle *et al.*, 1987). In fact, it turns out that this effect has a *power threshold*

$$P_c \simeq 17 \left(\frac{\omega_0}{\omega_p}\right)^2 \text{ GW}, \tag{4.98}$$

which we will proceed to derive via two different methods shortly. For the typical densities $n_e \sim 10^{17}$–10^{19} cm^{-3} accessible in a field-ionized plasma (Sec. 2.2), this translates into a power requirement of 2–10 TW.

Short pulse Terawatt lasers first became available in the early to mid-1990s, which is why the *experimental* investigation of self-focusing — also known as optical guiding or channeling — has only recently been widely pursued. Intensity amplification apart, the extraordinary interest in this effect is driven by the possibility of diffraction-free propagation in tenuous plasmas, an important prerequisite for benchtop particle acceleration and novel x-ray laser schemes (see Secs. 7.1, 7.2).

4.6.1 *Geometric optics picture of self-focusing*

A rough estimate of the threshold power needed to make a light beam self-focus due to relativistic effects can be obtained from geometrical arguments. This picture is over-simplified, but demonstrates the essence of this nonlinearity in a physically intuitive way[2]. Consider a laser beam with a radial profile $a(r) = a_0 \exp(-r^2/2\sigma_0^2)$ which has been focused to a spot size σ_0 just inside a region of uniform, underdense plasma, see Fig. 4.11.

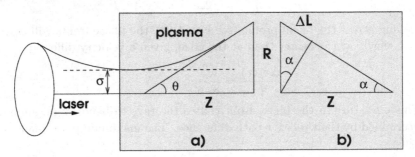

Fig. 4.11   Geometrical view of a) diffraction and b) self-focusing of a Gaussian laser beam.

In the absence of nonlinear effects, the beam will diffract with a divergence angle (Siegman, 1986)

$$\theta_d = \frac{dR}{dZ} = \frac{\sigma_0}{Z_R} = \frac{1}{k\sigma_0}, \tag{4.99}$$

where Z_R is the Rayleigh length (half the confocal parameter), defined here

---

[2]I am indebted to P. Mora for this derivation.

by:

$$Z_R = \frac{2\pi\sigma_0^2}{\lambda}. \tag{4.100}$$

At high intensities, we saw in the earlier discussion on instabilities that the dispersion relation of the wave is altered due to the effective relativistic mass increase of the electrons, and is given by Eq. (4.73). The corresponding refractive index is:

$$\eta(r) \equiv \frac{ck}{\omega} = \left\{ 1 - \frac{\omega_p^2}{\omega^2 \left[1 + a(r)^2/2 \right]^{1/2}} \right\}^{1/2}. \tag{4.101}$$

For a beam with a profile $a(r)$ as above, $\eta(r)$ is *peaked* on axis $(d\eta/dr < 0)$, which in optics terminology represents a 'positive', or focusing, lens. This should be contrasted with the divergent refractive index $(d\eta/dr > 0)$ created by field ionization discussed in, see Sec. 2.4. The phase velocity of the wave fronts passing through the focusing medium described by Eq. (4.101) can be approximated thus:

$$\frac{v_p(r)}{c} = \frac{1}{\eta} \simeq 1 + \frac{\omega_p^2}{2\omega^2} \left(1 - \frac{a^2(r)}{4} \right). \tag{4.102}$$

Looking across the beam profile (see Fig. 4.12), the phase fronts will travel more slowly at the center than at the edge, given a velocity difference:

$$\frac{\Delta v_p(r)}{c} = \frac{\omega_p^2}{8\omega^2} a_0^2 e^{-r^2/\sigma_0^2}.$$

This curvature in the phase front causes the rays to bend by an amount determined by their relative path difference. The maximum path difference is just

$$\Delta L = \mid \Delta v_p \mid_{\max} t = \left| \frac{\Delta v_p}{c} \right|_{\max} Z = \alpha R.$$

Thus the maximum focusing angle of the beam is given by

$$\alpha^2 = \frac{\omega_p^2 a_0^2}{8\omega^2}. \tag{4.103}$$

Beam spreading due to diffraction will therefore be canceled by self-focusing effects if $\theta = \alpha$, or setting Eq. (4.99) and Eq. (4.103) equal,

$$a_0^2 \left(\frac{\omega_p \sigma_0}{c} \right)^2 \geq 8. \tag{4.104}$$

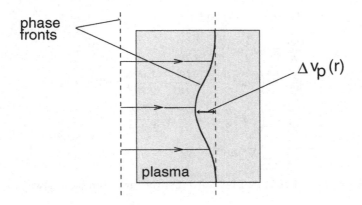

Fig. 4.12 Phase front bending due to mass drag effects.

This represents a power threshold, since the laser power $P_L \propto a_0^2 \sigma_0^2$.

4.6.2 *The nonlinear Schrödinger equation*

Back in Sec. 4.3.3, we derived the fully nonlinear wakefield equations in a window moving with the laser pulse. Recall that to justify the quasistatic approximation for the wakefield, it was argued that the EM pulse envelope had to vary slowly compared to the laser (or plasma) frequency. In what follows this envelope model will be developed more rigorously. We start from the normalized wave equation used as the basis for the stability analysis in Sec. 4.4, here denoting A as the dimensionless vector potential:

$$\frac{\partial^2 A}{\partial t^2} - c^2 \nabla^2 A = -\omega_p^2 \frac{nA}{\gamma}. \qquad (4.105)$$

Since we are interested in global beam behavior, it makes sense here to separate the rapid variations in the laser field phase, $\psi = \omega t - kz$, from its amplitude $a(r, z, t)$, by applying the slowly-varying envelope approximation (see Appendix B):

$$A = \frac{1}{2}\left(ae^{i\psi} + a^*e^{-i\psi}\right). \qquad (4.106)$$

The time- and space-derivatives are easily obtained by observing that

$\partial\psi/\partial t = i\omega$ and $\partial\psi/\partial z = -ik$, so that:

$$\frac{\partial A}{\partial t} = i\omega\frac{1}{2}ae^{i\psi} + \frac{1}{2}\frac{\partial a}{\partial t}e^{i\psi} + c.c.$$

$$\frac{\partial^2 A}{\partial t^2} = \left(-\omega^2\frac{a}{2} + i\omega\frac{\partial a}{\partial t} + \frac{1}{2}\frac{\partial^2 a}{\partial t^2}\right)e^{i\psi} + c.c.$$

$$\frac{\partial^2 A}{\partial z^2} = \left(-k^2\frac{a}{2} - ik\frac{\partial a}{\partial z} + \frac{1}{2}\frac{\partial^2 a}{\partial z^2}\right)e^{i\psi} + c.c.$$

$$\nabla_\perp^2 A = \frac{1}{2}\nabla_\perp^2 ae^{i\psi} + c.c.$$

The RHS of Eq. (4.105) can be conveniently rewritten as follows:

$$\omega_p^2\frac{nA}{\gamma} = \omega_p^2\frac{na}{2\gamma}e^{i\psi} + c.c.$$

$$= \frac{1}{2}\left[\omega_p^2(n-\gamma)\frac{a}{\gamma} + \omega_p^2 a\right]e^{i\psi} + c.c. \qquad (4.107)$$

Substituting the above expressions into the wave equation, neglecting 2nd derivatives of the (slowly varying) envelope amplitude a, and applying the linear dispersion relation $\omega^2 = \omega_p^2 + c^2k^2$ to cancel three other terms, we finally get:

$$i\omega\frac{\partial a}{\partial t} + ic^2k\frac{\partial a}{\partial z} - \frac{c^2}{2}\nabla_\perp^2 a - \frac{\omega_p^2}{2}(1 - \frac{n}{\gamma})a = 0. \qquad (4.108)$$

Transforming time- and space-variables to a window moving with the group velocity of the pulse, $v_g = c^2k/\omega$ (i.e. letting $\tau = t, \xi = z - v_g t$), simplifies the envelope equation to:

$$i\omega\frac{\partial a}{\partial\tau} = c^2\nabla_\perp^2\frac{a}{2} + \omega_p^2(1 - \frac{n}{\gamma})\frac{a}{2}.$$

This can be tidied up a little further by using the normalizations:

$$\tilde{\tau} = \frac{\omega_p^2\tau}{\omega} \quad \text{and} \quad \tilde{r} = k_p r,$$

for the time and space variables respectively. The envelope (or paraxial) equation in its commonly quoted dimensionless form (dropping the tildes) then becomes:

$$\frac{\partial a}{\partial\tau} + \frac{i}{2}\nabla_\perp^2 a + \frac{i}{2}\left(1 - \frac{n}{\gamma}\right)a = 0. \qquad (4.109)$$

Eq. (4.109) describes the evolution of the electromagnetic beam *envelope* $a(r, \xi, \tau)$ as the pulse progresses through the plasma. To derive a focusing threshold, we assume that the plasma density is homogeneous, and for the time-being stays that way, so that we can set $n = 1$. For a weakly nonlinear, circularly polarized pump, we can expand the relativistic factor (see page 88):

$$\gamma^{-1} \simeq 1 - \frac{|a|^2}{4}.$$

For linear polarization, the analysis is slightly more complex because we have to include the 2nd harmonic density response (Max *et al.*, 1974; Mori *et al.*, 1988; McKinstrie and Russell, 1988), which also leads to a nonlinear term $O(|a|^3)$. However, the overall result is qualitatively the same, so I will stick to circular light here for simplicity. The resulting equation is commonly referred to as the *nonlinear Schrödinger equation* (NLSE):

$$\frac{\partial a}{\partial t} = -\frac{i}{2}\nabla_\perp^2 a - \frac{i}{8}|a|^2 a. \qquad (4.110)$$

This equation, which has been studied exhaustively in the literature (see, for instance (Decoster, 1978) or (Shukla *et al.*, 1986)) exhibits a number of conservation properties, which can be derived directly by taking spatial moments. For example, multiplying both sides by a^* and adding the complex conjugate of the result $(a \times \partial/\partial a^*)$ yields:

$$\frac{\partial}{\partial t}|a|^2 = \frac{i}{2}a\nabla_\perp^2 a^* - \frac{i}{2}a^*\nabla_\perp^2 a.$$

Integrating of the focal area of the pulse in cylindrical geometry, we find:

$$\int_0^\infty \frac{\partial}{\partial t}|a|^2\, d^2r = -2\pi\frac{i}{2}\int (a^*\nabla_\perp^2 a - a\nabla_\perp^2 a^*)r\,dr$$

$$= -\pi i\int\left[a^*\left(\frac{\partial^2}{\partial r^2} + \frac{1}{r}\frac{\partial}{\partial r}\right)ar - c.c.\right]dr$$

$$= \pi i\int\left(r\left|\frac{\partial a}{\partial r}\right|^2 - c.c.\right)dr = 0.$$

We identify

$$P \equiv \int |a|^2\, d^2r \qquad (4.111)$$

variously as the wave action, photon number, or normalized beam power; which according to the proof above, is *conserved*. Many other conservation relations can be similarly derived, but the one of most interest here concerns the *Hamiltonian*,

$$\mathfrak{H} \equiv \frac{1}{2} \mid \nabla_\perp a \mid^2 - \frac{1}{16} \mid a \mid^4, \qquad (4.112)$$

which is also globally conserved, namely:

$$\frac{\partial}{\partial t} \int_0^\infty \mathfrak{H} d^2 r \equiv \frac{\partial H}{\partial t} = 0,$$

just as for the wave action. (See McKinstrie and Russell (1988) or Vidal and Johnston (1996) for a discussion of these identities and their correspondence with classical field theory). To determine whether a beam will focus, we need to take the r^2 moment of Eq. (4.110). In other words, we determine the 'acceleration' of the RMS beam width

$$< \delta r^2 > \equiv < r^2 > - < r >^2 .$$

If the beam does not deviate from the axis (i.e. *hosing* does not occur, see, for example, Shvets and Wurtele (1994)) then we can put $\partial < r > / \partial t = 0$, so that

$$\frac{\partial^2}{\partial t^2} < \delta r^2 > = \frac{\partial^2}{\partial t^2} < r^2 >$$

$$= \frac{1}{P} \int_0^\infty r^2 \frac{\partial \mid a \mid^2}{\partial t^2} d^2 r$$

$$= \frac{4H}{P}. \qquad (4.113)$$

Integrating twice then gives:

$$< r^2 > = \frac{2H}{P} t^2 + Ct + D, \qquad (4.114)$$

where C and D are constants which depend on the focusing optics and the initial spot size respectively. Now consider the situation where the wavefronts are initially parallel ($C = 0$), which will be approximately the case for a Gaussian beam focused to its diffraction limit just inside a region of plasma, Fig. 4.13. Before coming to the relativistic case, it is helpful to make a connection with classical Gaussian optics. For non-relativistic pump strengths at the focus $a \ll 1$, we can set $\mathfrak{H} = \frac{1}{2} \mid \nabla a \mid^2$ in Eq. (4.114), so that for a Gaussian beam profile $a(r) = a_0 \exp(-r^2/2\sigma_0^2)$, where σ_0 is the nominal focal spot size in vacuum; the radius at which the intensity is

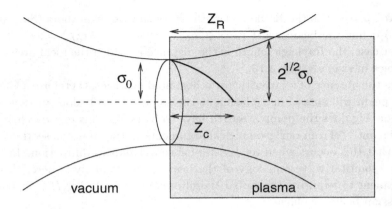

Fig. 4.13   Geometry for relativistic whole-beam focusing.

$1/e$ times its peak value. We then find for the normalized beam power and global Hamiltonian:

$$P = \pi a_0^2 \sigma_0^2; \quad H = \frac{\pi a_0^2}{2}, \qquad (4.115)$$

so that

$$
\begin{aligned}
< r^2 > (t = 0) &= P^{-1} \int r^2 \mid a \mid^2 d^2 r \\
&= \frac{2\pi}{P} \int r^3 a_0^2 e^{-r^2/\sigma_0^2} dr \\
&= \sigma_0^2.
\end{aligned}
$$

Thus our constant D in Eq. (4.114) is just σ_0^2, and the expression for the beam width simplifies to

$$< r^2 > = \sigma_0^2 + \frac{t^2}{\sigma_0^2}. \qquad (4.116)$$

Remembering that our time and space variable were normalized to ω_0/ω_p^2 and ω_p/c respectively, we substitute back $t \to \omega_p^2/\omega_0 t; r \to \omega_p r/c$ and write $z = ct$ to recover the usual the usual expression for a beam diffracting in vacuum:

$$\sigma(z) \equiv \sqrt{< r(z)^2 >} = \sigma_0 \left(1 + \frac{z^2}{z_R^2} \right)^{1/2}, \qquad (4.117)$$

where $z_R = k\sigma_0^2$ is the Rayleigh length. Note that some authors (Siegman, 1986) define the beam waist $r_0 = \sqrt{2}\sigma_0$, so that $z_R = kr_0^2/2 = \pi r_0^2/\lambda$. In either case, the Rayleigh length is the distance at which the focal area $\pi\sigma^2$ doubles in size, see Fig. 4.13.

In the absence of relativistic (or other) nonlinearities, therefore, a Gaussian beam will spread in a plasma at the same rate as it does in vacuum. So how big does the pump strength have to be before this picture changes significantly? From our geometrical arguments in the previous section, we saw that this occurs when relativistic focusing balances diffraction. In the more quantitative terminology of the beam-moment analysis here, this is equivalent to requiring the global Hamiltonian to vanish; i.e. $H = 0$. For a Gaussian beam, we have

$$H = \frac{\pi a_0^2}{2}\left(1 - \frac{a_0^2\sigma_0^2}{16}\right) = 0,$$

which is satisfied for $a_0^2\sigma_0^2 = 16$. As before, this represents a power threshold, and in terms of the definitions given by Eq. (4.115), we can express this critical power in normalized units as

$$P_c = 16\pi. \tag{4.118}$$

To express this in a more practical form, we recall the expression for the beam power P_L in watts:

$$P_L = \int_S I d^2r$$
$$= \pi c \epsilon_0 \int_0^\infty E^2(r) r dr.$$

The electric field E is related to the pump strength $a(r)$ by

$$E(r) = \frac{m\omega c}{e}a(r),$$

so we can write P_L in terms of the normalized power

$$\tilde{P} = \int_0^\infty a^2(r)d^2r,$$

thus:

$$P_L = \left(\frac{m\omega c}{e}\right)^2 \left(\frac{c}{\omega_p}\right)^2 \frac{c\epsilon_0}{2} \int_0^\infty 2\pi r a^2(r) dr$$

$$= \frac{1}{2}\left(\frac{m}{e}\right)^2 c^5\epsilon_0 \left(\frac{\omega}{\omega_p}\right)^2 \tilde{P}$$

$$\simeq 0.35 \left(\frac{\omega}{\omega_p}\right)^2 \tilde{P} \text{ GW}.$$

The critical power $\tilde{P}_c = 16\pi$ thus corresponds to a physical threshold:

$$P_c \simeq 17.5 \left(\frac{\omega}{\omega_p}\right)^2 \text{ GW}, \qquad (4.119)$$

which is the usual value quoted in the literature (Litvak, 1970; Max *et al.*, 1974; Sprangle *et al.*, 1988).

The beam radius defined by Eq. (4.114) can also be written

$$< r^2 > = \sigma_0^2 + \frac{t^2}{\sigma_0^2}\left(1 - \frac{P}{P_c}\right)$$

$$= \sigma_0^2\left[1 + \frac{z^2}{z_R^2}\left(1 - \frac{P}{P_c}\right)\right]. \qquad (4.120)$$

This expression points to three distinct regimes depending on the ratio P/P_c. For $P/P_c < 1$, the beam spread angle (see Fig. 4.11) $\theta \equiv < r^2 >^{1/2}/z$ is reduced by an amount $(1 - P/P_c)^{1/2}$ (Shen, 1984) — an effect which can in fact be detected experimentally by looking at the radial beam profile in the 'far-field' after it has passed through the plasma (Monot *et al.*, 1995a). For $P = P_c$, the beam should, in the absence of other losses or instabilities, propagate indefinitely with constant radius: a dream scenario, of course, for designers of plasma-based particle accelerators or ultrafast x-ray laser schemes, see Chapter 7. Finally, for $P > P_c$, Eq. (4.120) evidently predicts that the beam will collapse to zero radius in a distance:

$$z_c = \frac{z_R}{(P/P_c - 1)^{1/2}}. \qquad (4.121)$$

This is actually borne out by numerical solutions of the NLSE (4.110) containing a simple cubic nonlinearity (McKinstrie and Russell, 1988; Castillo-Herrera and Johnston, 1993). We will come these numerical studies of relativistic beam propagation shortly, but before we do that, we need to consider the effect that the transverse beam dynamics can have on the propagation medium, which we have assumed to remain unchanged up to now.

4.6.3 *Ponderomotive channel formation*

If the pulse length is long compared to the plasma period, $\omega_p \tau \gg 1$, we can still consider the electron fluid response to be adiabatic, or 'frozen' on the timescale of the EM envelope. The 1D equations for the plasma response to a short laser pulse derived in Sec. 4.3 can in fact be generalized for multi-dimensional wakes (Sprangle *et al.*, 1992), as we will see later when we discuss magnetic field generation. For our present purposes, it suffices to write down Eq. (4.49) in vector form to include both longitudinal and transverse plasma velocity components:

$$\frac{d}{dt}(\gamma \boldsymbol{v}) = c\nabla\phi - \frac{c}{2\gamma}\nabla a^2, \qquad (4.122)$$

where we now assume that v varies much more slowly than the laser field $a\cos\omega_0 t$. In the adiabatic limit, we let $\boldsymbol{v} \to 0$, and take the divergence of Eq. (4.122), to get:

$$\nabla^2\phi = \frac{1}{2\gamma}\nabla^2 a^2 = \nabla^2\gamma.$$

The relativistic factor is then just a function of the pulse amplitude: $\gamma = (1 + a^2)^{1/2}$. In this context, we are interested in the transverse response of the plasma. If the pulse is cigar-shaped, we can take $\nabla = \nabla_\perp$, and apply Poisson's equation (ϕ and n normalized as before)

$$\nabla_\perp^2\phi = k_p^2(n - 1),$$

to obtain an appealingly simple expression for the density perturbation:

$$n = 1 + k_p^{-2}\nabla_\perp^2\gamma. \qquad (4.123)$$

The 2nd term in Eq. (4.123) arises from the *transverse* ponderomotive force exerted by the laser pulse, and leads to in a reduction of the electron density along the propagation axis, see Fig. 4.14. An important prediction of Eq. (4.123) is that if the laser pulse is sufficiently intense and tightly focused — either by the optics or as a result of self-focusing — then complete expulsion of electrons from the beam center will occur. As an example, consider the Gaussian pulse profile $a(r) = a_0\exp(-r^2/2\sigma^2)$ we had previously. After time-averaging over the laser period, the density depression

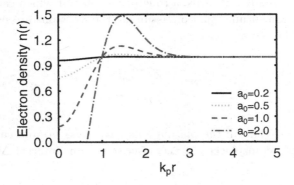

Fig. 4.14   Cavitation: ponderomotive expulsion of electrons from the beam axis for a normalized spot size $k_p\sigma = 1$ and four different pump strengths a_0.

term in cylindrical coordinates can be written:

$$\nabla_\perp^2 \gamma = \frac{1}{4\gamma} \nabla_\perp^2 a^2 = \frac{1}{4\gamma} \frac{4a_0^2}{\sigma^2} \left(\frac{r^2}{\sigma^2} - 1 \right) \exp(-r^2/\sigma^2).$$

The deepest depression is on the laser axis at $r = 0$, so we obtain a 'cavitation' condition, $n = 0$, for:

$$\frac{a_0^2}{k_p^2 \sigma_0^2} > 1,$$

or, in more practical units:

$$I_{18} \lambda_\mu^2 > \frac{1}{20} n_{18} \sigma_\mu^2, \qquad (4.124)$$

where σ_μ and λ_μ are the spot size and laser wavelength in μm; I_{18} and n_{18} are the laser intensity and plasma density in units of 10^{18} Wcm^{-2} and 10^{18} cm^{-3} respectively. This condition is actually quite easy to fulfill with TW lasers; easier in fact, than the matching condition $\omega_p \tau \sim 1$ required for wakefield generation. Indeed, the first experiments on propagation of high-intensity, femtosecond pulses could more or less ignore the plasma wave generation, and take the electron response to be transverse and adiabatic.

Expulsion of electrons from the beam axis will be sustained for the duration of the pulse provided the ions remain immobile. One has to remember, however, that pushing the electrons out will leave a huge, unneutralized space-charge over an area $\sim \pi \sigma_0^2$. The resulting *Coulomb explosion* will drive the ions radially outwards, a rough timescale for which can be found from the ion momentum equation assuming the electrons are ponderomotively force-balanced as above:

$$m_i \frac{\partial v_i}{\partial t} = Ze\nabla_\perp \phi$$
$$= Zm_e c^2 \nabla_\perp \gamma. \qquad (4.125)$$

Here, we assume that all the plasma ions have mass m_i and carry charge $q_i = +Ze$. The ion displacement Δx_i after some time interval Δt_i is therefore:

$$\Delta x_i = \frac{Zm_e}{m_i} \Delta t_i^2 c^2 \nabla_\perp \gamma. \qquad (4.126)$$

Depending on the laser intensity, the ponderomotive force term can be simplified to:

$$\nabla_\perp \gamma \simeq \begin{cases} \dfrac{a_0^2}{4r_c}, & a_0 \ll 1 \\[2ex] \dfrac{a_0}{\sqrt{2}r_c}, & a_0 \gg 1, \end{cases}$$

where r_c is the channel radius. This may of course be smaller than the vacuum spot size σ_0 (with a corresponding enhancement of a_0) if self-focusing has occured. In the high-intensity limit, we suppose that the ions undergo acceleration until $\Delta x_i \sim r_c$, which will take a time:

$$\Delta t_i \simeq \left(\frac{m_i}{Zm_e} \right)^{1/2} \frac{2^{1/4} r_c}{c a_0^{1/2}}. \qquad (4.127)$$

For a Ti:sapphire laser focused to an intensity $I = 10^{19}$ Wcm^{-2} over a spot size $\sigma_0 = 3$ μm$\simeq r_c$ into a hydrogen gas jet ($Z = 1, m_i/m_e = 1836$), this gives an acceleration time $\Delta t_i \simeq 430$ fs. A sub-100 fs pulse will therefore not experience much ion channeling in this case. However, thanks to their much higher inertia, the ions will continue to coast outwards after the pulse has passed through, leaving a deeper channel which can be detected and/or exploited by a 2nd probe beam, see Sec. 4.8.

The maximum velocity acquired by the ions will be roughly $v_i \simeq$

$\Delta x_i / \Delta t_i$, giving an kinetic energy gain of:

$$\frac{1}{2} m_i v_i^2 \sim \frac{1}{2} m_i \left(\frac{Z m_e}{m_i} \right)^2 \frac{c^4 a_0^2}{2 r_c^2} \Delta t_i^2$$

$$= Z m_e c^2 \frac{a_0}{2\sqrt{2}}, \tag{4.128}$$

which is roughly the energy gained by a charge $+Ze$ sliding down the ponderomotive potential. (A more rigorous calculation gives $U_{\text{pond}} = Z m_e c^2 (\gamma - 1)$.) For $a_0 = 4$, this is already in the MeV range, providing a readily detectable signature of channel formation.

4.7 Dynamics of Relativistic Self-Focusing

The moment equations in Sec. 4.6.2 are rather useful in predicting whether an intense pulse will focus or not, but they can only be used up to a certain point. For example, we saw earlier that Eq. (4.120) predicts that for $P \gg P_c$, the beam will collapse to zero width in a distance $z \sim z_R (P/P_c)^{-1/2}$, which since power is conserved, implies that the intensity will become infinite. In reality, the physics rescues us from this uncomfortable predicament: either additional defocusing forces come into play (consider, for instance the $a_0 \gg 1$ or $n \to 0$ limits of Eq. (4.109), or the paraxial approximation breaks down for some reason. The latter will occur, for example, if the spot size becomes smaller than the laser wavelength (i.e. $k_0 \sigma < 1$), or if parametric instabilities develop to such an extent that strong electron heating occurs. These scenarios eventually render the NLSE invalid, and one has to either allow the wave vectors a finite transverse component (Feit and Fleck Jr., 1988) or resort to a fully kinetic treatment including the full set of Maxwell equations (Mori *et al.*, 1988).

4.7.1 *Numerical solutions of the wave-envelope equation*

This limitation notwithstanding, it is still possible to gain valuable insight into the dynamical behavior of self-focusing by numerical integration of the NLSE Eq. (4.109) coupled with adiabatic electron response Eq. (4.123) (Borisov *et al.*, 1990; Gibbon *et al.*, 1995). The technical details of how this is done are deferred as usual to Chapter 6: we consider a few examples here to illustrate some of the typical signatures of self-channeling. The first and most obvious point to consider is to check the predictions of the moment equation (4.113) for the beam radius. Remember that this theorem can only predict *global* beam behavior up to the first collapse: effects such as filamentation (transverse beam break-up) or strong cavitation are simply not catered for. To verify the virial theorem implied by Eq. (4.109) therefore, we set up an initial radial Gaussian beam profile with $\sigma_0 = 7.5 \ \mu$m and pump strengths a_0 chosen such that the beam powers P/P_c in each case are 0.5, 1.0 and 3.0 respectively. The resulting beam behavior in Fig. 4.15 shows how the rms width changes as expected according to the predictions. The solution at the highest power is self-trapped because of cavitation, see Fig. 4.15 b): the beam first self-focuses to a diameter $O(5 \ c/\omega_p)$, whereupon the enhanced ponderomotive force creates a density channel deep enough to prevent the pulse from diffracting out again.

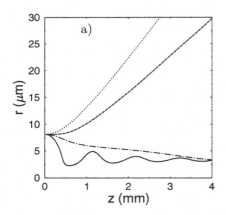 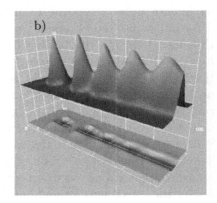

Fig. 4.15   Numerical simulation of self-focusing using the paraxial wave equation: a) beam radii for powers $P \ll 1$ (vacuum solution, dotted), $P = 0.5$ (dashed), 1.0 (dash-dotted) and 3.0 (solid) times the critical power P_c respectively; b) intensity (top) and electron density (bottom) surfaces for the $P/P_c = 3$ case.

4.7.2   *Guiding of short pulses*

So far in our treatment of nonlinear propagation effects, we have ignored plasma wave excitation completely, concentrating instead on the asymptotic behavior of a long pulse with $\omega_p \tau \gg 1$. In this case (see Fig. 4.12), we have seen that self-focusing occurs when the refractive index has a maximum on axis, or when $\partial \eta / \partial r < 0$. When a plasma wakefield is generated, however, the refractive index of the light wave can be modified not only by the electron quiver motion, but also by the longitudinal bunching of the plasma density. We can include this effect in the optical picture by generalizing the refractive index described by Eq. (4.101) thus:

$$\eta_r \simeq 1 - \frac{\omega_p^2}{2\omega^2} \left(1 - \frac{a_0^2}{4} + \frac{n}{n_0} \right), \qquad (4.129)$$

where n is the perturbed plasma wakefield density given by Eq. (4.38) for moderate pump strengths. In the short pulse limit, $\xi_L \ll \pi/k_p$, we may drop the second term in this equation and approximate

$$\frac{n}{n_0} \sim \frac{<a^2>}{2} = \frac{a_0^2}{4},$$

which cancels the relativistic term in Eq. (4.129)! Actually, the cancelation is not exact: a more precise expression for η_r may be determined by using some of the identities within the quasistatic approximation — Eqs. (4.57,4.58) — which we came across earlier, namely

$$\frac{n_e}{\gamma} = \frac{n_0}{1+\phi},$$

where ϕ is the wake potential. Thus, the refractive index may be written:

$$\eta_r = 1 - \frac{\omega_p^2}{2\omega^2} \frac{n_e}{n_0 \gamma} = 1 - \frac{\omega_p^2}{2\omega^2} \frac{1}{1+\phi}.$$

The solution for ϕ can be found from Eq. (4.42) in both long and short pulse limits. For long pulses, we have $\phi \simeq a^2/2$ and we recover the previous result in Eq. (4.129). For short pulses though, we have $\phi \simeq k_p^2 a_0^2 \xi^2 / 8$ and

$$\eta_r = 1 - \frac{\omega_p^2}{2\omega^2} \left(1 - \frac{k_p^2 a_0^2 \xi^2}{8} \right).$$

Physically this means that the critical power requirement for focusing depends on the position along the length of the pulse, and is increased overall by a factor $k_p^2 \xi_L^2 / 4 = \pi^2 \xi_L^2 / \lambda_p^2$.

This important result, that pulses shorter than a plasma wavelength cannot be relativistically guided, was first pointed out by Sprangle *et al.*, (1990b). Since then, there have been numerous studies on wakefield generation based on numerical solutions of the two-dimensional fluid equations

(Sprangle *et al.*, 1992; Andreev *et al.*, 1992; Antonsen Jr. and Mora, 1993) to specifically investigate the influence of the plasma wakefield on the propagation behavior of the pulse. Technically speaking, these models are a step up in complexity from the adiabatic approach described in Sec. 4.7.1, combining the fully space-time-dependent wave-envelope equation (4.109) with a *dynamic* equation for the plasma wakefield, such as Eq. (4.62).

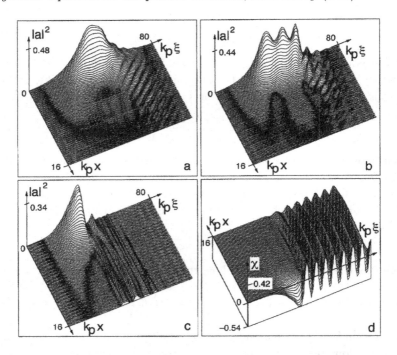

Fig. 4.16 Beam break-up by self-modulational instability, simulated using a two-dimensional fluid model: a), b) and c) snapshots of laser intensity at different times, d) refractive index corresponding to c). *Courtesy:* P. Mora, École Polytechnique, Palaiseau; Fig. 1 from Antonsen Jr. and Mora (1992). ©1992, American Physical Society.

As well as confirming the above prediction concerning guiding, this approach explicitly demonstrates the full implications of the self-modulational instability for high-intensity pulse propagation. This causes breakup of the pulse into beamlets of length $\lambda_p/2$, accompanied by amplification of the plasma wakefield (Sprangle *et al.*, 1992; Andreev *et al.*, 1992; Abramyan *et al.*, 1992), see Fig. 4.16. Further analyses by various authors (Esarey *et al.*, 1994; Antonsen Jr. and Mora, 1992), showed that this effect could be exploited to relax the conditions needed for guided wakefield generation of $\omega_p \tau_L \sim \pi$ and $P > P_c$, which if simultaneously satisfied, translate to a power requirement:

$$P_L \simeq 17 \left(\frac{\omega \tau_L}{\pi}\right)^2 \text{ GW}$$

$$= 60\lambda_\mu^{-2} \left(\frac{\tau_L}{100\text{fs}}\right)^2 \text{ TW}, \tag{4.130}$$

currently accessible only at Petawatt laser facilities. The self-modulated wakefield generation relies on the instability to break a longer pulse up into smaller resonant pump beamlets, so that higher densities can be used, and hence lower laser powers.

This all sounds very promising until one realizes that the parameter regime of interest is precisely where one expects significant forward Raman scattering to occur. Indeed, it is well known that FRS has a sister instability RMI (relativistic modulational instability), which produces exactly the same effect as the wakefield modulation, but in 1D. Thus, the argument boils down to a competition between longitudinal and radial transport of pump photons (Decker *et al.*, 1994). An excellent clarifying discussion of the whole issue of beam breakup can be found in the review by Mori (1997).

4.7.3 *Particle-in-cell simulations*

Although the fluid models described in the previous section can efficiently tackle most aspects of electromagnetic wave propagation and accompanying effects such as wakefield excitation, they are completely unable to cope when things become kinetic (e.g. if wavebreaking or particle-trapping occurs). This is a serious deficiency considering that one of the primary aims of this research is to trap and accelerate particles to extreme (>GeV) energies. As we have seen in Sec. 4.4, scattering instabilities tend to generate large amplitude plasma waves, which in turn will either break, releasing thermal energy; or pick up and accelerate thermal electrons to relativistic energies. Such a scenario can only be handled self-consistently with a kinetic model such as a Vlasov or Particle-in-Cell (PIC) code, see Chapter 6. On the other hand, a 2D electromagnetic PIC code includes all of the physics which we have encountered so far in this chapter: the inexperienced code-user can quickly become swamped in graphical diagnostics, most of which is usually of little relevance to the particular problem at hand. Despite this tendency to overkill, PIC simulation has become a vitally important tool in this field, and is frequently used to perform numerical experiments ahead of actual (more expensive) laboratory investigations using real lasers and targets.

Some of the first self-consistent 2D simulations of high-intensity electromagnetic wave propagation in this context were performed during a Los Alamos–UCLA collaboration by Forslund *et al.*, (1985) with the PIC code WAVE. These simulations, while confirming many of the predictions of fluid theory on whole-beam self-focusing and beat-wave generation (see Sec. 4.3.4), also revealed a wealth of new effects related to plasma wave excitation and fast particle generation. For example, Raman scattering (backward, forward and side) and its consequent electron heating was shown to be more-or-less inevitable in the parameter regime where one would hope to observe relativistic whole-beam focusing — a feature which seems to have been borne out by recent experiments. Forslund *et al.* also pointed out the presence of large stationary (DC) magnetic fields generated by heated electrons comoving along the laser beam axis; an effect which is increasingly pronounced as the intensity is increased, as demonstrated in a more recent study by Pukhov and Meyer-ter-Vehn (1996).

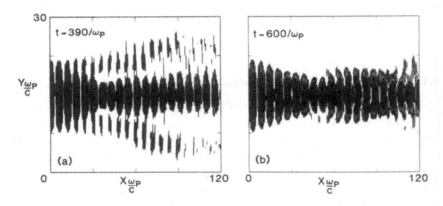

Fig. 4.17   Two-dimensional PIC simulation of relativistic self-focusing after Mori *et al.* (1988). At the earlier time ($390/\omega_p$), only the central portion of the pulse is focused: the wings are scattered away from the axis. At $600/\omega_p$, after the ions have been pushed away from the center to form a channel, nearly all the light is focused, subsequently exhibiting radial oscillations. *Courtesy:* W. B. Mori, UCLA; Fig. 4 from Mori *et al.*, (1988). ©1988, American Physical Society.

The first major study of relativistic self-focusing using the Los Alamos WAVE code was made by Mori *et al.*, (1988). This and subsequent work revealed a number of departures with the predictions of fluid models. Recall that the fluid model in cylindrical geometry predicts a collapse of the beam accompanied by complete ponderomotive explusion of electrons from the beam axis after the first focus — Fig. 4.15. In a PIC simulation using cartesian slab geometry, this behavior may be less pronounced for various reasons: first, because the focusing power is weaker and the beam collapse is saturated sooner; second, because of Raman scattering out of the focal cone — a point to be examined shortly. On the other hand, a channel *can* eventually form if the ions are allowed to move, so that qualitatively similar behavior to the (fixed-ion) fluid model is obtained after a sufficiently long time, see Fig. 4.17.

Since these early numerical studies there has been a relentless increase in available computing power, with the result that three-dimenstional PIC simulations can now be performed with laser and plasma parameters close to those actually used in experiments. Notable contributions to this topic have been made by the UCLA and MPQ-Garching groups — albeit with differing emphasis. For example, 3D simulations by Pukhov and Meyer-ter-Vehn (1996) confirm a long-standing prediction of fluid theory (Sec. 4.6.2): that self-focusing is much stronger in 3D than in 2D slab geometry — see Exercise 7 and the analysis by Chen and Sudan (1993a). Moreover, these authors stress the potential importance of the copropagating electron beam formed at the front end of the laser pulse, which contributes to the creation of a light channel by the combination of magnetic pinching and relativistic modification of the local refractive index. By contrast, Tzeng and Mori (1996; 1998) show that electron cavitation (channel formation) tends to be suppressed by a combination of strong plasma heating and Raman scattering of light out of the focal cone *before* it can be focused.

4.8 Experiments on Relativistic Laser Propagation

Before femtosecond lasers arrived, the field of relativistic wave propagation in plasmas provided a boundless arena for the speculative and untestable ideas described in this chapter. At the beginning of the 1990s, laser intensities were approaching the watershed mark of 10^{18} Wcm^{-2}, which for 1 μm wavelength light meant that electron quiver motion was at last well and truly relativistic.

Considering the vast amount of literature devoted to self-focusing, the first burning question for many was simply: did relativistic focusing in underdense plasmas really take place, and if so, how could one demonstrate the fact? Obviously the first requirement is to satisfy the power threshold given by Eq. (4.98), which for typical densities generated in gaseous media (10^{18}-10^{19} cm^{-3}) implies laser powers in the Terawatt range.

4.8.1 *Homogeneous plasmas*

The first serious attempt to observe relativistic channeling was made at the University of Chicago by an international team led by C. K. Rhodes (Borisov *et al.*, 1992a). These experiments were performed with a 500 fs KrF* laser ($\lambda = 248$ nm), delivering a beam with peak power of 0.3 TW, focused to 8×10^{17} Wcm^{-2} in a chamber filled with various gases. For these laser parameters, the electron density needed to trigger self-focusing follows from Eq. (4.98), and comes to $n_e \simeq 10^{21}$ cm^{-3}. For this reason, the team was forced to consider heavier gases than hydrogen, which, for a given fill pressure, yield higher electron densities thanks to multiple field ionization. The disadvantage of such media (e.g. N_2) is that sharp density gradients tend to be produced wherever the laser intensity crosses the threshold for a particular ionization stage. As we saw in Sec. 2.4, this can lead to strong *de*focusing of the laser beam, often preventing it reaching its maximum focus and clamping the density to lower values.

These misgivings aside, images of light diffracted at an angle to the primary propagation axis appeared to show propagation features qualitatively consistent with relativistic beam dynamics. Further experiments by the Chicago group claimed to demonstrate self-channeling over 100 Rayleigh-lengths, or 3 mm for the wavelength used. Critics of these experiments pointed out that the resolution of the imaging system was no better than 40 μm, so that focusing down to the expected equilibrium beam diameter of 1–2 μm could not have been verified directly. In fact, a similar image with the same Rayleigh-length would have been produced by a beam focused to 10 μm, an effect which could easily arise from ionization-induced refraction.

A less ambiguous imaging system was employed by the Berkeley group (Sullivan *et al.*, 1994), who studied the Thomson-scattered radiation at 90° to the laser axis, spectrally filtered at the laser wavelength. This technique provided a spatial resolution of 8 μm, enough to detect differences between ideal focusing behavior in vacuum and nonlinear effects in the plasma medium. Although some nonlinear optical effects were identified in this experiment — such as filamentation and spectral blue-shifting (p. 24) — they could be more readily explained by ionization dynamics rather than relativistic effects. No clear-cut evidence for channeling beyond the critical power was seen.

The same $90°$ Thomson-scattering technique was adopted by the Saclay group, who in addition used a pulsed hydrogen gas-jet to provide a relatively homogeneous, well-characterized plasma medium (Auguste *et al.*, 1994). Their resulting side-on images of the beam path revealed an unmistakable lengthening of the propagation channel with laser power; extending over many Rayleigh lengths and ultimately limited by the diameter of the gas jet (Monot *et al.*, 1995b). Moreover, consistent qualitative agreement was found between the experimental images and Thomson images reconstructed from a paraxial propagation model (Sec. 6.5), see Fig. 4.18. Very similar

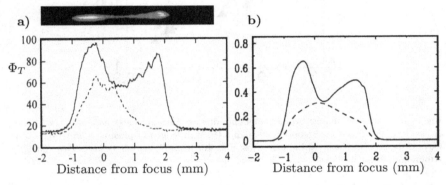

Fig. 4.18   a) Experimental Thomson scattering of relativistic self-channeling at two laser powers $P/P_c = 2$ (CCD image and solid lines), and $P/P_c = 0.125$ (dashed lines). The curves in b) are theoretical predictions from a wave-envelope model. The Rayleigh length, $Z_R = 300$ μm. *Courtesy:* T. Auguste and P. Monot, CEA-Saclay; Figs. 2 and 6 from Gibbon *et al.*, (1996). ©IEEE, 1996.

propagation behavior was subsequently reported by the NRL and Michigan groups (Krushelnick *et al.*, 1997; Wagner *et al.*, 1997), who verified the dependence of the channel length on the ratio P/P_c by varying both the laser power and the plasma density independently.

A contrasting picture of sub-ps laser propagation emerged from experiments by the LLNL and UCLA groups, who concentrated on the light *transmitted* through a length of underdense plasma (Coverdale *et al.*, 1995). They found that the percentage of laser power collected within the focal cone angle of the beam decreased with increasing plasma length, resulting in losses of up to 50% for a 1 mm propagation distance. This implies that rather than being channeled, most of the light is scattered out of the focal cone by instabilities such as Raman forward and side-scattering (Sec. 4.4). These conclusions were supported both by 2D PIC simulations and by measurements showing an anti-correlation between the anti-Stokes (Raman-scattered) line in the spectrum and the fraction of transmitted light, see Fig. 4.19.

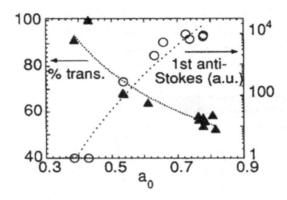

Fig. 4.19   Growth of anti-Stokes sideband due to SRS (circles) associated with decrease in transmitted laser energy through a 0.8 mm length of plasma. *Courtesy:* C. A. Coverdale, Sandia National Laboratories; Fig. 1c) from Coverdale *et al.* (1995). ©1995, American Physical Society.

The Saclay and Livermore experiments are not necessarily contradictory, since they emphasize different aspects of the laser propagation. Just because one sees a channel does not mean that the whole beam is guided: more recent experiments have in fact verified this (Chen *et al.*, 1998a). Ideally one should perform this experiment using all these diagnostics simultaneously, along with plenty of laser power to give a large range of P/P_c. With just one or two main diagnostics at hand, it is difficult to judge the 'quality' of self-channeling. In particular, there is no way of measuring the laser intensity amplification in a plasma directly: this has either to be inferred from a model, or from secondary photon scattering or particle emission. For target media with multiple ionization stages, one could in principle detect the ion emission and deduce the intensity from the highest charge state found (see Sec. 2.2). However, this will not give spatial information and one could not distinguish between isolated high-intensity filaments and whole-beam focusing.

The additional presence of longitudinal plasma waves — with their ability to accelerate electrons to multi-MeV energies — provides another indirect means of detecting channeling effects (Ting *et al.*, 1996). In a wakefield experiment by the Michigan group (Umstadter *et al.*, 1996), direct evidence for the so-called modulational instability (112) was obtained by measuring the number of accelerated electrons as a function of laser power. The electron yield increased dramatically as soon as the laser power exceeded P_c, suggesting that self-focusing took place. Moreover, under certain conditions, far-field images of the electron beam revealed a 'hole' at its center, a feature attributed to a lack of electrons (i.e. cavitation) along the laser axis.

Evidence of additional ion channeling, in which a long-lived waveguide is created via ponderomotive expulsion of both electrons and *ions* (Sec. 4.6.3), has come from several groups using a variety of complementary diagnostics. The NRL team used a second probe beam to test the quality of the plasma waveguide created by the ion blowout (Krushelnick *et al.*, 1997), whereas the Michigan group were able to create interferograms of ponderomotively produced channels several ps after the passage of a TW pump laser (Chen *et al.*, 1998a). In an extensive series of investigations on ion acceleration in underdense plasmas, Krushelnick *et al.* (1999) have exploited the 'Coulomb explosion' scenario of MeV ion production to infer enhancements in laser intensity brought about by self-focusing.

4.8.2 *Preformed, inhomogeneous plasmas*

One way of avoiding the undesirable defocusing effects that accompany multiphoton-ionized plasmas, while at the same time producing higher densities, is to ablate the plasma from a solid target using a long pulse before sending in the high-intensity short pulse. In this way, plasmas with densities in excess of 10^{20} cm^{-3} can be created, having scalelengths of a few hundred microns. Although the propagation length of an incident TW laser pulse up the plasma gradient may be less than a Rayleigh length, the critical power P_c is reduced to 100-200 GW owing to the higher density. Thus, initial focusing and channeling will occur much more readily than in a gas jet (Mackinnon *et al.*, 1995; Borghesi *et al.*, 1997; Malka *et al.*, 1997; Fuchs *et al.*, 1998a), and can also lead to curious turbulent effects such as 'post-soliton' formation (Naumova *et al.*, 2001). Of particular note here are the experimental campaigns by O. Willi's group at Imperial College using the VULCAN laser at the Rutherford-Appleton Laboratory (RAL)

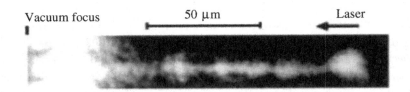

Fig. 4.20   Self-emission at 0.527 μm from a TW-laser produced channel in a preformed plasma created from a plastic film situated to the left of the vacuum focus. *Courtesy:* M. Borghesi, Queen's University Belfast; Fig. 3 from Borghesi *et al.*, (1997). ©1997, American Physical Society.

and by a joint Limeil-LULI team with the P102 TW laser at Limeil. In the investigation by Borghesi *et al.*, (1997), evidence of channeling was provided by interferograms (showing density fluctuations) and 2nd harmonic self-emission from the plasma. Taken together, these diagnostics implied the formation of a channel 130 μm long with a diameter of 5 μm. Further support for these findings came from 3D PIC simulations, revealing apparently similar focusing behavior of the pulse under conditions close to those in the experiment.

Similar results using plastic foil targets were found at Limeil (Malka *et al.*, 1997; Fuchs *et al.*, 1998b), who also inferred light amplification from the electron spectra taken during the interaction. Their interferograms showed density depressions of up to 20% at the center of the channel.

An intriguing variation on this experiment has been performed by the Garching group, who detected *neutrons* generated by fusion reactions of energetic (several hundreds of keV) deuterium ions originating from the channel (Pretzler *et al.*, 1998). This effectively opens up a new field for tabletop lasers, namely photonuclear physics — more of which we will encounter later in Chapter 7.

4.9   Magnetic Field Generation

One of the more esoteric features of high intensity laser-matter interaction is its potential for the creation of extremely high magnetic fields in the megagauss to gigagauss range. To put these numbers into perspective, consider that an electromagnet found in a well-equipped low-temperature physics lab might deliver up to 1 Tesla, or 10^4 Gauss on a good day. When we generate B-fields in a plasma, we do so by stirring up electron currents.

These will have roughly the dimension of the laser spot size, and persist until collisions or turbulence (from wavebreaking) disturb the coherence of the electron motion.

In an underdense plasma, a stationary (DC) magnetic field is generated in the plasma wake behind the laser pulse. At first sight, the presence of a B-field seems contradictory: in arriving at our 1D solution for the wakefield given by Eq. (4.43), we assumed that the plasma wave was purely longitudinal, or curl-free ($\nabla \times \mathbf{E_p} = \mathbf{0}$). This is quite correct: recall that in Sec. 4.3.1 we linearized the fluid equations (apart from the laser source term), discarding terms $O(E_z^2)$ and higher. To this order, the wake *is* curl-free, and one may easily verify from Eqs. (4.46) and (4.47) that $\frac{\partial E_z}{\partial r} = \frac{\partial E_r}{\partial \xi}$. An electrostatic plasma wave may contain both longitudinal *and* radial field components without necessarily inducing a significant B-field. Therefore, magnetic fields, if they occur, must be produced via higher order fluid nonlinearities.

We have already met one such plasma wave nonlinearity: the relativistic drag effect, or 'mass increase', in Sec. 4.3.4. It turns out that there are several more: electrostatic waves can exhibit nonlinear behavior (such as steepening and frequency shifts) due to geometrical effects — a feature first pointed out by Dawson (1959) in an analytical study of cylindrical Langmuir oscillations.

Magnetic nonlinearities are more subtle because they are inherently multi-dimensional, but can be illustrated by simple physical arguments. Consider a longitudinal plasma wake with wavenumber k_p. Recall that to lowest order, in can be described by a harmonic density perturbation $n_1 = n_0 \sin k_p \xi$, corresponding to the electric fields derived in Sec. 4.3.1. The fluid velocity acquired by the electrons is given by: $v_1 = (v_{z0} \sin k_p \xi, -v_{r0} \cos k_p \xi)$. Thus, to next order, the plasma wave will drive a nonlinear current $\mathbf{J}_2 = n_1 \mathbf{v}_1$, which will *change direction* during a plasma period, due to the fact that v_z and v_r are 90° out of phase. In other words, we generate a current loop, which will in turn generate a DC magnetic field, see Fig. 4.21.

From Ampere's law we can deduce that the B-field created by the current loop in Fig. 4.21 will also be 2nd order, containing both a stationary component and an oscillating one at $2\omega_p$. An excellent discussion of the physics behind this mechanism can be found in the article by Gorbunov *et al.*, (1997). To analyze the magnetic field more quantitatively, we need a nonlinear, 3D fluid model. We now proceed to develop this in a manner similar to the Akhiezer-Polovin approach for plane waves described at the

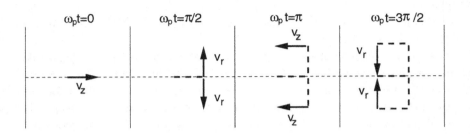

Fig. 4.21   Schematic picture of magnetic field generation in a plasma wake. At successive $\frac{1}{4}$-periods of the cycle, electrons see alternating longitudinal and radial fields, leading to the formation of current loops on either side of the beam axis.

beginning of the chapter, but this time making no assumption about the type of oscillation. Returning to the Lorentz-Maxwell equations (4.1–4.5), we take the curl of Eq. (4.1) and eliminate E using Faraday's law to get:

$$\nabla \times \frac{\partial \boldsymbol{p}}{\partial t} + \nabla \times [(\boldsymbol{v}.\nabla)\boldsymbol{p}] = \frac{e}{c}\frac{\partial \boldsymbol{B}}{\partial t} - \frac{e}{c}\nabla \times (\boldsymbol{v} \times \boldsymbol{B}). \qquad (4.131)$$

Using the vector identity

$$\boldsymbol{p} \times (\nabla \times \boldsymbol{p}) = \nabla \frac{\boldsymbol{p}^2}{2} - (\boldsymbol{p}.\nabla)\boldsymbol{p},$$

we can rewrite the 2nd term on the LHS of Eq. (4.131) as:

$$(\boldsymbol{v}.\nabla)\boldsymbol{p} = mc^2\nabla\gamma - \boldsymbol{v} \times (\nabla \times \boldsymbol{p}),$$

where $\gamma = (1 + p^2/m^2c^2)^{1/2}$. Inserting this expression into Eq. (4.131), we finally obtain:

$$\frac{\partial \boldsymbol{\Omega}}{\partial t} = \nabla \times (\boldsymbol{v} \times \boldsymbol{\Omega}), \qquad (4.132)$$

where

$$\boldsymbol{\Omega} \equiv \frac{1}{m}\nabla \times \boldsymbol{p} - \frac{e\boldsymbol{B}}{mc}$$

can be identified as a generalized *fluid vorticity*. Since $\boldsymbol{B} = \nabla \times \boldsymbol{A}$, we may also write $\boldsymbol{\Omega} = m^{-1}\nabla \times \boldsymbol{p}_c$, where $\boldsymbol{p}_c \equiv \boldsymbol{p} - e\boldsymbol{A}/c$ is the canonical momentum.

If, before the arrival of the laser pulse, we have $\boldsymbol{p}_c = 0$ (or $\boldsymbol{p} = \boldsymbol{A} = 0$), then $\boldsymbol{\Omega}(t = 0) = 0$. Eq. (4.132) then tells us that $\boldsymbol{\Omega} = 0$ *for all time*, so we can immediately write

$$B = \frac{c}{e}\nabla \times p. \tag{4.133}$$

This is a pretty useful result, because it immediately gives us the magnetic field provided we know the fluid momentum. To get the latter, we return to the Lorentz equation of motion, first noting that $v = p/m\gamma$ and make use of Eq. (4.133) to combine the 2nd terms from both sides of Eq. (4.1) thus:

$$\begin{aligned}
\frac{e}{c}v \times B + (v.\nabla)p &= \frac{e}{c}\frac{p \times B}{m\gamma} + \frac{(p.\nabla)}{m\gamma}p \\
&= \frac{1}{m\gamma}[p \times (\nabla \times p) + (p.\nabla)p] \\
&= \frac{1}{m\gamma}\nabla\frac{p^2}{2} \\
&= mc^2\nabla\gamma.
\end{aligned}$$

The equation of motion can thus be written:

$$\frac{\partial p}{\partial t} = -eE - mc^2\nabla\gamma, \tag{4.134}$$

or, since $E = -\nabla\phi - c^{-1}\partial A/\partial t$,

$$\frac{\partial}{\partial t}\left(p - \frac{eA}{c}\right) = \nabla(e\phi - mc^2\gamma). \tag{4.135}$$

Eqs. (4.134) and (4.135) are quite useful forms of the Lorentz equation — particularly in hybrid fluid (or kinetic) models which do not follow the full electromagnetic spatial and temporal detail of the problem (Sprangle *et al.*, 1992; Chen and Sudan, 1993b). For our purposes however, we need to go a bit further and eliminate E. Taking the time-derivative of Eq. (4.134) and combining Eq. (4.133) with Ampère's law, we obtain a wave equation for p:

$$\frac{\partial^2 p}{\partial t^2} + c^2\nabla \times (\nabla \times p) = -\frac{4\pi e^2}{m}\frac{n_e p}{\gamma} - mc^2\nabla\frac{\partial\gamma}{\partial t}. \tag{4.136}$$

The electron density is found by combing Eq. (4.134) with Poisson's equation Eq. (4.2), to get

$$4\pi e^2 n_e = 4\pi e^2 n_0 + \nabla.\frac{\partial p}{\partial t} + mc^2\nabla^2\gamma.$$

Inserting this into Eq. (4.136), we finally obtain

$$\frac{\partial^2 \boldsymbol{p}}{\partial t^2} + c^2 \nabla \times (\nabla \times \boldsymbol{p}) + \frac{\omega_p^2}{\gamma} \boldsymbol{p} = -mc^2 \nabla \frac{\partial \gamma}{\partial t}$$

$$-\frac{\boldsymbol{p}}{m\gamma} \nabla \cdot (\frac{\partial \boldsymbol{p}}{\partial t} + mc^2 \nabla \gamma). \quad (4.137)$$

This is an exact, nonlinear equation for the fluid momentum in arbitrary geometry. Once solved, we can get the magnetic field from Eq. (4.133). Alternatively, we can also formulate an analogous wave equation for $\boldsymbol{B}$:

$$\frac{\partial^2 \boldsymbol{B}}{\partial t^2} + c^2 \nabla \times (\nabla \times \boldsymbol{B}) + \frac{\omega_p^2}{\gamma} \boldsymbol{B} = -\frac{4\pi ec}{m} \nabla \left(\frac{n_e}{\gamma} \right) \times \boldsymbol{p}, \quad (4.138)$$

which explicitly shows the nonlinear source term for magnetic field generation discussed earlier. Equations like (4.137) and (4.138) have been studied in the context of plasma wakefield generation by a number of authors (Bell and Gibbon, 1988; Miano and de Angelis, 1989; Gorbunov *et al.*, 1996; Sheng *et al.*, 1998). Although the detailed mathematical approach in these studies varies, they all basically agree that: i) the magnetic field has both static and oscillatory components, ii) the magnitude of the B-field scales as:

$$B_0 \sim \begin{cases} \dfrac{a_0^4}{\tilde{\sigma}_L}, & a_0 \ll 1 \\[2mm] \dfrac{a_0^2}{\tilde{\sigma}_L}, & a_0 \gg 1, \end{cases} \quad (4.139)$$

in the low and high-intensity limits respectively, where $\tilde{\sigma}_L$ is the laser spot size in units of c/ω_p. Numerically, it turns out that magnetic fields $\gg 1$ MG can be generated for *narrow* plasma wakes of $O(c/\omega_p)$ in diameter with pump strengths $a_0 > 1$.

Apart from their potential use as a source of large, stationary magnetic fields, 3-dimensional wakefields have also provoked interest in their radiative properties. The oscillatory $2\omega_p$-components of Eq. (4.137) and Eq. (4.138) can lead to significant *Terahertz* emission from the plasma (Hamster *et al.*, 1993), which, among other things, is of interest to the condensed matter community for following dynamic processes, and for imaging of organic matter such as in the diagnosis of skin cancer (Pickwell *et al.*, 2004).

Exercises

(1) Complete the derivation of the exact nonlinear wave equation for longitudinal plasma oscillations Eq. (4.23) on p. 60.

(2) The wake potential produced by a short pulse laser has the general form

$$\phi(\xi) = -\frac{k_p}{4} \int_{\xi}^{\infty} d\xi' \mid a(\xi') \mid^2 \sin[k_p(\xi - \xi')].$$

a) Determine the corresponding wake field-strength and density and show that these satisfy their respective differential equations (4.41) and (4.40).

b) Calculate the potential and longitudinal field of a plasma wake driven by a square laser pulse:

$$a^2(\xi) = a_0^2\{\theta(\xi) - \theta(\xi - l)\},$$

where $\theta(\xi)$ is the Heaviside unit step function. Show that the wakefield is maximized for pulse lengths $l = \lambda_p/2$. Why are slightly longer pulse lengths needed for $\sin^2$ or Gaussian pulse shapes?

(3) Show that linearly or circularly polarized electromagnetic waves, the relativistic γ-factor for the fluid motion can be factorized into transverse and longitudinal components, as expressed by the identity (4.50) on p. 73.

(4) Show that in the limit $\beta_g \to 1$, the nonlinear wake potential in Eq. (4.62) reduces to the more familiar and palatable form:

$$\frac{\partial^2 \phi}{\partial \xi^2} = \frac{k_p^2}{2} \left[\frac{(1 + a^2)}{(1 + \phi)^2} - 1 \right].$$

This can also be obtained directly from Poisson's equation and the continuity equation $n(1 - u) = 1$ within the same limit. Solve this numerically to find the equivalent resonance curve to Fig. 4.5 for a relativistic pump ($a_0 \gg 1$).

(5) Charge conservation is an inherent property of Maxwell's equations. Prove this by deriving the continuity equation (4.51) using Poisson's

equation and Ampère's law (p. 57).

(6) Integrate the beat-wave envelope equations, Eq. (4.68), and show that the maximum amplitude is given by Eq. (4.69).

(7) a) Show that the expression for the root-mean-square width of a Gaussian laser beam Eq. (4.117) follows from Eq. (4.116).

b) The nonlinear Schrödinger equation for relativistic laser propagation in slab geometry is:

$$\frac{\partial a}{\partial t} + \frac{i}{2}\frac{\partial^2 a}{\partial x^2} + \frac{i}{8}\mid a \mid^2 a = 0.$$

Rederive the RMS beam width $< x^2 >$ in this 2D case and compare with the 3D result in a). Comment on the implications for self-focusing and cavitation in this geometry.

Chapter 5

Interaction with Solids: Overdense Plasmas

In the previous three chapters the interaction process has for the most part been treated as a reactive response of the matter to the light passing through it. In other words, one either considers how the properties of the pulse change as it propagates, or tries to find out what it leaves behind: ahead of the pulse, the material can be assumed to be undisturbed. For solids, the picture is radically different. A metal target, for example, will initially *reflect* the laser pulse like a mirror, sending the beam back specularly. Even a glass target, although initially transparent, will be rapidly ionized by the intense electric field, and a reflective plasma wall will be created within a few laser cycles at the surface of the target.

The subsequent interaction physics depends sensitively on the laser intensity, and to make matters worse, the last fifteen years have seen femtosecond-solid experiments escalate across an intensity range of *seven* orders of magnitude. Unsurprisingly then, it is not easy to decide what physics to include and what to omit when constructing theoretical models of such interactions. Coupling of the laser energy to the target material posed the first such dilemma, and still remains one of the more hotly disputed issues in the field. This is because more than one physical picture is possible depending on whether the material is treated as a dense conductor, or a 'sandwich' of cold solid plus a hot, thin or extended layer of plasma in the region of the laser's focal spot. To date, there is no single model which can adequately describe all the main pieces of absorption physics, not to mention the numerous other effects — mass and energy transport, nonlinear propagation, fast particle generation, and so on — which can also take place.

Departures from the common wisdom which prevailed for nanosecond interactions (see, for example Kruer, 1988) were predicted towards the end of the 1980s by a number of authors anticipating the first experiments with sub-picosecond lasers (Gamaliy and Tikhonchuk, 1988; More *et al.*, 1988a; Mulser *et al.*, 1989; Milchberg and Freeman, 1989). They pointed out several ways in which the traditional laser-plasma physics would not apply to short-pulse interactions.

First, field ionization over the first few laser cycles rapidly creates a surface plasma layer with a density many times the critical density n_c. As we saw from the linear dispersion relation (4.33) in the previous chapter, this is the density at which the plasma becomes opaque for an electromagnetic wave with frequency ω, and is defined such that

$$\omega^2 = \frac{4\pi e^2 n_c}{m}, \tag{5.1}$$

where e and m are the electron charge and mass respectively. In practical units, this translates to

$$n_c \simeq 1.1 \times 10^{21} \left(\frac{\lambda}{\mu m}\right) \ cm^{-3}. \tag{5.2}$$

For example, aluminium (a favourite material in early experiments) already has 3 valence electrons; another 6 can be stripped for a few hundred eV. The electron density created in this ionization state is given by:

$$n_e = Z^* n_i = \frac{Z^* N_A \rho}{A}, \tag{5.3}$$

which, taking the effective ion charge $Z^* = 9$, the Avogadro number $N_A = 6.02 \times 10^{23}$, density $\rho = \rho_{\text{solid}} = 1.9 \ g \ cm^{-3}$ and the atomic number $A = 26$, gives $n_e = 4 \times 10^{23} \ cm^{-3}$. For a pulse with a FWHM of 100 fs and a peak intensity of $10^{18} \ Wcm^{-2}$, this is the density which will be created by the leading edge of the pulse, a good 200 fs or so prior to its maximum.

Second, the short pulse duration means that there is not enough time for a substantial region of coronal plasma to form in front of the target during the interaction. Just as in long-pulse ICF interactions, this underdense region is created by ablation. The plasma pressure created during heating causes matter to blow off at roughly the sound speed:

$$c_s = \left(\frac{Z^* k_B T_e}{m_i}\right)^{1/2}$$

$$\simeq 3.1 \times 10^7 \left(\frac{T_e}{\text{keV}}\right)^{1/2} \left(\frac{Z^*}{A}\right)^{1/2} \ cm \ s^{-1}, \tag{5.4}$$

where k_B is the Boltzmann constant, T_e the electron temperature and m_i the ion mass. Assuming the plasma expands isothermally (Kruer, 1988), the density profile will assume an exponentially decreasing form with a well-defined scale length

$$L = c_s \tau_L$$
$$\simeq 3 \left(\frac{T_e}{\mathrm{keV}}\right)^{1/2} \left(\frac{Z^*}{A}\right)^{1/2} \tau_{\mathrm{fs}} \text{Å}. \tag{5.5}$$

Thus, if a 100 fs Ti:sapphire laser pulse heats the target to a few hundred eV, we can expect a preformed plasma with scale-length in the region $L/\lambda=$ 0.01–0.1, which is pretty steep compared to the 100s-of-micron plasma coronas typical of ICF experiments. This of course supposes that the *contrast* of the laser system — the level of the main pulse intensity to the pedestal or *prepulse* created by amplified spontaneous emission (ASE) — is sufficiently high that the surface remains unperturbed until the main pulse arrives.

Finally, thanks to this steep density gradient, the laser pulse interacts directly with the solid-density plasma which has just formed: laser energy can be deposited at much higher densities than in nanosecond interactions, where it is absorbed at or below the critical density n_c. The initial situation we are faced with then, is an intense, electromagnetic wave impinging on a highly overdense, mirror-like wall of plasma. In the absence of absorption, the electromagnetic field will form a standing wave pattern in front of the target, augmented by an evanescent component penetrating into the overdense region to a characteristic *skin depth* $l_s = c/\omega_p$, where $\omega_p = (4\pi e^2 n_e/m)^{1/2}$ is the plasma frequency, see Fig. 5.3.

Throughout this chapter we will encounter many variations and extensions of this interaction scenario, which together form a rich but highly complex physical basis for interpreting contemporary experiments with solids or grainy, solid-like targets, such as liquid droplets and clusters.

5.1 Ionization

Determination of the ionic charge-state distribution in a plasma subjected to the rapidly changing conditions during short-pulse laser interactions is no easy task: indeed, the atomic physics of dense, non-equilibrium plasmas is really a separate discipline in itself (Salzmann, 1998). After all, without precise knowledge of the ionization degree, Z^*, one cannot even begin to determine basic plasma properties like the electron density, equation of state, transport coefficients and so on.

For high density, optically thick plasmas, the radiative and absorptive processes in the constituent plasma ions are balanced, so that a local thermal equilibrium (LTE) is reached, and the ionization state can be determined by statistical means. Under these circumstances, the relative ion populations are related by the well-known Saha-Boltzmann equation (see, for example, Hutchinson, 1987, Chapter 6):

$$\frac{n_e n_{Z+1}}{n_Z} = \frac{g_{Z+1}}{g_Z} \frac{2m^3}{h^3} \left(\frac{2\pi T_e}{m}\right)^{3/2} \exp(-\Delta E_Z/T_e), \qquad (5.6)$$

where n_Z, n_{Z+1} are the ion densities corresponding to ionization states Z and $Z + 1$; g_Z, g_{Z+1} are the respective statistical weights of these levels (taking electron degeneracy into account), and ΔE_Z is the energy difference between the two states. This equation is subject to the constraints:

$$\sum n_Z = n_0; \quad \sum Z n_Z = n_e. \qquad (5.7)$$

For a given element at atomic density n_0, the Saha equation gives the relative proportions of ions — from singly charged up to fully stripped (hydrogen-like) — plus the net electron density as a function of temperature T_e.

While the Saha equation is often a good starting point, short pulse lasers tend to produce optically thin plasmas which span many orders of magnitude in density and temperature all at once. By optically thin, one means that radiation from recombining ions can escape the plasma completely; a property which of course forms the basis of plasma spectroscopy (Gauthier, 1988). This more typical situation is known as *non*-LTE, and usually requires the solution of time-dependent atomic rate equations in order to determine the charge distribution. These take the following general form (Lee, 1987):

$$\frac{dn_Z}{dt} = n_e n_{Z-1} S(Z-1) - n_e n_Z [S(Z) + \alpha(Z)] + n_e n_{Z+1} \alpha(Z+1), \quad (5.8)$$

where $S(Z)$ and $\alpha(Z)$ are the ionization and recombination rates of the ion with charge state Z, respectively. The recombination rate generally comprises a number of separate processes, such as radiative recombination, 3-body collisional recombination and dielectronic recombination.

For high-atomic-number elements, solving Eq. (5.8) quickly becomes an expensive or even intractable proposition, because the number of excited states can run into thousands. Worse still, for many important transitions the atomic data needed to compute the coefficients $S(Z)$ or $\alpha(Z)$ is simply not available. To get round this, mixed models are often employed which reduce the number of equations needed by grouping together excited states into 'superlevels' (Edwards and Rose, 1993; Town *et al.*, 1995; Djaoui, 1999).

I conclude this section by emphasizing that this is very much an on-going area of research with many grey areas. For example, at near-solid densities and/or low temperatures, the electrons become degenerate and are more correctly described by Fermi-Dirac rather than Boltzmann statistics. This is effectively restricts electrons making transitions into low free energy states, leading among many other effects, to a reduction in ionization degree (Rose, 1988).

In this case, at least for LTE-plasmas, so-called Thomas-Fermi models can be used to model the ionization (More, 1979; Pfalzner, 1991; Fromy *et al.*, 1992; Salzmann, 1998). An additional complication in the presence of very strong laser fields is that field ionization will take place too — at least early on before sufficient numbers of free electrons are released for collisional ionization to take over. Although one can estimate the initial rates using the formulae in Sec. 2.2, these have to be treated with caution because tunneling is also affected by the presence of neighbouring ions (Pfalzner, 1992).

5.2 Collisional Absorption

In view of the scenario for femtosecond-solid interactions outlined at the start of the chapter, it is not surprising that the first theoretical works appearing at the end of the 1980s concentrated on the issue of laser absorption in step-like, highly overdense plasma profiles. Intensities at that time were quite low by today's standards, reaching 10^{12}–10^{14} Wcm^{-2}, so collisions were justifiably assumed to dominate the absorption physics. The dominant absorption mechanism in this intensity regime depends on the density profile: from ICF work we know that nanosecond pulses are absorbed mainly by inverse-bremsstrahlung in the underdense, coronal plasma extending out many hundreds of microns from the ablator, with typical scale-lengths $L/\lambda \sim 100$ (Kruer, 1988; Lindl, 1997).

On the sub-picosecond timescale, we have already seen that we can expect scale-lengths more like $L/\lambda \leq 0.1$, so the standard formulae (which we will come to shortly) are unlikely to be valid in this case. On the other hand, a steep profile allows the laser field to access much higher densities, where the plasma is highly collisional. It is therefore still a fair question to ask how much laser energy can be absorbed by a solid-density target in this way.

To calculate the absorption coefficient, we can go to the other extreme and assume that the plasma has not expanded at all, so that the density is a perfect step-profile. In this case the Fresnel equations of metal optics can be applied (Born and Wolf, 1980; Godwin, 1994). Alternatively, one can solve the Helmholtz wave equations directly for an arbitrarily shaped profile. The second method is of course more general and ultimately more useful, since it can be used as part of a more complete physical interaction model. In the $L/\lambda \to 0$ limit one should recover the characteristic Fresnel behavior in any case.

5.2.1   *Helmholtz equations*

The Helmholtz equations for electromagnetic wave propagation in an inhomogeneous plasma are well known and comprehensive treatments can be found, for example, in the books by Ginzburg (1964) and Kruer (1988). Nonetheless, it will prove instructive to see how they are obtained in order to assess their range of applicability in short pulse interactions later on. Just as for wave propagation in the underdense, homogeneous plasmas considered in the previous chapter, we start from Maxwell's equations. This time however, we initially consider only small field amplitudes and a non-relativistic fluid response. The appropriate equation of motion here is the Lorentz equation of motion for the electrons, including collisional damping:

$$m\frac{\partial \boldsymbol{v}}{\partial t} = -e(\boldsymbol{E} + \frac{\boldsymbol{v}}{c}\times\boldsymbol{B}) - m\nu_{ei}\boldsymbol{v}, \qquad (5.9)$$

where ν_{ei} is the electron-ion collision frequency, given by (Dendy, 1993; Kruer, 1988):

$$\nu_{ei} = \frac{4(2\pi)^{1/2}}{3}\frac{n_e Z e^4}{m^2 v_{te}^3}\ln\Lambda$$

$$\simeq 2.91 \times 10^{-6} Z n_e T_e^{-3/2}\ln\Lambda \ \text{s}^{-1}. \qquad (5.10)$$

Here Z is the number of free electrons per atom, n_e is the electron density in cm^{-3}, T_e is the temperature in eV and $\ln\Lambda$ is the Coulomb logarithm, accounting for the usual limits, $b_{\min}$ and $b_{\max}$, of the electron-ion scattering cross-section. Here, these are determined by the classical distance of closest approach and the Debye length respectively, so that:

$$\Lambda = \frac{b_{max}}{b_{min}} = \lambda_D \cdot \frac{k_B T_e}{Z e^2} = \frac{9 N_D}{Z}, \qquad (5.11)$$

where

$$\lambda_D = \left(\frac{k_B T_e}{4\pi n_e e^2} \right)^{1/2} = \frac{v_{te}}{\omega_p}, \qquad (5.12)$$

and

$$N_D = \frac{4\pi}{3} \lambda_D^3 n_e$$

is the number of particles in a Debye sphere. Physically, binary collisions result in a frictional drag on the electron motion, taken into account by the damping term in Eq. (5.9).

The relevant EM wave equations for $\boldsymbol{E}$ and $\boldsymbol{B}$ are obtained in the usual way by taking the curl of the Faraday and Ampère equations (4.3, 4.4) respectively, to give:

$$\nabla^2 \boldsymbol{E} - \frac{1}{c^2} \frac{\partial^2 \boldsymbol{E}}{\partial t^2} = \frac{4\pi}{c^2} \frac{\partial \boldsymbol{J}}{\partial t} + \nabla(\nabla \cdot \boldsymbol{E}), \qquad (5.13)$$

$$\nabla^2 \boldsymbol{B} - \frac{1}{c^2} \frac{\partial^2 \boldsymbol{B}}{\partial t^2} = -\frac{4\pi}{c} \nabla \times \boldsymbol{J}. \qquad (5.14)$$

The right-hand sides of each equation represent the source terms of the EM waves in the plasma. Next, we linearize Eqs. (5.9–5.14) and assume that all field and fluid quantities have a harmonic time-dependence $\exp(-i\omega t)$, where ω is the laser frequency. Formally, we make an expansion

$$f(\boldsymbol{x}, t) = f_0(\boldsymbol{x}) + f_1(\boldsymbol{x}) e^{-i\omega t} + f_2(\boldsymbol{x}) e^{-2i\omega t} + ...,$$

which effectively results in the following simplifications:

$$\frac{\partial}{\partial t} \rightarrow -i\omega$$
$$n_e \rightarrow n_o + n_1$$
$$\boldsymbol{J} \rightarrow -e n_0 \boldsymbol{v}_1$$
$$(\boldsymbol{E} + \boldsymbol{v} \times \boldsymbol{B}) \rightarrow \boldsymbol{E}_1.$$

Inserting these approximations into the Lorentz equation (5.9) allows us to solve for $\boldsymbol{v}_1$, namely:

$$\boldsymbol{v}_1 = \frac{-i}{\omega + i\nu_{ei}} \frac{e \boldsymbol{E}_1}{m}.$$

This immediately gives us the induced current

$$\boldsymbol{J}_1 = -en_o\boldsymbol{v}_1 = \sigma_e\boldsymbol{E}_1, \tag{5.15}$$

where σ_e, the alternating current (AC) electrical conductivity of the plasma, is given by

$$\sigma_e = \frac{i\omega_p^2}{4\pi\omega(1+i\tilde{\nu})}. \tag{5.16}$$

We have also written $\tilde{\nu} = \nu_{ei}/\omega$ for convenience. Equations 5.15 and 5.16 can be thought of as a form of Ohm's law for a plasma subjected to an oscillating electric field. Substituting expression (5.15) for $\boldsymbol{J}_1$ into the RHS of the wave equation (5.13) for $\boldsymbol{E}_1$ gives:

$$\nabla^2\boldsymbol{E}_1 + \frac{\omega^2}{c^2}\boldsymbol{E}_1 = \frac{\omega_p^2}{c^2}\frac{\boldsymbol{E}_1}{1+i\tilde{\nu}} + \nabla(\nabla\cdot\boldsymbol{E}_1). \tag{5.17}$$

For a planar, transverse EM wave propagating in a uniform plasma we have $\nabla \to i\boldsymbol{k}$, and $\boldsymbol{E}_1$ perpendicular to $\boldsymbol{k}$, so that $\nabla\cdot\boldsymbol{E}_1 = 0$. In this limit we recover the standard linear dispersion relation:

$$-k^2 + \frac{\omega^2}{c^2}\left(1 - \frac{\omega_p^2}{\omega^2(1+i\tilde{\nu})}\right) = 0. \tag{5.18}$$

From this we identify the dielectric constant of the propagation medium

$$\varepsilon \equiv \frac{k^2c^2}{\omega^2} = 1 - \frac{\omega_p^2}{\omega^2(1+i\tilde{\nu})} = 1 + \frac{4\pi i\sigma_e}{\omega},$$

which can be readily generalized to a non-uniform plasma by defining a local permittivity, and allowing it to vary in space. In the remaining analysis we will consider a plasma density with a gradient in one direction, so we write

$$\varepsilon(x) \equiv n^2(x) = 1 - \frac{n_0(x)/n_c}{(1+i\tilde{\nu}(x))}, \tag{5.19}$$

where $n(x)$ is the local refractive index, n_0 the equilibrium electron density and n_c the critical density of the EM wave.

A simplification of the wave equation (5.17) is possible for an inhomogeneous plasma if restrict ourselves to a plane wave incident at some fixed angle to the density gradient, polarized out of the propagation plane, see Fig. 5.1. In this case the wave has a *periodicity* in y given by:

$$\boldsymbol{E}_1 = (0,0,E_z)e^{iky\sin\theta}.$$

Thus, we can write the gradient operator $\nabla = (\partial/\partial x, ik\sin\theta, 0)$ and again verify that $\nabla \cdot \boldsymbol{E}_1 = 0$. Making use of Eq. (5.19), the wave equation reduces to the Helmholtz equation for the electric field:

$$\frac{\partial^2 E_z}{\partial x^2} + k^2(\varepsilon - \sin^2\theta)E_z = 0. \tag{5.20}$$

Note that the wave modes involved here are purely transverse: within the linear approximation used, there is no coupling between an s-polarized EM wave and electrostatic modes (or Langmuir waves), owing to the fact that $n_1 = \nabla \cdot \boldsymbol{E}_1 = 0$.

Consider now a p-polarized wave $\boldsymbol{E}_1 = (E_x, E_y, 0)$, see Fig. 5.1. In this case, we clearly have $\nabla \cdot \boldsymbol{E}_1 \neq 0$; a component of the laser field lies along the density gradient, so we can expect to drive plasma waves at some point along the light path. The equation for the electric field in this case

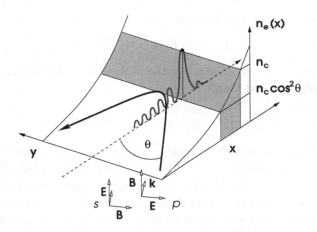

Fig. 5.1 Geometry of plane-wave incident on a plasma density profile for p-polarized light (E-field in the plane of ∇n_e), and for s-polarized light, (E-field in the z-direction). In the latter case the resonance is absent.

is somewhat involved (albeit still solvable), since it includes both EM and ES components — see, for example the tutorial by Mulser (1979). In fact it is more convenient to first solve for B_z instead, and then obtain E from Ampère's law, which after substituting $\boldsymbol{J}_1 = \sigma_e\boldsymbol{E}_1$, becomes (Kruer, 1988)

$$\nabla \times \boldsymbol{B}_1 = -\frac{i\omega\varepsilon}{c}\boldsymbol{E}_1. \tag{5.21}$$

In an analogous fashion, Faraday's law can be written:

$$\nabla \times \boldsymbol{E}_1 = \frac{i\omega}{c} \boldsymbol{B}_1. \tag{5.22}$$

As with the electric field, we substitute the expression for the current Eq. (5.15) into the magnetic field wave equation (5.14), and then use Eq. (5.21) and Eq. (5.22) to eliminate $\boldsymbol{E}_1$ from the RHS:

$$\nabla^2 \boldsymbol{B}_1 + \frac{\omega^2}{c^2} \boldsymbol{B}_1 = -\frac{4\pi}{c} \nabla \times (\sigma_e \boldsymbol{E}_1)$$

$$= \frac{4\pi i \omega \sigma_e}{c^2} \boldsymbol{B}_1 - \frac{4\pi i}{\omega} \frac{\nabla \sigma_e}{\varepsilon} \times (\nabla \times \boldsymbol{B}_1),$$

which, after eliminating σ_e using Eq. (5.19) and some rearranging, gives:

$$\nabla^2 \boldsymbol{B}_1 + \frac{\omega^2}{c^2} \boldsymbol{B}_1 + \frac{\nabla \varepsilon}{\varepsilon} \times (\nabla \times \boldsymbol{B}_1) = 0. \tag{5.23}$$

Applying the same oblique-incidence ansatz as before

$$\boldsymbol{B}_1 = (0, 0, B_z) e^{iky \sin \theta},$$

and assuming a density gradient in the x-direction only, we finally arrive at the Helmholtz equation for B:

$$\frac{\partial^2 B_z}{\partial x^2} - \frac{1}{\varepsilon} \frac{\partial \varepsilon}{\partial x} \frac{\partial B_z}{\partial x} + k^2 (\varepsilon - \sin^2 \theta) B_z = 0. \tag{5.24}$$

If the density gradient L is gentle enough, so that $\varepsilon(x)$ is slowly varying over a laser wavelength, i.e. $L^{-1} \sim \varepsilon^{-1} \partial \varepsilon / \partial x \ll k$, or $kL \gg 1$, then Eqs. (5.20) and (5.24) can be solved within the so-called Wentzel-Kramers-Brillouin (WKB) approximation (Ginzburg, 1964; Kruer, 1988), which for s-polarized light yields a solution in the form of an Airy function. In the context of femtosecond interactions, however, it is the *opposite* limit, $kL \ll 1$, which is usually more relevant. In this case, WKB fails, and one must generally resort to numerical methods, which is just what a number of workers did in order to analyze the early absorption experiments with short pulse lasers (Milchberg and Freeman, 1989; Kieffer *et al.*, 1989a; Fedosejevs *et al.*, 1990b; Rae and Burnett, 1991).

A method for solving the Helmholtz equations (5.20) and (5.24) is given in Sec. 6.6. This type of *wave-solver* code is an integral part of many workhorse fluid models for simulation of femtosecond interactions currently in use (Davis *et al.*, 1995; Teubner *et al.*, 1996b). This model is employed here to illustrate some of the main features of collisional absorption found in the literature. The three principle parameters which one can vary are: the scale-length, the angle of incidence and the collision frequency. For a given value of kL, the reflectivity (or absorption) shows a characteristic angular dependence. Milchberg and Freeman (1989) investigated this behavior for a range of parameters close to those used in their experiments at that time (Milchberg *et al.*, 1988). For long scale-lengths ($kL \gg 1$), we recover either the WKB result for *s*-polarized light, or classical resonant absorption for *p*-polarized light, see also Sec. 5.5.1.

For *s*-light, the absorption coefficient in this limit is given for an exponential profile by (Kruer, 1988):

$$\eta_{\mathrm{WKB}} = 1 - \exp\left(-\frac{8\nu_{ei}L}{3c}\cos^3\theta\right). \tag{5.25}$$

For *p*-light, the analysis is a little more involved (Ginzburg, 1964; Kruer, 1988; Kull, 1983), but with moderate damping, one obtains a maximum absorption of about 60% at an optimum angle of incidence given by

$$\sin\theta_{\mathrm{opt}} = 0.8\,(kL)^{-1/3}. \tag{5.26}$$

These features can be found in Fig. 5.2, which shows the absorption fraction of both *s*- and *p*-light for three different scale-lengths. These curves represent a small subset of the detailed analyses by Milchberg and Freeman (1989), Fedosejevs *et al.* (1990b) and Kieffer *et al.* (1989a). Note the shift of the absorption peak for *p*-light to large angles as the scale-length is reduced, as predicted by Eq. (5.26). Actually provided the wave amplitude is small, the Ginzburg result is a fair approximation down to about $L/\lambda = 0.1$; below $L/\lambda = 0.08$, Eq. (5.26) has no solution, and we effectively enter the *skin heating* regime where absorption takes place at over-critical densities (Kieffer *et al.*, 1989a), see also Exercise 2b) on p. 212.

5.2.2 *Normal skin effect*

In the limit $L \to 0$, we have a perfect density step, and we should recover the Fresnel-like absorption behavior of metal-optics. This situation is the 'ideal' for many applications in short-pulse laser-solid interactions, so it is worth reviewing the field solutions in this case. Consider first *s*-polarized light. Starting from the Helmholtz equation 5.20 for the electric field, we represent the density by a Heaviside step function:

$$n_0(x) = n_0\Theta(x),$$

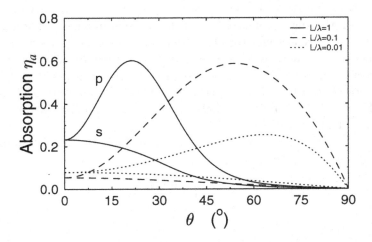

Fig. 5.2 Angular absorption dependence for various density scale-lengths: L/λ=1 (solid curves), L/λ=0.1 (dashed) and L/λ= 0.01 (dotted). Optimum absorption for p-polarized light occurs at an angle of incidence which increases as the density gradient becomes steeper.

and for the time-being neglect collisions in the dielectric constant, so that Eq. (5.19) reduces to:

$$\varepsilon(x) = 1 - \frac{\omega_p^2}{\omega^2}\Theta(x).$$

In the vacuum region ($x < 0$), the electric field thus has the solution

$$E_z = 2E_0 \sin(kx\cos\theta + \phi), \tag{5.27}$$

where $k = \omega/c$, E_0 is the amplitude of the laser field and ϕ a phase factor still to be determined. In the overdense region, the field is evanescent:

$$E_z = E(0)\exp(-x/l_s), \tag{5.28}$$

where

$$l_s = \frac{c}{\omega_p}\left(1 - \frac{\omega^2}{\omega_p^2}\cos^2\theta\right)^{-1/2}. \tag{5.29}$$

The decay length, l_s, is identified as the collisionless skin-depth: in the highly overdense limit, $n_0/n_c \gg 1$, we have $l_s \simeq c/\omega_p$. To complete our solution, we match up Eq. (5.27) and Eq. (5.28) together with their derivatives at the boundary $x = 0$. After a little algebra, this procedure gives

$$E(0) = 2E_0 \frac{\omega}{\omega_p} \cos \theta$$

$$\tan \phi = -l_s \frac{\omega}{c} \cos \theta.$$

An example of this solution is shown in Fig. 5.3 for an s-polarized wave normally incident onto a density step of modest height ($n_0/n_c = 5$) to emphasize the field structure in the skin layer. Note the discontinuity in the magnetic field gradient: this arises because the B-field changes sign at the boundary.

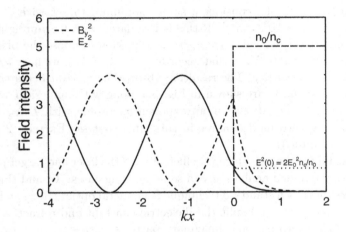

Fig. 5.3 Normally incident electromagnetic fields in an overdense plasma skin-layer with $n_0/n_c = 5, E_0 = 1$.

The effect of electron-ion collisions under these conditions will be to make the plasma behave like a metal surface with finite conductivity. The calculation of reflectivity under these conditions is covered in many texts, such as Born and Wolf (1980), so one needs only quote the results here, which are the Fresnel equations:

$$R_s = \left| \frac{\sin(\theta - \theta_t)}{\sin(\theta + \theta_t)} \right|^2, \qquad \text{for } s\text{-light} \qquad (5.30)$$

and

$$R_p = \left| \frac{\tan(\theta - \theta_t)}{\tan(\theta + \theta_t)} \right|^2, \qquad \text{for } p\text{-light} \qquad (5.31)$$

where θ is the angle of incidence as before, and

$$\theta_t = \sin^{-1} \left\{ \frac{\sin \theta}{n} \right\}$$

is the generalized, complex angle of the transmitted light rays (from Snell's law). The refractive index $n = \sqrt{\varepsilon}$ can be obtained from Eq. (5.19) as before, though this time the density is just taken to be equal to the solid density; a scenario sometimes referred to as the *Drude model* (see Sec. 1.4).

By way of example, consider a solid aluminium target with $Z^* = 3$, such that $n_e \simeq 2 \times 10^{23}$ cm^{-3}. If this is irradiated with 0.8 μm light from a Ti:sapphire laser, we will have $n_e/n_c \simeq 100$. Now suppose the plasma is initially heated to 120 eV, so that according to Eq. (5.10), we have $\nu/\omega = 5$ at the maximum density. The resulting absorption curves calculated from Eqs. (5.30) and (5.31) are shown in Fig. 5.4 along with numerical solution of the Helmholtz equations for an exponential profile with $L/\lambda = 0.001$. Evidently these two models agree in this limit (but see Exercise 2 at the end of the chapter.)

The scaling of the absorption coefficient with the laser and target parameters can be obtained analytically in some limiting cases. To find this scaling, we first borrow a standard technique from metal optics (Born and Wolf, 1980; Godwin, 1994) and split the dielectric constant and refractive index from Eq. (5.19) into real and imaginary parts: $\varepsilon = \varepsilon_r + i\varepsilon_i$, $n = n_r + in_i$. Since $\varepsilon = n^2$, we may write:

$$\varepsilon_r = 1 - \frac{\tilde{\omega}_p^2}{1 + \nu^2} = n_r^2 - n_i^2 \qquad (5.32)$$

$$\varepsilon_i = \frac{\nu\tilde{\omega}_p^2}{1 + \nu^2} = 2n_r n_i, \qquad (5.33)$$

where $\tilde{\omega}_p = \omega_p/\omega$ and $\nu = \nu_{ei}/\omega$. At normal incidence, Eqs. (5.30 and 5.31) reduce to

$$R_s = R_p = \left| \frac{1 - n}{1 + n} \right|^2,$$

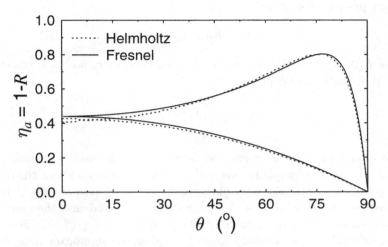

Fig. 5.4 Angular absorption dependence for a step-profile with $n_e/n_c = 100$ and $\nu/\omega = 5$ calculated from the Fresnel equations (5.30,5.31). The dotted curves were computed numerically for similar conditions from the Helmholtz equations.

giving an absorption fraction

$$\eta = 1 - R_s = \frac{4n_r}{(1 + n_r)^2 + n_i^2}. \qquad (5.34)$$

The problem reduces to solving for n_r and n_i from Eqs. (5.32) and (5.33). There are two cases of interest here corresponding to light ($\nu \ll 1$) or heavy ($\nu > 1$) damping respectively. We also assume that the plasma is highly overdense: $\tilde{\omega}_p^2 \gg 1$. In the first limiting case, ($\nu \ll 1$), we can write: $n_i \simeq \nu\tilde{\omega}_p^2/2n_r$, and solve for n_r from Eq. (5.32) to obtain $n_r = \nu^2\tilde{\omega}_p^2/2$. The same procedure can be followed in the ($\nu > 1$) limit to give $n_r = n_i = \tilde{\omega}_p^2/2\nu$. Substituting these values for the refractive index into Eq. (5.34), we finally obtain the absorption coefficient for the normal skin effect (More *et al.*, 1988a; Gamaliy, 1994):

$$\eta_{\mathrm{nse}} = \begin{cases} \dfrac{2\nu_{ei}}{\omega_p}, & \nu_{ei} \ll \omega_0 \\[3mm] \dfrac{2\omega_0}{\omega_p}\left(\nu_{ei}/\omega_0\right)^{1/2}, & \nu_{ei} > \omega_0. \end{cases} \qquad (5.35)$$

As for any conducting material, the electric field in the overdense plasma decays exponentially (the wave-vector is imaginary), so for a slab of semi-infinite plasma we may write:

$$E(x) = E(0) \exp(-x/\delta_s),$$

where $E(0)$ is the field at the plasma-vacuum boundary and δ_s is the effective (collisional) skin-depth:

$$\delta_s = \frac{c}{\omega_p} \mid 1 + i\nu \mid^{1/2} . \tag{5.36}$$

We conclude this discussion on collisional absorption by noting that Eq. (5.35) can in principle be inverted to yield information about the transport properties of high density plasmas. If the absorption or reflectivity is determined experimentally with a high degree of confidence, then one can deduce ν_{ei} and compare it to the Spitzer expression Eq. (5.10). For very dense and cold — or *strongly coupled* — plasmas, significant departures from the classical value can be expected. This threshold is usually denoted by the ion coupling parameter (ratio of potential to thermal energy):

$$\Gamma = \frac{(Ze)^2}{k_B T_e a_i}, \tag{5.37}$$

where $a_i = (\frac{4\pi}{3}n_i)^{-1/3}$ is the average ion spacing.

Classical theory breaks down when Γ approaches unity, a condition easily satisfied by solid-density plasmas at temperatures of a few eV; precisely the regime which high-contrast, femtosecond lasers can access. Consequently, a number of theoretical (Pfalzner and Gibbon, 1998; Millat *et al.*, 2003; Schlanges *et al.*, 2003) and experimental (Shepherd *et al.*, 1988; Theobald *et al.*, 1999; Riley *et al.*, 2002) works have appeared recently which attempt to explore this unusual state of matter.

5.3   Heating and Thermal Conduction

It is clear that absorption of laser energy will inevitably result in heating of the target material. On the other hand, increasing the temperature of the plasma will influence the rate at which it can convert oscillation energy into thermal energy. Thus, heating and absorption are really coupled and should be treated accordingly in a self-consistent way. This is indeed how the first models of this process were put together, for example by More *et al.*, (1988a) and Rozmus and Tikhonchuk (1990). Because the absorption issue

is complex (not to say controversial), we will first tackle the heating problem
on its own and then see how to take absorption into account afterwards.

In the absence of hydrodynamic motion, our overdense, step-like plasma
slab provides us with a textbook transport problem: propagation of a
nonlinear heat-wave into a cold, conducting, semi-infinite medium with
a continuous heat source at one boundary, see Fig. 5.5. Mathematically,

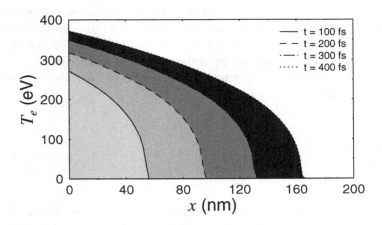

Fig. 5.5 Nonlinear heat-wave advancing into a semi-infinite, solid-density plasma. The
curves are obtained from the numerical solution of the Spitzer heat flow equation for
constant laser absorption at the target surface (left boundary).

this is described by the energy transport equation for a collisional plasma
(Nicholas, 1982)

$$\frac{\partial \epsilon}{\partial t} + \nabla.(\boldsymbol{q} + \boldsymbol{\Phi}_a) = 0, \tag{5.38}$$

where ϵ is the energy density, $\boldsymbol{q}$ is the heat flow and $\boldsymbol{\Phi}_a = \eta_a \boldsymbol{\Phi}_L$ is the
absorbed laser flux. Equations like Eq. (5.38) and their formal derivation
from kinetic theory can be found in most plasma physics texts: a more
complete version used in hydro codes is given later, see Eq. (5.59) in Sec. 5.4.

The simplest possible model one can apply for femtosecond pulses is
volume heating. If the penetration depth of the heat wave l_h over the pulse
duration is less than the skin depth, $l_s = c/\omega_p$, then the thermal transport
can be neglected. This is the laser-plasma equivalent of putting a pot of
water on a stove: a volume $V \simeq l_s \pi \sigma^2$ will be heated *en bloc*. Setting

$\epsilon = \frac{3}{2} n_e k_B T_e$ and $\nabla.\Phi_a \sim \Phi_a/l_s$, we have

$$\frac{dT_e}{dt} \simeq \frac{\Phi_a}{n_e l_s}, \tag{5.39}$$

or

$$\frac{d}{dt}(k_B T_e) \simeq 4 \frac{\Phi_a}{\text{Wcm}^{-2}} \left(\frac{n_e}{\text{cm}^{-3}}\right)^{-1} \left(\frac{l_s}{\text{cm}}\right)^{-1} \text{keV fs}^{-1}. \tag{5.40}$$

Thus, a pulse with peak intensity of 10^{15} Wcm^{-2} will produce an initial heating rate of around 1 keV fs^{-1} near the surface of a solid-density plasma. After a few femtoseconds, thermal transport inevitably sets in because of the huge temperature gradients generated, and heat is carried away from the surface into the colder target material according to Eq. (5.38). For ideal plasmas, we write

$$\epsilon = \frac{3}{2} n_e k_B T_e$$

as before, and

$$q(x) = -\kappa_e \frac{\partial T_e}{\partial x}, \tag{5.41}$$

which is the usual Spitzer-Härm heat-flow (Spitzer, Jr. and Härm, 1952; Spitzer Jr., 1962). The constant κ_e is known as the Spitzer thermal conductivity and is given by:

$$\kappa_e = 32 \left(\frac{2}{\pi}\right)^{1/2} \frac{n_e}{\nu_0 m^{5/2}} T_e^{5/2}, \tag{5.42}$$

where

$$\nu_0 = \frac{2\pi n_e Z e^4 \log \Lambda}{m^2}.$$

Substituting for ϵ and q in Eq. (5.38) and restricting ourselves to one spatial dimension by letting $\nabla = (\partial/\partial x, 0, 0)$, gives a diffusion equation for T_e:

$$\frac{3}{2} n_e k_B \frac{\partial T_e}{\partial t} = \frac{\partial}{\partial x} \left(\kappa_e \frac{\partial T_e}{\partial x}\right) + \frac{\partial \Phi_L}{\partial x}. \tag{5.43}$$

Despite the nonlinear $T_e^{5/2}$-dependence of the diffusion coefficient, equation (5.43) has a well-known self-similar solution (Zel'dovich and Raizer, 1966). By this we mean that the *shape* of the heatwave remains the same while it penetrates the target, even though the surface temperature may be increasing. In the example depicted in Fig. 5.5, the heat front initially advances at a phenomenal 5×10^7 cm s^{-1}, comfortably justifying the label 'ultrafast' often used to describe short-pulse heatflow.

It proves instructive to derive this self-similar solution to illustrate some of the essential aspects of plasma heating in the femtosecond regime. Following Rozmus and Tikhonchuk (1990), we rewrite all the dependent and independent variables in dimensionless form thus: $\tau = t/t_0$; $\xi = x/(v_0 t_0)$; $\Theta = T_e/(mv_0^2)$, where t_0 and v_0 represent time and velocity constants to be determined later. Substituting these expressions into Eq. (5.43), we can write:

$$\frac{\partial \Theta}{\partial \tau} = \frac{\partial}{\partial \xi} \left(\Theta^{5/2} \frac{\partial \Theta}{\partial \xi} \right), \tag{5.44}$$

provided we choose

$$t_0 = \frac{64}{3} \left(\frac{2}{\pi} \right)^{1/2} \frac{v_0^3}{\nu_0}. \tag{5.45}$$

To determine v_0 we apply the boundary condition at the edge of the target ($x = 0$), namely, that the heat flow into the plasma is equal to the incoming laser flux:

$$q|_{x=0} = \eta_a I_0, \tag{5.46}$$

where η_a is the absorption coefficient. For the moment, we will assume that this is constant, and does not depend on the temperature. This is actually a reasonably accurate assumption for most of the collisionless absorption mechanisms which we will meet in Sec. 5.5, but of course less realistic for collisional absorption. This second, more complex situation can also be handled within this model by modifying the above boundary condition. Substituting Eq. (5.41) into Eq. (5.46) we get:

$$-\frac{3}{2} n_e m v_0^3 \Theta^{5/2} \frac{\partial \Theta}{\partial \xi} \bigg|_{x=0} = \eta_a I_0, \tag{5.47}$$

which after choosing

$$v_0 = \left(\frac{2 \eta_a I_0}{3 n_e m} \right)^{1/3}, \tag{5.48}$$

reduces to the simple form:

$$\Theta^{5/2} \frac{\partial \Theta}{\partial \xi} \bigg|_{\xi=0} = -1. \tag{5.49}$$

We now try and find a self-similar variable for the dimensionless equations (5.44) and (5.49). A more physical justification behind the choice of variable based on dimensional analysis can be found in Zel'dovich and Raizer (1966). Basically, we seek a combination $\zeta = \xi/\tau^\alpha$, so that the derivatives become:

$$\frac{\partial}{\partial \tau} = \frac{-\alpha\xi}{\tau^{\alpha+1}}\frac{d}{d\zeta}; \qquad \frac{\partial}{\partial \xi} = \tau^{-\alpha}\frac{d}{d\zeta}.$$

At the same time, we let $\Theta(\xi,\tau) = \Psi(\zeta)\tau^\beta$, so that the boundary condition Eq. (5.49) now reads:

$$\Psi^{5/2}\frac{\partial \Psi}{\partial \zeta}\bigg|_{\zeta=0} = -\tau^{\alpha-7\beta/2}.$$

By definition, our new variable Ψ should only depend on ζ and not on τ or xi. Thus, for the τ-dependence to vanish, we require

$$\alpha = \frac{7\beta}{2}.$$

Making the same substitutions in Eq. (5.44) we find:

$$-\alpha\zeta\frac{d\Psi}{d\zeta} + \beta\Psi = \tau^{5\beta/2-2\alpha+1}\frac{d}{d\zeta}\left(\Psi^{5/2}\frac{\partial \Psi}{\partial \zeta}\right).$$

Again, for the τ-dependence on the RHS to vanish, we require

$$\frac{5\beta}{2} - 2\alpha + 1 = 0.$$

Solving these two algebraic equations for α and β gives $\alpha = 7/9, \beta = 2/9$, so that our heat flow equation is reduced to a nonlinear *ordinary* differential equation:

$$\frac{d}{d\zeta}\left(\Psi^{5/2}\frac{\partial \Psi}{\partial \zeta}\right) - \frac{2\Psi}{9} + \frac{7\zeta}{9}\frac{\partial \Psi}{\partial \zeta} = 0, \tag{5.50}$$

with the boundary condition

$$\Psi^{5/2}\frac{\partial \Psi}{\partial \zeta}\bigg|_{\zeta=0} = -1. \tag{5.51}$$

Eq. (5.50) has no exact analytical solution, but can be solved numerically by a 'shooting' method; an integration from the left boundary, subject to the global requirement that the total thermal energy increases by the rate at which laser energy is absorbed. Defining the normalized thermal energy

$W = \int_0^\infty \Theta d\xi$, we can express its rate of increase in terms of normalized variables as

$$\frac{\partial W}{\partial \tau} = \int_0^\infty \frac{\partial \Theta}{\partial \tau} d\xi'$$

$$= \int_0^\infty \frac{\partial}{\partial \xi} \left(\Theta^{5/2} \frac{\partial \Theta}{\partial \xi} \right) d\xi'$$

$$= - \Theta^{5/2} \frac{\partial \Theta}{\partial \xi} \bigg|_{\xi=0} = 1. \tag{5.52}$$

In terms of the self-similar variable, we have $W = \tau \int_0^\infty \Psi d\zeta$, so the boundary condition reduces to a normalization condition for Ψ:

$$\frac{\partial W}{\partial \tau} = \int_0^\infty \Psi d\zeta = 1. \tag{5.53}$$

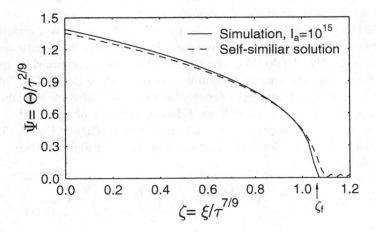

Fig. 5.6   Self-similar heat-flow for constant laser absorption at $\zeta = 0$. The solid curve is obtained by renormalizing one of the later curves in Fig. 5.5.

The solution, shown in Fig. 5.6, is characterized by the two boundary values $\Psi(0) = 1.35$ and $\zeta_f \simeq 1.05$. To verify that this self-similar solution really does describe the heat front penetration, the curves of Fig. 5.5 have been cast into normalized form and superimposed. Re-expressing the self-similar variable in terms of actual laser and target parameters, the condition

$\Psi(0) = 1.35$ physically corresponds to a surface temperature of

$$T_e(0) = 250 \left(\frac{n_e}{10^{23} \text{ cm}^{-3}}\right)^{-2/9} Z^{2/9} \left(\frac{I_a}{10^{15} \text{ Wcm}^{-2}}\right)^{4/9} \left(\frac{t}{100\text{fs}}\right)^{2/9} \text{ eV.}$$

(5.54)

Likewise, the heat front position is simply given by

$$x_f = v_0 t_0 \zeta_f \tau^{7/9}$$
$$= 1.05 v_0 t_0^{2/9} t^{7/9}$$
$$\simeq 65 \left(\frac{n_e}{10^{23} \text{ cm}^{-3}}\right)^{-7/9} Z^{-2/9} \left(\frac{I_a}{10^{15} \text{ Wcm}^{-2}}\right)^{5/9} \left(\frac{t}{100\text{fs}}\right)^{7/9} \text{ nm.}$$

(5.55)

These are essentially the expressions obtained by Rosen (1990) using simple physical arguments based on classical heat-flow and constant absorbed intensity I_a. Treating the heated plasma layer as a blackbody radiator, Rosen then applied these results to estimate the soft x-ray yield from femtosecond laser-produced plasmas.

Although appealingly simple, the formula (5.54) tends to overestimate the temperature because the absorption has not been taken into account self-consistently. To do this, we need to substitute an appropriate expression for η_a into the boundary condition at the surface Eq. (5.47). According to the temperature-dependence which results, different self-similarity scalings are obtained, as well as different versions of the ODE (5.50). For example, inserting the collisional absorption coefficient Eq. (5.35) into Eq. (5.47) ultimately leads to a surface temperature scaling of (Rozmus and Tikhonchuk, 1990):

$$T_e(0) = 119 \left(\frac{n_e}{10^{23} \text{ cm}^{-3}}\right)^{1/12} Z^{1/12} \left(\frac{I_a}{10^{15} \text{ Wcm}^{-2}}\right)^{1/3} \left(\frac{t}{100\text{fs}}\right)^{1/6} \text{ eV,}$$

(5.56)

which we notice has a much weaker intensity- and time-dependence than in Eq. (5.54).

This type of heat-flow model can be made more sophisticated by using a Fokker-Planck treatment of collisions, which explicitly takes into account non-local transport (Rozmus and Tikhonchuk, 1990; Gamaliy, 1994). Even with this approach, a self-similar solution can still be found which describes the heat front penetration into the target, and an upper limit set on the pulse duration above which the plasma pressure becomes large enough to drive hydrodynamic motion.

5.4 Hydrodynamics

A discussion of hydrodynamics in the context of femtosecond interactions may at first sight seem somewhat superfluous. After all, one of the *raisons d'être* of short pulse technology was to try and avoid plasma expansion. As noted at the beginning of this chapter, however, it is not possible to keep the ions completely immobile, even with the best possible laser contrast ratio. Rapidly increasing thermal plasma pressure will cause the material to expand sooner or later. Moreover, there are a number of instances — for example in hard x-ray generation — where a *controlled*, finite density gradient is desirable because it enhances the collective absorption component into hot electrons (see Sec. 5.5). This can be created either by a well-characterized pedestal, or, preferably, by using a femtosecond prepulse timed to arrive at a predetermined moment before the main pulse.

Whether the plasma expansion is good or bad, it is still important to be able to model the hydrodynamics accurately. Once again, this is an area where the field of short pulse interactions has benefitted directly from earlier ICF research: many of the hydro-codes in use today have been adapted from earlier versions developed to model implosions. The main ingredients of these codes are: ionization; collisional absorption; thermal transport; equation-of-state; hydrodynamics; plus a number of optional bits of physics such as nuclear burn rates, soft x-ray emission and lasing. The first three pieces of this jigsaw have already been discussed in previous sections of this chapter, so we will concentrate on the actual hydro-part in what follows.

The starting point for any hydrodynamics code is basically the *two-fluid* plasma model. This comprises a set of continuity, momentum and energy equations for both electrons *and* ions. These can be rigorously derived from the Vlasov-Boltzmann kinetic equation for each species — a standard but tedious procedure which we need not repeat here; the reader is referred instead to the many excellent treatments of kinetic theory available in the literature (Kruer, 1988; Dendy, 1993; Krall and Trivelpiece, 1986). An additional assumption is usually made that for a quasi-neutral plasma ($n_e = Zn_i$), the mass and momentum transport is essentially dominated by the ions, whose mass $M = Am_p \gg m$, where m_p and m are the proton and electron mass respectively; A is the atomic number. One may therefore define an average fluid density ρ and velocity u thus:

$$\rho \equiv n_i M + n_e m \simeq n_i M,$$

$$u \equiv \frac{1}{\rho}(n_i M u_i + n_e m u_e) \simeq u_i.$$

This permits us to reduce the two-fluid description to a single-fluid one, but we usually retain both energy equations to allow for thermodynamic imbalance between particle species ($T_e \neq T_i$). The basis set of equations for our laser-plasma hydro-code is then (Nicholas, 1982):

$$\frac{\partial \rho}{\partial t} + \nabla\cdot(\rho u) = 0 \qquad (5.57)$$

$$\frac{\partial \rho u}{\partial t} + \nabla\cdot(\rho u u) + \nabla P - f_p = 0 \qquad (5.58)$$

$$\frac{\partial \epsilon_e}{\partial t} + \nabla\cdot\left[u(\epsilon_e + P_e) - \kappa_e \nabla T_e - \frac{Q_{ei}}{\gamma_e - 1} - \Phi_a\right] = 0 \qquad (5.59)$$

$$\frac{\partial \epsilon_i}{\partial t} + \nabla\cdot\left[u(\epsilon_i + P_i) - \kappa_i \nabla T_i + \frac{Q_{ei}}{\gamma_i - 1}\right] = 0, \qquad (5.60)$$

where the energy density ϵ_α is the sum of internal and kinetic fluid energies:

$$\epsilon_\alpha = \frac{P_\alpha}{\gamma_\alpha - 1} + \frac{1}{2}\rho u^2,$$

with γ_α defined as the number of degrees of freedom for the electrons and ions respectively. Q_{ei} is the electron-ion equipartition rate (Huba, 1994)

$$Q_{ei} = \frac{2m}{M}\frac{n_e k_B (T_e - T_i)}{\tau_{ei}} \qquad (5.61)$$

and $\tau_{ei} = \nu_{ei}^{-1}$ is the usual electron-ion collision time given by Eq. (5.10). The thermal conductivities κ_e and κ_i have the form given by Eq. (5.42), but are in general anisotropic (Huba, 1994).

Coupling to the laser is included via the absorption term Φ_a in the electron energy equation (5.59) and via the ponderomotive force in the momentum equation (5.58)

$$f_p = -\frac{\omega_p^2}{16\pi\omega^2}\nabla E_L^2,$$

which was derived in Sec. 3.2 and Exercise 3.3. The latter essentially acts only on the electrons, the ions having too much inertia to respond effectively to the laser field. The force nevertheless still manifests itself in the fluid momentum because of the electric field set up when electrons are displaced by the ponderomotive potential, see Sec. 5.7 and Chapter 10 of Kruer (1988).

The pressures and temperatures of the fluid components are connected via an equation of state. This can be calculated using the ideal gas law $P_\alpha = k_B n_\alpha T_\alpha$ at low densities, or some more sophisticated model, such as the Thomas-Fermi statistical model at high densities (Pfalzner, 1991). Further refinements can be made by including non-equilibrium and low-temperature effects (More *et al.*, 1988b). As usual, we defer the technical details of the model to Sec. 6.8 and show here some examples of its use in femtosecond interaction studies.

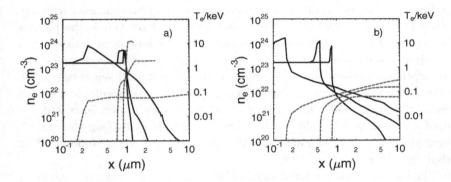

Fig. 5.7 Density (solid line) and temperature (dashed line) profiles from hydrodynamic simulations of two laser pulses with pulse durations of a) 100 fs and b) 100ps. The laser impinges from the right onto an initially cold Al target extending to $x = 1\ \mu m$. In a), the three snapshots are taken at times 0 fs, 300 fs and 5 ps after the peak of the pulse; in b) these snapshots are at -150 ps, -100 ps and 0 ps relative to the peak.

First, the hydro-code presents us with a means to illustrate just what is meant by a 'short' laser pulse by considering two cases with the same laser energy but different pulse lengths. To choose parameters which typify the experiments performed in the late 1980s — now routinely used for *prepulses* — we take a laser energy of 20 mJ focused to a 25 μm spot size ($1/e$ intensity). For a pulse duration of 100 fs this gives an intensity on target of 10^{16} Wcm^{-2}. The same pulse stretched to 100 ps will likewise have a flux of 10^{13} Wcm^{-2}. The results of hydro-simulations with these two sets of parameters are depicted in Fig. 5.7, which shows the density and temperature profiles at the peak of the laser pulse. Note the stark contrast in the density profiles: for the long pulse, a substantial underdense plasma corona has formed in which efficient inverse-bremsstrahlung (IB) heating can take place, eventually resulting in an overall energy absorption of 70%. For the short pulse there is almost *no* collisional absorption, partly due to the rapid temperature rise, causing the collision rate to fall; but in this case also due to inadequate modeling of IB in the steep gradient.

This particular code (MEDUSA) has no wave-solver of the type described in Sec. 5.2.1; only a WKB absorption model (Eq. 5.25), which as we saw, severely underestimates the absorption in step-profiles. In the example shown, however, we have imposed an 'anomalous' absorption component to match the total absorption fraction in the long pulse simulation. Even with this cosmetic patch to make the absorbed energy the same for both pulses, the density profile in the short-pulse interaction remains extremely steep $(L/\lambda < 0.1)$ while the laser is incident. At later times, the plasma does expand with ablation velocities similar to the long pulse interaction.

Hydro-codes equipped with wave-solvers can in principle compute the absorption, transport and subsequent hydrodynamics all self-consistently. One of the first to do this was Delettrez (1988), whose results in Fig. 5.8 using the LILAC code (Richardson *et al.*, 1986) clearly demonstrated the inadequacy of purely collisional absorption models to explain the experimental results at laser intensities 10^{14}–10^{16} Wcm^{-2} being announced at that time (Kieffer *et al.*, 1989b).

These ingredients - ionization, collisions, wave propagation, thermal transport and hydrodynamics — form the basis of many standard short pulse simulation codes now in common use (More *et al.*, 1988a; Ng *et al.*, 1994; Davis *et al.*, 1995; Teubner *et al.*, 1996b) and have also been extended to two- and three-dimensional models. More sophisticated transport codes also exist which solve the Fokker-Planck (FP) equation in order to include non-local heat flow (Town *et al.*, 1994; Matte *et al.*, 1994; Town *et al.*, 1995; Limpouch *et al.*, 1994). The self consistent treatment of ionization dynamics and FP heat flow leads to a strongly non-Maxwellian electron distribution function, which modifies the heat flow penetration into the target compared to the classical Spitzer-Härm theory of thermal transport. These workhorse models are depicted schematically in Fig. 5.9.

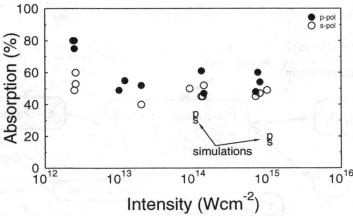

Fig. 5.8 Comparison between absorption measurements using a carbon target made by Meyerhofer *et al.* at LLE and rates calculated from the LILAC hydrocode incorporating a purely collisional absorption model. *Courtesy:* J. Delettrez, Laboratory for Laser Energetics; (unpublished, 1988).

5.5 Collisionless Absorption

Early experiments performed at modest intensities (Milchberg *et al.*, 1988; Kieffer *et al.*, 1989b; Murnane *et al.*, 1989; Landen *et al.*, 1989; Fedosejevs *et al.*, 1990a) were in good agreement with hydrodynamic models, both in terms of the measured reflectivity, and the typical plasma parameters inferred from atomic x-ray spectra. However, a number of discrepancies between theory and experiment soon became apparent as lasers were upgraded and intensities increased. First, for intensities above 10^{15} Wcm^{-2} or so, the plasma temperature rises sufficiently fast that collisions become ineffective during the interaction. Recall that from Eq. (5.54), we have a temperature scaling:

$$T_e \propto I_a^{4/9} t^{2/9},$$

or a 400 eV temperature near the critical surface after just 10 fs for an absorbed intensity $I_a = 10^{15}$ Wcm^{-2}. According to Eq. (5.10), the collision frequency scales as $T_e^{-3/2}$, which therefore implies a scaling falling off as:

$$\nu_{ei} \sim I_a^{-2/3} t^{-1/3}.$$

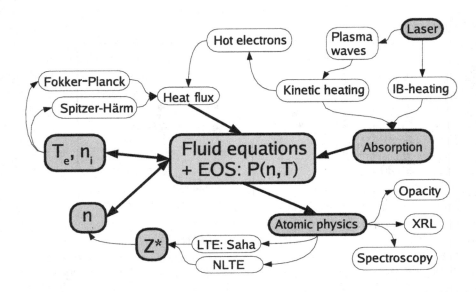

Fig. 5.9 Building blocks of a hydrodynamic laser-plasma simulation code.

This effect is actually taken care of in the self-consistent models of Rozmus and Tikhonchuk (1990), as described at the end of Sec. 5.3.

The second limiting effect occurs when the electron quiver velocity becomes comparable to the thermal velocity, thus reducing the effective collision frequency further (Silin, 1965; Pert, 1995):

$$\nu_{\text{eff}} \simeq \nu_{ei} \frac{v_{te}^3}{(v_{os}^2 + v_{te}^2)^{3/2}}. \tag{5.62}$$

A temperature of 1 keV corresponds to a thermal velocity $v_{te} \simeq$ 0.05, so collisional absorption starts to turn off for irradiances $I\lambda^2 \geq$ 10^{15} Wcm$^{-2}\mu$m^2, and therefore could not account for the high absorption observed in experiments in this regime, for example, by Chaker *et al.*, (1991) and by Meyerhofer *et al.*, (1993). Given this discrepancy, alternative absorption mechanisms have been sought which do not rely on collisions between electrons and ions.

In fact, there are a number of collisionless processes which can couple laser energy to the plasma. The best-known of these is resonance absorption, which was studied extensively in the 1970s and '80s with 2D PIC codes in order to understand the origin of fast electrons generated in nanosecond laser-plasma interactions (Estabrook *et al.*, 1975; Forslund *et al.*, 1977; Estabrook and Kruer, 1978; Adam and Héron, 1988; Kruer, 1988). Ironically, fast electron generation is highly *undesirable* in the conventional ICF context because the long stopping range of these particles leads to preheat of the fuel core (see Fig. 1.2), thus preventing DT pellets from being compressed to the necessary densities for high thermonuclear gain. With the advent of short pulses, however, fast electrons are very much back in favour, both as an integral element of the fast ignitor scheme, and because they can be used to produce hard x-rays in high-Z targets beyond the hot surface plasma — applications which will be examined in detail in Chapter 7.

5.5.1 *Resonance absorption*

It is not immediately clear how effective resonance absorption can be in the steep density gradients expected during short pulse interactions. In the standard picture (Ginzburg, 1964), a p-polarized light wave tunnels through to the critical surface ($n_e = n_c$), where it drives up a plasma wave, see Fig. 5.1. This wave grows over a number of laser periods and is eventually damped either by collisions at low intensities or by particle trapping and wave breaking at high intensities (Kruer, 1988).

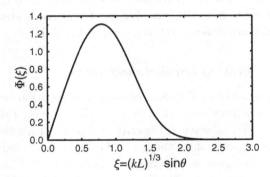

Fig. 5.10 The Denisov function: self-similar behavior of resonance absorption for long density scale-lengths.

For long density scale-lengths — usually defined by the inequality $kL \gg 1$, where $k = 2\pi/\lambda$ is the laser wave vector and $L^{-1} \equiv \mid \frac{d}{dx} \log N_e \mid_{x=x_c}$ — the absorption fraction has a self-similar dependence on the parameter

$\xi = (kL)^{1/3} \sin \theta$, where θ is the usual angle of incidence (Denisov, 1957; Ginzburg, 1964, and see also Exercise 4). In this limit, the angular absorption behavior is given by the curve $\phi(\xi)$ shown in Fig. 5.10. To a good approximation,

$$\phi(\xi) \simeq 2.3\xi \exp(-2\xi^3/3), \qquad (5.63)$$

and the fractional absorption is given by:

$$\eta_{\text{ra}} = \frac{1}{2}\Phi^2(\xi).$$

This behavior is more-or-less independent of the damping mechanism provided the pump amplitude is small. Collisionless resonance absorption can therefore be modeled in a hydro-code even if the collision rate is too small to give significant inverse-bremsstrahlung (Rae and Burnett, 1991). This is usually done by introducing a phenomenological collision frequency in the vicinity of the critical surface (Forslund *et al.*, 1975a) such that one recovers the $\sim 60\%$ maximum absorption rate in the long scale-length limit at an angle given by Eq. (5.26).

Although the overall energy balance will be taken care of in this manner, the way in which it is divided up into thermal and suprathermal electron heating can only be determined self-consistently using a *kinetic* approach such as particle-in-cell (PIC) (Birdsall and Langdon, 1985) or Vlasov simulation. An example of a PIC simulation of resonance absorption for a relatively long density scale-length is shown in Fig. 5.11.

5.5.2 *Vacuum heating (Brunel mechanism)*

As indicated previously in Fig. 5.2, resonance absorption ceases to work in its usual form in very steep density gradients. To see this, consider a resonantly driven plasma wave at the critical density with a field amplitude E_p. In a sharp-edged profile, there will be little field swelling, and E_p will be roughly the same as the incident laser field E_L. Electrons will therefore undergo oscillations along the density gradient with an amplitude $x_p \simeq eE_L/m_e\omega^2 = v_{\text{os}}/\omega$. The resonance breaks down if this amplitude exceeds the density scale length L, i.e. if $v_{\text{os}}/\omega > L$.

The inability of the plasma to support Langmuir waves near the critical surface would seem to rule out any form of mode conversion normally necessary to couple electromagnetic wave energy to the plasma. On the other hand, one can intuitively see that electrons near the edge of an abrupt

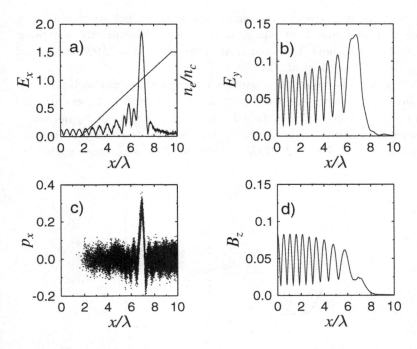

Fig. 5.11 PIC simulation of resonance absorption for parameters $\theta = 9^o, v_{os}/c = 0.07, L/\lambda = 5\,(kL = 10\pi)$, and $n_e^{max}/n_c = 1.5$: a) Density profile and electric field normal to the target, b) electric field parallel to the target, c) particle phase-space (momentum vs position), d) laser magnetic field.

change in plasma-vacuum interface will be directly exposed to the laser field. Indeed, a thermal electron arriving near the edge at the right moment in the laser cycle may be dragged out violently into the vacuum well beyond the thermal Debye sheath, $\lambda_D = v_{te}/\omega_p$. As the field reverses its direction, this same electron will be turned around and accelerated back into the plasma. Because the latter is highly overdense, the electric field only penetrates to a skin depth $\sim c/\omega_p$, so the electron can travel virtually unhindered into the target, where it is eventually absorbed through collisions.

This simple but important mechanism was first pointed out by Brunel (1987), motivated in part by a rather different but geometrically similar problem; namely, the losses suffered by an electromagnetic pump wave in a laser-based particle accelerator scheme. For this 'not-so-resonant, resonant absorption', Brunel developed an analytical model, which we reproduce in a much simplified guise here: similar analyses have subsequently been made by a number of authors (Bonnaud *et al.*, 1991; Kato *et al.*, 1993; Wilks and Kruer, 1997; Greschik *et al.*, 2000).

The model is based on the *capacitor approximation*, in which the magnetic field of the wave is ignored (i.e. the $v \times B$ term is neglected), and assumes that the laser electric field E_L has some component E_d normal to the target surface which pulls electrons back and forth across their equilibrium positions, see Fig.5.12. If the wave is incident with some angle θ on

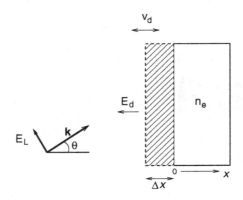

Fig. 5.12   Capacitor model of the Brunel heating mechanism.

an initially smooth, mirror-like surface, a standing wave will be set up so that the driving electric field is given by

$$E_d = 2E_L \sin \theta. \qquad (5.64)$$

Suppose this field pulls a sheet of electrons out to a distance Δx from its initial position. The surface number density of this sheet is $\Sigma = n_e \Delta x$ (just as for capacitor plates), so the electric field created between $x = -\Delta x$ and $x = 0$ is

$$\Delta E = 4\pi e \Sigma.$$

Equating this to the driving field above, we can then solve for Σ:

$$\Sigma = \frac{2E_L \sin \theta}{4\pi e}. \qquad (5.65)$$

When the charge sheet returns to its original position, it will have acquired a velocity $v_d \simeq 2v_{os} \sin\theta$, where v_{os} is the usual electron quiver velocity in the laser field, as defined by Eq. (2.19). Assuming these electrons are all 'lost' to the solid, then the average energy density absorbed per laser cycle is given by:

$$P_a = \frac{\Sigma}{\tau} \frac{mv_d^2}{2}$$
$$\simeq \frac{1}{16\pi^2} \frac{e}{m\omega} E_d^3.$$

Comparing this to the incoming laser power $P_L = cE_L^2 \cos\theta/8\pi$ and substituting Eq. (5.64), we obtain the fractional absorption rate:

$$\eta_a \equiv \frac{P_a}{P_L} = \frac{4}{\pi} a_0 \frac{\sin^3\theta}{\cos\theta}, \tag{5.66}$$

where $a_0 = v_{os}/c$.

According to Eq. (5.66), we expect more absorption at large angles of incidence and higher laser irradiance, $I\lambda^2 \propto a_0^2$. In fact, this expression predicts over 100% absorption if either of these parameters is large enough. Fortunately, there are two corrections which we can immediately make to improve the model. First, we take into account the reduced driver field amplitude due to imperfect reflectivity, replacing Eq. (5.64) by

$$E_d = [1 + (1 - \eta_a)^{1/2}]E_L \sin\theta. \tag{5.67}$$

Second, anticipating that the return velocities of the electrons may become relativistic at intensities in excess of 10^{18} Wcm^{-2}, we replace their kinetic energy with $U_k = (\gamma - 1)mc^2$ in the expression for the absorbed power, with $\gamma = (1 + v_d^2/c^2)^{1/2}$. Using both these corrections, we arrive at an *implicit* expression:

$$\eta_B = \frac{1}{\pi a_0} f \left[(1 + f^2 a_0^2 \sin^2\theta)^{1/2} - 1 \right] \frac{\sin\theta}{\cos\theta}, \tag{5.68}$$

where $f = 1 + (1 - \eta_a)^{1/2}$ is the field amplification factor. This is essentially Brunel's original result (Brunel, 1987), apart from a numerical factor $O(1)$ arising from a different estimate for the density of heated electrons.[1].

There are two limiting cases which allow us to progress further with the analysis: $a_0 \ll 1$ and $a_0 \gg 1$. In the low-intensity limit, we obtain

---

[1]N.B.: In Eq. (10) of Brunel's paper, v_{osc} corresponds to our v_d; v_L to our v_{os}.

something similar to the previous result in Eq. (5.66), except that pump depletion is now included:

$$\eta_{\text{vh}}^{\text{low}} = \frac{a_0}{2\pi} f^3 \alpha(\theta) \tag{5.69}$$

where $\alpha(\theta) = \sin^3\theta / \cos\theta$. By writing $\eta_a = 1 - (f-1)^2$ and letting $\beta = a_0\alpha/2\pi$, Eq. (5.69) reduces to a simple quadratic for f:

$$\beta f^2 + f - 2 = 0,$$

which has one physically meaningful root:

$$f = \frac{(1+8\beta)^{1/2} - 1}{2\beta}.$$

In the strongly relativistic limit (strictly speaking, $fa_0\sin\theta \gg 1$), we find:

$$\eta_a = \frac{f^2}{\pi} \frac{\sin^2\theta}{\cos\theta},$$

which, we note, is *independent* of a_0, and can also be solved similarly to Eq. (5.69). Letting $\alpha' = \sin^2\theta / \cos\theta$, we find $f = 2(\frac{\alpha'}{\pi} + 1)^{-1}$, so that finally:

$$\eta_{\text{vh}}^{\text{rel}} = \frac{4\pi\alpha'}{(\pi + \alpha')^2}. \tag{5.70}$$

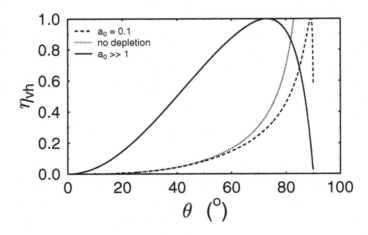

Fig. 5.13   Angular dependence of vacuum heating predicted by Brunel model.

Both these limiting cases are shown in Fig. 5.13 and reveal behavior reminiscent of the normal skin effect in Sec. 5.2.2. As with Fresnel-type absorption, the pump-depletion effect prevents the rate becoming unphysically large at grazing incidence, as one can see by comparing the two curves with and without this correction for the low-intensity case: the peak shifts to lower angles as a_0 is increased. In the extreme relativistic case, the absorption curve is much broader, and inspection of Eq. (5.70) shows that it peaks at $\alpha' = \pi$, corresponding to $\theta_{\mathrm{opt}} = 73.06^\circ$.

Just as for collisional absorption (Sec. 5.2), the situation becomes more complicated for more realistic density profiles with finite gradients. This case was first considered by Gibbon and Bell (1992), whose simulations revealed a complex transition between resonance absorption and vacuum heating[2] depending on the irradiance and scale-length.

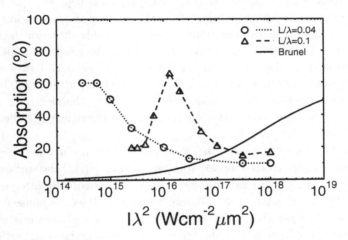

Fig. 5.14 Absorption fraction vs. irradiance according to Brunel — Eq. (5.68) with $\theta = 45^\circ$ (solid line); and Gibbon & Bell (1992) for density scale-lengths $L/\lambda=0.1$ (dashed) and $L/\lambda=0.04$ (dotted) and $n_e/n_c = 2$.

For high irradiances and short scale-lengths, the absorption saturates at around 10–15%, but for intermediate values (e.g. $I\lambda^2 = 10^{16}$ Wcm$^{-2}\mu$m^2, $L/\lambda \sim 0.1$), the absorption can still be as high as 70%, see Fig. 5.14. The

---

[2]This regrettable misnomer was coined by the present author to distinguish between the EM and ES versions of Brunel's mechanism, exemplified by Fig. 5.14. Unfortunately, the phrase appears to have stuck.

simulation results for steep profiles exhibit clear departures from the simple capacitor model above: some additional mechanism must limit the absorption efficiency well before $v_{os}/c \simeq 1$. Also, the angular absorption for modest intensities and densities tends to peak near $45°$ (Gibbon and Bell, 1992; Andreev *et al.*, 1994), rather than the grazing incidence angles suggested by Fig. 5.13.

The main reason for this is that the electrostatic model is inadequate to describe an obliquely incident plane wave because of the DC currents set up along the surface of the target. These create an additional magnetic field which can deflect the fast electrons and inhibit them from returning to the plasma. Brunel reasoned that this limitation could be overcome by using two geometrically opposed laser pulses at $\pm\theta$. This way, the surface currents cancel to a large extent, and the formula Eq. (5.68) should be reasonably valid. This idea was confirmed in 2-dimensional PIC simulations (Brunel, 1988; Estabrook and Kruer, 1986), but still begs the question of how to improve the analytical model for the single-pump scenario typical of most experimental investigations. Progress in this direction has been made by Yu *et al.*, (1998), who take into account the effects of a DC electron cloud outside the target surface for normally incident light; and by Andreev *et al.*, (1999), Ruhl *et al.*, (1999) and Sheng *et al.*, (2000), who showed independently that at relativistic intensities, the electrons are directed along the laser axis, rather than parallel to the density gradient, see Exercise 5 on p. 213.

In simulations with *mobile* ions, a rather different picture emerges. The strong space-charge created by electrons circulating outside the surface pulls out a supersonic, underdense ion shelf which can drastically alter the absorption physics (Brunel, 1988; Gibbon, 1994) as well as the phase content of the reflected light (Kingham *et al.*, 2001). After a characteristic time $t_{\text{shelf}} \simeq 100\sqrt{A/Z}\lambda\mu\text{m}$ fs, the absorption and hot electron distribution resemble that seen in early simulations of resonance absorption (Forslund *et al.*, 1977; Estabrook and Kruer, 1978); the main difference being that the pressure balance assumed and imposed in long pulse (ns) interactions is unlikely to be achieved for femtosecond pulses. This lack of hydrodynamic equilibrium can strongly influence the hot electron fraction and temperature scaling. For extreme intensities ($I > 10^{20}$ Wcm$^{-2}\mu$m^2) at normal incidence, energy can also be transferred directly to the ions via charge imbalance both in the low density shelf region and in the overdense plasma, where a collisionless shock is formed (Denavit, 1992), see Secs. 5.7 and 5.8.

5.5.3 *Anomalous skin effect*

A third collisionless mechanism which is closely related to vacuum heating is the *anomalous* skin effect. This is actually a well-known effect in solid-state physics (Ziman, 1969), and was originally studied for step-like vacuum-plasma interfaces by Weibel (1967). Its potential significance for short pulse interaction was therefore recognized quite early by a number of workers (Gamaliy and Tikhonchuk, 1988; Mulser *et al.*, 1989; Rozmus and Tikhonchuk, 1990; Andreev *et al.*, 1992).

Physically, the anomalous skin effect is not as mysterious as it sounds. Consider first the situation for the normal skin effect. Electrons within the skin layer $l_s = c/\omega_p$ oscillate in the laser field and dissipate energy through collisions with ions (inverse-bremsstrahlung). This is in effect, the $L = 0$, or Fresnel limit considered earlier in Sec. 5.2.2. The oscillation energy is thus locally thermalized provided that the electron mean-free-path, $\lambda_{mfp} = v_{te}/\nu_{ei}$ is smaller than the skin depth.

Now suppose that the temperature is increased so that $\lambda_{mfp} > l_s$, and that the mean thermal excursion length $v_{te}/\omega > l_s$. Under these conditions the laser field is carried further into the plasma and the relationship between induced current and electric field becomes *nonlocal*. The effective collision frequency is now given by the excursion time in this enlarged skin layer, l_a, i.e. $\nu_{eff} = v_{te}/l_a$. To estimate l_a, we can modify the general expression (5.36) for the skin depth found earlier by replacing ν_{ei} with ν_{eff}, so that for $\nu_{eff}/\omega > 1$, we have

$$l_a \simeq \frac{c}{\omega_p} \left(\frac{\nu_{eff}}{\omega} \right)^{1/2} = \frac{c}{\omega_p} \left(\frac{v_{te}}{\omega l_a} \right)^{1/2}.$$

Solving for l_a, we find (Weibel, 1967):

$$l_a \simeq \left(\frac{c^2 v_{te}}{\omega \omega_p^2} \right)^{1/3}.$$

A more accurate value for l_a can be found from kinetic theory, in which the nonlocal nature of the electron current is taken into account self-consistently via an integral equation involving the velocity distribution function. From this 'quasi-linear' analysis (Rozmus and Tikhonchuk, 1990; Andreev *et al.*, 1992; Yang *et al.*, 1995), one finds:

$$\frac{1}{l_a^3} = \frac{2\pi^2 \omega_p^2 \omega}{c^2} \int_0^\infty dv \, v f_0(v), \tag{5.71}$$

where f_0 is the lowest order term in an expansion of the slowly varying part of the distribution function, usually taken to be Maxwellian:

$$f_0(v) = f_M(v) \equiv \frac{1}{(2\pi v_{te}^2)^{3/2}} \exp(-v^2/2v_{te}^2). \tag{5.72}$$

Substituting Eq. (5.72) into Eq. (5.71), we then find:

$$l_a = \left(\frac{2}{\pi}\right)^{1/6} \left(\frac{c^2 v_{te}}{\omega \omega_p^2}\right)^{1/3}, \tag{5.73}$$

and the corresponding absorption rate is (Rozmus and Tikhonchuk, 1990; Andreev *et al.*, 1992):

$$\eta_{\text{ase}} = \frac{8\omega l_a}{3\sqrt{3}c} \simeq \left(\frac{T_e}{511 \text{ keV}}\right)^{1/6} \left(\frac{n_c}{n_e}\right)^{1/3}. \tag{5.74}$$

In more sophisticated studies of the anomalous skin effect, the full Vlasov equation is solved numerically in order to correctly describe the nonlocal relationship between the current and the electric field (Matte and Aguenaou, 1992; Andreev *et al.*, 1992; Ruhl and Mulser, 1995). Nonetheless, just as with resonance absorption, the effect can also be included in a hydrodynamic model by replacing ν_{ei} with the effective collision frequency $\nu_{\text{eff}} = v_{te}/l_a$.

The Fresnel equations can again be used to obtain the angular absorption dependence in the step-profile limit (Andreev *et al.*, 1992). The maximum absorption for p-polarized light is nominally $\eta_{\text{ase}} \sim 2/3$ at grazing incidence angles, *independent* of density and temperature, but can be enhanced further if the distribution function is anisotropic.

5.5.4 *Sheath inverse-bremsstrahlung*

The phrase sheath inverse-bremsstrahlung (SIB) was coined by Catto and More (1977) to describe yet another absorption mechanism which can occur in overdense plasma layers. Unlike in the anomalous skin effect (ASE) regime however, the electron transit time through the skin depth is *longer* than a laser period. The energy transfer takes place via a series of irreversible kicks received by electrons from the laser field as they are turned around at the plasma-vacuum interface; hence the analogy with inverse-bremsstrahlung for small v_{os}/c.

In a comprehensive analysis of this effect, Yang *et al.*, (1995) extended the theory of Catto & More by including the $\mathbf{v} \times \mathbf{B}$ term in the equation of motion, and showed that SIB and ASE are essentially two limits of the same absorption mechanism. In the SIB limit, i.e. $\omega l_s / v_{te} \gg 1$, the absorption rate can also be expressed in closed analytical form:

$$\eta_{\text{sib}} = \frac{8}{\sqrt{2\pi}} \frac{v_{te}}{c} \frac{a[(a+1)\exp(a)E_1(a) - 1]}{1/2 + \frac{\sqrt{\pi a}}{2}\exp(a)\text{erfc}(a^{1/2})}, \tag{5.75}$$

where $a = \omega^2 l_s^2 / 2 v_{te}^2$, $E_1(a) = \int_a^\infty \exp(-t^2)/t \, dt$ and erfc is the complementary error function. We see that after multiplying by c/v_{te}, this expression depends only on the parameter a. One can also find an exact expression for the absorption, valid for all values of a, which can be evaluated numerically (Yang *et al.*, 1995).

For most practical purposes, the absorption rate for normal incidence is not huge. For example, consider two plasma layers both with a temperature of 5 keV and with electron densities n_e/n_c of 5 and 400 respectively. The value of the normalized transit time is then 0.05 and 4 respectively, so we can say that SIB will dominate in the first case; ASE in the second. The absorption in either case is only 5%, however, but as we noted before in Sec. 5.5.3, absorption can be increased by going to oblique incidence. A corresponding generalisation for *s*- and *p*-polarized light was presented in a follow-up study by Yang *et al.*, (1996). The authors describe this oblique-incidence variation of SIB as *sheath-transit* absorption, which basically sums up the physical situation of electrons being inelastically reflected from the Debye sheath near the surface of the plasma.

Both SIB and ASE are complementary to the Brunel's vacuum-heating effect discussed in Sec. 5.5.2 in that they are important for steep gradients $L/\lambda \ll 1$ when the light pressure P_L is *less* than the plasma pressure P_e. Quantitatively, this implies that $v_{\text{os}}^2/v_{te}^2 < n_e/n_c$ or:

$$\frac{P_L}{P_e} = \frac{2I_0/c}{n_e k_B T_e} \simeq \frac{660 I_{18}}{160 n_{23} T_{\text{keV}}} < 1, \tag{5.76}$$

where I_{18} is the laser intensity in units of 10^{18} Wcm^{-2}, n_{23} the electron density in units of 10^{23} cm^{-3}, and T_{keV} the temperature in keV. As this threshold is crossed, we can expect a transition from SIB/ASE to resonance absorption/vacuum heating: a feature confirmed in PIC simulations by Yang *et al.*, (1996) and which probably also accounts for some of the erratic changes in absorption in the intensity-L/λ plane originally reported by Gibbon and Bell (1992).

5.5.5   *Relativistic j × B heating*

Despite its exotic label, this mechanism actually dates back to the pre-femtosecond era. Its potential importance for light absorption was originally pointed out by Kruer and Estabrook (1985). Physically, it is very similar to the Brunel mechanism, in that electrons are directly accelerated by a laser field incident on a step-like density profile. The main difference is that the driving term is the high-frequency $v \times B$ component of the Lorentz force, which oscillates at twice the laser frequency. A *linearly* polarized wave $E = E_0(x)\hat{y}\sin\omega t$ gives rise to a longitudinal force term:

$$f_x = -\frac{m}{4}\frac{\partial v_{os}^2(x)}{\partial x}(1 - \cos 2\omega t). \qquad (5.77)$$

The first term on the RHS is the usual DC ponderomotive force (Sec. 3.2), which in this context tends to push the electron density profile inwards (see Sec. 5.8). The second, high-frequency component leads to heating in an analogous fashion to the component of a p-polarized electric field parallel to the density gradient. This '$j \times B$' term, which works for *any* polarization apart from circular, is most efficient for normal incidence and becomes significant at relativistic quiver velocities.

Using fully 2D PIC simulations, Wilks *et al.*, (1992) demonstrated the potential importance of this heating mechanism along with a host of other processes, such as hole boring and magnetic field generation. The absorption fraction is small, but increases with laser irradiance (Kruer and Estabrook, 1985) and tends to produce highly anisotropic distribution functions Gamaliy and Dragila (1990).

5.5.6   *Absorption measurements*

From the preceding sections it should be clear that there are plenty of ways of coupling laser energy to high density plasmas, even on the femtosecond timescale. For many of the applications described later in Chapter 7, this is good news; for others a mixed blessing, depending on where the energy actually goes: i.e. thermal electrons, hot electrons, or ions. Because of this complexity, it is quite difficult to predict which coupling processes will prevail for a given set of laser and target parameters, so the question is whether one can unambiguously detect signatures of a particular absorption mechanism. Many early absorption measurements in the late 1980s/early 1990s set out to do just this, although the main concern at that time was whether one would see any absorption *at all*, given that collisional effects start to fall off at higher intensities. With hindsight we know that we *do*: a glance at Fig. 5.16 should be enough to quash any doubts that high fractional absorption is routinely achieved even at the highest intensities currently available.

Absorption can be measured in a variety of ways, the simplest of which is to monitor the amount of light specularly reflected from the surface of the target; the reflectivity R. Assuming no other scattering processes occur, then the absorption fraction is just given by $\eta_a = 1 - R$. The presence of an underdense region of plasma may alter this mirror-like behavior to one where light is diffusely scattered in a 2π solid angle, or even back-scattered via parametric instabilities such as RBS, see Sec. 4.4.2.

A schematic diagram of a typical experimental setup for absorption measurements is shown in Fig. 5.15. The double dielectric mirror (DM) arrangement allows backscattered light to be neatly collected with the same optics used to focus the pulse itself onto the target. An Ulbricht sphere (US) collects the diffusely reflected light, the amount of which is registered by the photodiode (PD). A further energy detector (ED) measures the specularly reflected light inside the focal cone determined by the f-number of the off-axis parabolic mirror.

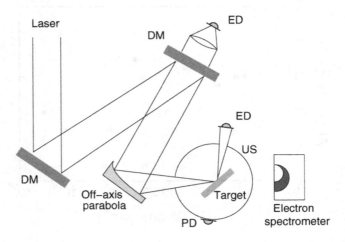

Fig. 5.15 Experimental setup for absorption and hot electron measurements from short-pulse laser interaction with a solid target.

Perhaps the most telling signature of an absorption mechanism is its dependence on the angle of incidence and polarization of the laser electric field relative to the target normal. As we have seen in preceding sections, an absorption maximum at low incidence angles (i.e. close to the target normal, say $\theta = 5$–$15°$) points to resonance absorption in a relatively long scale-length preplasma. High absorption at near grazing incidence ($\theta = 70$–$80°$) would suggest that some form of skin effect or vacuum heating is dominating the interaction physics. Experimentally, the angular dependence is normally found by rotating the target with respect to the focal axis, adjusting the diagnostics accordingly. This tends to restrict data points to 'octagonal' values like $22.5°, 45°, 67.5°$, and so on, because the target is usually placed inside a vacuum chamber with a limited number of port-holes.

Despite these constraints, this procedure has been very successful, with a number of groups finding clear evidence of resonance absorption during earlier 'low-contrast' interactions; experiments in which a long pedestal (or deliberate prepulse) from the amplifier chain created a substantial region of underdense plasma tens of microns thick in front of the target (Kieffer *et al.*, 1989b; Fedosejevs *et al.*, 1990a; Teubner *et al.*, 1993; Bastiani *et al.*, 1997).

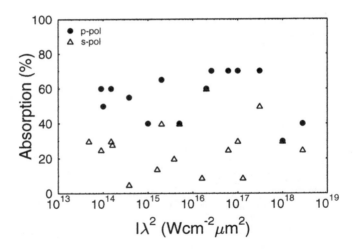

Fig. 5.16   Absorption measurements of short pulse laser-solid interactions since 1988 using *p*-polarized light (filled circles) and *s*-polarized light (triangles). For details of the experimental conditions, see the accompanying table on p. 170.

As the collection of measurements in Fig. 5.16 and Table 5.1 shows, the overall picture of absorption physics is still rather muddy: large discrepancies are found in absorption fractions measured at the same intensity with different lasers. Nevertheless, there is an unmistakably higher absorption for p-polarized light than s-polarized, which certainly supports the thesis developed in the previous sections that nonlinear, collective effects will dominate the physics in this regime.

Experimental evidence for particular mechanisms has also been reported by a number of groups. For example, Grimes *et al.*, (1999) observe an angular absorption dependence at low intensities consistent with vacuum heating. Teubner *et al.*, (1996a) have seen Fresnel-type behavior for KrF* pulses at intensities in excess of 10^{18} Wcm^{-2} suggestive of the anomalous skin effect. At relativistic intensities ($> 10^{19}$ Wcm^{-2} @ 1 μm), Santala *et al.*, (2000) measured a change in the *direction* of hot electrons (see next section) as the density scale-length of a preplasma was increased, suggesting a transition from vacuum heating to either $j \times B$ or Raman instabilities. Currently there is a general notion (although still not really proven) that $j \times B$ heating must dominate the absorption physics at relativistic intensities. For this reason a large number of experiments are now performed at normal incidence, rather than at a finite angle to the target. This raises an intriguing yet unanswered question: do resonance absorption and vacuum heating switch off or do they simply manifest themselves differently because of momentum conservation (see p. 162)?

5.6 Hot Electron Generation

There are at least three tell-tale signatures of high-intensity collective effects in laser-plasma interactions: high, angular-dependent absorption into hot electrons, hard x-ray emission and fast ion generation. Over the last decade, many experiments have been performed which consistently verified this picture of sub-ps interactions. As seen in the previous section, absorption in excess of 50% for p-polarized light has been measured for intensities of 10^{16} Wcm^{-2} and above (Kieffer *et al.*, 1989b; Klem *et al.*, 1993; Meyerhofer *et al.*, 1993; Teubner *et al.*, 1993; Sauerbrey *et al.*, 1994), x-rays in the keV-MeV range have been detected (Kmetec *et al.*, 1992; Klem *et al.*, 1993; Chen *et al.*, 1993; Rousse *et al.*, 1994; Schnürer *et al.*, 1995; Teubner *et al.*, 1996a), and fast ion blow-off was also seen quite early on (Meyerhofer *et al.*, 1993; Fews *et al.*, 1994).

All of the collisionless absorption mechanisms discussed in Sec. 5.5 will,

Table 5.1 Absorption measurements with short pulse lasers since 1988, including information on the laser system and target materials used. The 'ASE' column gives the quoted contrast ratio of the amplified spontaneous emission or prepulse from the laser relative to the main pulse intensity. Where available, fractional absorption results (percentages) for p- and s-polarized light are quoted separately as η_p and η_s respectively. Further details can be found in the original source cited in the last column.

Laboratory	Energy	λ/nm	τ_L/fs	I_L/ Wcm^{-2}	ASE	Target(s)	θ_i/°	η_p	η_s	Reference
Bell Labs	7 mJ	308	400	10^{15}	10^{-3}	Al	45	60	25	Milchberg (1988)
Berkeley	5 mJ	616	160	10^{16}	10^{-4}	Si	–		20–30	Murnane (1989)
LLE/INRS-E	300 mJ	1053	1000	10^{16}	10^{-3}	Al, Cu	0–45	40	40	Kieffer (1989b)
MPIBC	20 mJ	248	150	2.5×10^{16}	$< 10^{-7}$	Al, Au	0–75	60	30	Fedosejevs (1990a)
LLE/INRS-E	300 mJ	1053	1000	10^{14}	10^{-3}	Al, Cu, Ta	45	55	35	Chaker (1991)
MPIBC	30 mJ	248	400	10^{17}	10^{-10}	Al	0–67.5	65	40	Teubner (1993)
LLE	60 mJ	1053	1000	2×10^{16}	$< 3 \times 10^{-6}$	Al, Si	0–70	63	28	Meyerhofer (1993)
LLNL	5 J	1053	800	2×10^{16}	10^{-3}	CH, Al, Ta	22.5	60	60	Klem (1993)
				10^{17}			22.5	30	30	
MPIBC/Rice	30 mJ	248	400	5×10^{17}	10^{-10}	Al, C	45	70	25	Sauerbrey (1994)
VSOI	500 mJ	1053	1500	10^{18}	10^{-6}	Al	0–67.5	70	30	Andreev (1994)
LLNL (USP)	200 mJ	400	120	2×10^{14}	10^{-7}	Al.Ta	0		30	Price (1995)
				10^{18}			0		9	
RAL/Jena	250 mJ	248	400	5×10^{18}	10^{-7}	Al	0–80	70	50	Teubner (1996a)
LULI/LOA	60 mJ	800	120	4×10^{16}	10^{-8}	SiO$_2$	45	70		Bastiani (1997)
Limeil/Jena	5 J	528	350	4×10^{18}	10^{-12}	Al	45	40	25	Feurer (1997)
Austin		620	120	10^{15}	10^{-3}	Al, Fe	20–65	55	5	Grimes (1999)

one way or another, result in the *superheating* of some fraction of the electrons to energies much higher than the initial bulk plasma temperature T_e. More often than not, this suprathermal electron component has a Maxwellian form with a characteristic temperature $T_h \gg T_e$, see Fig. 5.17. This is perhaps quite surprising, given that the hot electrons are accelerated by a *coherent* electric field — whether belonging to the laser itself or within the plasma: one might expect to see a more monochromatic beam-like tail with some thermal spread. However, as Bezzerides *et al.*, (1980) pointed out, the random, stochastic nature of the particle acceleration in standing-wave fields leads to strong cycle-to-cycle fluctuations in the trajectories and energies acquired by electrons. Averaging these single-particle distributions over time inevitably leads to a Maxwellian velocity distribution.

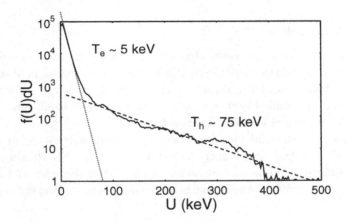

Fig. 5.17 Typical bi-Maxwellian electron distribution resulting from collisionless heating by a laser. This example resulted from a $1\frac{1}{2}$-D PIC simulation using the BOPS code (p. 242) with a laser irradiance 5×10^{16} Wcm$^{-2}\mu$m^2 incident at $45^\circ P$ onto a plasma with $n_e/n_c = 3$ and $L/\lambda = 0.2$.

This radical departure of the electron distribution function from a pure (single-temperature) Maxwellian is a hallmark of collective heating mechanisms. Contrast this with collisional heating, which leads to a broadening of $f(v)$ and corresponding increase in T_e. The determination of the hot electron temperature, T_h, and its associated hot electron *fraction* (that is, number of suprathermal electrons, n_h, often expressed relative to the critical density n_c) poses one of the most important physics issues in short-pulse

laser-solid interactions. Many applications — from femtosecond diffractometry to fast ignitor schemes — depend crucially on achieving the highest
possible conversion efficiency of laser energy into hot electrons of a specific
energy.

To date, this issue remains controversial because the interaction physics
consists of many competing effects, which vary in importance according to
the conditions: laser intensity, target material, contrast ratio, and so on.
Thus, as we have already seen, it is very difficult to isolate an absorption
mechanism, either experimentally or even in a simulation. In what follows,
the signatures of hot electron generation will be examined for the main
absorption processes already discussed, and compared to the available experimental data for femtosecond laser pulses collected over the last decade
or so.

5.6.1 *Long pulses*

Before we tackle the processes characteristic of short pulse interactions,
however, it is helpful to recall the established theories of resonant absorption
from Los Alamos and Livermore, since these will serve as a long pulse
benchmark. A detailed review of this work was made some time ago by
Haines (1979), which we freely borrow from here.

From the theoretical viewpoint, long laser pulses (that is, of several
picoseconds duration and longer) have the advantage that things tend to
be in equilibrium. Thus, one can safely assume that the laser and plasma
are in *pressure balance*, which, mathematically stated, means simply that:

$$P_e = n_e k_B T_e = \frac{I_0}{c} = \frac{E_0^2}{8\pi}. \tag{5.78}$$

Physically, this implies that the density profile near the critical point adjusts
according to the local laser pressure and temperature, that is:

$$n_e = \frac{I_0}{c k_B T_e} \simeq 200 \frac{I_{18} \lambda_\mu^2}{T_{\text{keV}}} n_c. \tag{5.79}$$

Contrast this with the short pulse case, where the electron density is essentially pinned to the solid density via the quasi-neutrality condition $n_e = Z n_i$
(p. 128). A relationship like Eq. (5.79) is appropriate if the ions have sufficient time to move in response to the laser pressure. Below the reflection
point, an underdense, outwardly streaming plasma 'shelf' will also form
(Kruer, 1988).

An equation for the equilibrium energy balance can be obtained by simply assuming that the absorbed laser flux is entirely carried away by a population of free-streaming hot electrons with a temperature T_h:

$$\eta_a I_0 = \beta n_h v_h \frac{m v_h^2}{2}, \tag{5.80}$$

where η_a is the usual absorption fraction, $v_h = (k_B T_h/m)^{1/2}$ and n_h are the mean hot electron velocity and number density respectively. The constant $\beta = \sqrt{2/\pi}$ arises from the assumption that the hot electrons form a 1-dimensional Maxwellian distribution. Thus, in this standard picture, the hot electrons determine the energy balance, whereas the cold bulk electrons are responsible for momentum (or pressure) balance — a situation which will prevail as long as $n_h < n_c \ll n_e$.

In particle-in-cell simulations carefully set up to ensure pressure balance from the outset, Forslund, Kindel and Lee (1977) (henceforth denoted FKL) argued that the hot electron temperature should scale like:

$$T_h^{\text{FKL}} \simeq 14(I_{16}\lambda_\mu^2)^{1/3} T_e^{1/3} \text{ keV}, \tag{5.81}$$

where I_{16} denotes the laser intensity in units of 10^{16} Wcm^{-2}, λ_μ is the laser wavelength in μm and T_e the temperature of the cold, bulk electrons in keV. This scaling law is based on the physically appealing ansatz that an electron which is resonant with a plasma wave will gain an energy:

$$T_h \sim e\Delta\phi = eE_p L,$$

where E_p is the local longitudinal electric field and L the half-width of the density perturbation (which in turn, is proportional to the density scale-length). Forslund *et al.*, (1977) found that in their simulations, L turned out to be the geometric mean of the plasma skin depth in the upper shelf region, the Debye length at the critical density, and the laser wavelength divided by 2π. In other words,

$$L \simeq \left(\frac{c^2 v_{te}}{\omega^2 \omega_p}\right)^{1/3}, \tag{5.82}$$

which is reminiscent of expression (5.73) for the anomalous skin depth, l_a. In a steepened, step-like profile, we can take $E_p \simeq E_0$, so that combining Eq. (5.79) with Eq. (5.82) to eliminate ω_p, one can recover the expression quoted above for T_h, Eq. (5.81), see Exercise 8.

A similar analysis of hot electron spectra resulting from resonance absorption was made independently by Estabrook and Kruer (1978). They found a much stronger inverse dependence of the steepened density scale-length on the intensity: $L \propto I^{-0.48}$ and deduced two empirical formulae for T_h by varying I and T_e in their 2D PIC simulations:

$$T_h^{\mathrm{EK1}} = 20T_e^{1/4}(I_{16}\lambda_\mu^2)^{0.39},$$
$$T_h^{\mathrm{EK2}} = T_e + 21T_e^{0.04}(I_{16}\lambda_\mu^2)^{0.42}. \qquad (5.83)$$

The units are the same as in Eq. (5.81). Again, Estabrook & Kruer appeal to the capacitor model to explain the weak scaling of T_h on laser intensity. With increasing intensity, higher plasma electric fields are generated but their resonance width is reduced, so that the net energy gain by an electron, $\Delta U \sim eE_pL$, will scale more weakly than $E_p \sim I^{1/2}$. These two model curves, depicted in Fig. 5.18, predict 50% lower temperatures, but very similar scaling, to data from nanosecond laser-plasma experiments over a wide range of parameters.

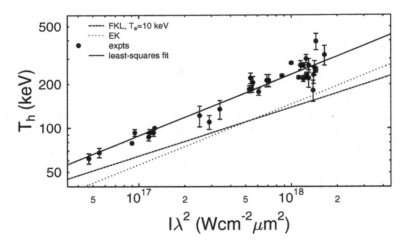

Fig. 5.18 Scaling of hot electron temperature with intensity according to steady-state theories (Eqs. 5.81 and 5.83) and experimental x-ray measurements made with the Los Alamos nanosecond CO_2 laser (Priedhorsky *et al.*, 1981). The least-squares fit is given by $T_h \simeq 89\,(I_{17}\lambda^2)^{0.42\pm0.12}$ keV.

5.6.2 *Short pulses*

Turning now to femtosecond laser-plasma interactions, the first point to note is that the steady-state picture above is unlikely to apply for two reasons: i) the much sharper density and field profiles and ii) the nonequilibrium nature of the interaction on this short time-scale. So what should we use instead? The variety of absorption mechanisms discussed previously makes it difficult to formulate a comprehensive theory of hot electron generation along the lines just described.

Nevertheless, one can at least construct approximate scaling-laws for T_h from the absorption models as they stand. Let us start with the Brunel model, in which electrons are directly accelerated by the laser field incident on a step-like density profile, see Sec. 5.5.2. According to the simple electrostatic version of this model, the electrons acquire velocities $v_d = 2v_{os} \sin \theta$, see Eq. (5.64). The hot electron temperature is therefore just

$$T_h^{\rm B} = \frac{mv_d^2}{2} \simeq 3.7 I_{16} \lambda_\mu^2, \qquad (5.84)$$

where, as before, I_{16} is the intensity in 10^{16} Wcm^{-2} and λ_μ the wavelength in microns. The above estimate does not allow for pump depletion or relativistic effects, which would reduce this scaling at higher intensities. Since $\eta_{\rm B} \sim I_0^{1/2}$, this can only be consistent with the energy balance relation (5.80) if n_h is independent of I_0.

PIC simulations support this scaling in a double-pump geometry (2 laser beams at $\pm 45°$), where surface currents are all but cancelled (Brunel, 1988). In the experimentally more relevant case of a single, obliquely incident pump onto a steep but finite density gradient ($L/\lambda \leq 0.1$), electromagnetic simulations with *fixed* ions give a much weaker scaling (Gibbon and Bell, 1992):

$$T_h^{\rm GB} \simeq 7 \left(I_{16} \lambda^2 \right)^{1/3}, \qquad (5.85)$$

which is similar to $T_h^{\rm FKL}$ for long-pulse resonance absorption in steepened density profiles, but numerically a factor of two or so lower depending on the density scale-length.

The weak scaling for small L/λ and fixed ions partly comes about because the absorption fraction *decreases* with intensity. Closer inspection of Fig. 5.14 on p. 161 reveals a scaling $\eta_{\rm vh} \sim a_0^{-4/3} \sim (I\lambda^2)^{-2/3}$ for this case.

Thus, by invoking the energy balance Eq. (5.80) again, one would have:

$$T_h \sim 7 \left(I\lambda^2\right)^{\frac{2}{9}} \left(\frac{n_h}{n_c}\right)^{-2/3} .$$

Comparing this expression to T_h^{GB} above would suggest that the hot electron fraction must also decrease weakly with laser intensity ($\sim I^{1/6}$) under these conditions. Analogous simulations with somewhat longer pulse lengths and mobile ions (Gibbon, 1994) yield temperatures more in line with T_h^{FKL}. Although there is still no pressure balance in this case, a quasi-stationary state is rapidly reached in which the bulk ions form a shock in the overdense region, but a small fraction is also pulled out, forming an underdense shelf. Under these conditions, the absorption is dramatically enhanced (reaching values of 80%) and more electrons are able to sample the full laser field, leading to higher temperatures.

Another school of thought on hot electron generation (Wilks *et al.*, 1992) argues that at 'relativistic' intensities, $T_h \propto e\Phi_p$, where Φ_p is the ponderomotive potential due to the standing wave field formed when the laser is reflected by the target surface. This appealingly simple expression arises naturally from the $j \times B$ mechanism (Sec. 5.5.5), which when generalized to relativistic intensities, predicts that

$$T_h^{\text{W}} \simeq mc^2(\gamma - 1)$$

$$= mc^2 \left[\left(1 + \frac{p_{os}^2}{m^2 c^2}\right)^{1/2} - 1 \right]$$

$$\simeq 511 \left[\left(1 + 0.73 I_{18}\lambda_\mu^2\right)^{1/2} - 1 \right] \text{ keV}. \qquad (5.86)$$

Basically, Eq. (5.86) is expected to apply at normal incidence for intensities above 10^{18} Wcm^{-2}, where the $j \times B$ mechanism becomes significant and other mechanisms are suppressed because the laser electric field vector is perpendicular to the density gradient. Sure enough, this is more-or-less what is found in 2D PIC simulations of laser interactions with step-like density profiles (Wilks *et al.*, 1992; Wilks, 1993) or extended density ramps (Pukhov and Meyer-ter-Vehn, 1997). However, these simulations were all performed with *mobile* ions, meaning that the absorption and hot electron generation was accompanied by hole boring (Sec. 5.8), adding further nuances to the interaction physics.

Hole boring, or any kind of surface deformation for that matter, modifies the interaction geometry in such a way that the laser field may couple directly (and usually more efficiently) to plasma oscillations parallel to the density gradient. If a hole or chanel is formed, Brunel-type absorption may take place along the sides; whereas the $j \times B$ mechanism may be enhanced at the back of the hole due to additional self-focusing of the light in the underdense plasma in front of the target (see Sec. 4.6). If this underdense region is very extensive (10s or 100s of microns), perhaps as a result of a prepulse or high-intensity pedestal, then a very intense pulse may drive electrons forward in a snow-plough manner before it even reaches the critical surface. Also in the underdense region, the pulse could undergo relativistic self-focusing (Sec. 4.6) or suffer Raman forward-, back- and side-scatter instabilities (Sec. 4.4).

5.6.3 *Measurements of hot electron temperature*

Overall, there may be many competing processes contributing to the final hot electron energy spectrum, depending on factors such as pulse intensity, duration, contrast ratio, angle of incidence, and so on. Rather than attempt to argue in favour of one or the other source of hot electrons in various limiting cases, we turn to the experimental data for assistance: can measurements of T_h provide some clues as to which mechanisms, if any, predominate for a particular intensity range?

A summary of hot electron measurements over the last decade or so is shown in Fig. 5.19. The labels refer to the experiments listed in Table 5.2. Qualification for this chart consists of a hot electron temperature measurement by one of three methods: electron spectrometer, K_α x-rays from layered targets, or bremsstrahlung radiation. Perhaps a little unfairly, fast ion measurements (Meyerhofer *et al.*, 1993) have been excluded here because of the complex relationship between hot electron temperature and ion expansion in the relativistic regime, see Sec. 5.7. A fairly loose definition of short pulse has been applied here — anything from 30 fs to 3 ps, in fact — but which is more than sufficient to distinguish from the long pulse experiments in, for example, Fig. 5.18.

Superimposed on the experimental points are four theoretical curves representing some of the models discussed previously. A number of general features are immediately apparent. First, despite the wide range of laser wavelengths (248 nm–1053 nm), target materials and contrast ratios used in these measurements, there is an indisputable general dependence of T_h on the product $I\lambda^2$, rather than intensity alone, as predicted by all theories, see Exercise 6 on p. 213.

Table 5.2 Hot electron temperature measurements using short pulse lasers.

Laboratory	λ/nm	t_L/fs	θ_i	ASE	Target	I/Wcm^{-2}	T_h/keV	Method	Reference
STA	807	120	30p	10^{-6}	Ta	3×10^{18}	300	γ	Kmetec (1992)
LLE	1050	1300	60p	10^{-5}	Al/Si	3×10^{15}	3	K$_\alpha$	Chen (1993)
LULI	620	100	7p		Al/Si	2×10^{16}	8.5	K$_\alpha$	Rousse (1994)
	800	120	45p	10^{-5}	Si	4×10^{16}	19	e$^-$, K$_\alpha$	Bastiani (1997)
IOQ-RAL	248	400	45p	10^{-7}	Al/Si	10^{18}	8	e$^-$	Teubner (1996a)
IOQ-CELV	528	350	45p	10^{-12}	Cu	10^{19}	420	e$^-$	Feurer (1997)
CELV	1056	300	0	10^{-8}	CH	2×10^{18}	370	e$^-$	Malka (1997)
						9×10^{18}	890	e$^-$	
MBI	1056	1500	40p			5×10^{17}	140	γ	Schnuerer (1995)
	1056	700	45p	10^{-6}	Ta	5×10^{17}	66	γ	Schnuerer (1997)
MBI-LOA	800	80	45p	10^{-5}	Ta	3×10^{18}	250	γ	Schnuerer (2000)
IC-RAL	1056	2500	30p			5×10^{18}	390	γ	Beg (1997)
	1056	1000	45p			10^{19}	1000	γ	Norreys (1999)
INRS	530	400	45p			5×10^{17}	25	γ	Yu (1999a)
						5×10^{18}	40	γ	
IOQ	248	500	25p		Al/Si	1.4×10^{16}	3.5	K$_\alpha$	Uschmann (1999)
	800	100	45p			5×10^{18}	300	γ	
	800	100	45p			2×10^{19}	1000	γ	
LLNL	1064	500	25p	10^{-7}		2×10^{19}	300	K$_\alpha$	Schwoerer (2001)
	1064	500	25p			10^{20}	4000	e$^-$	Wharton (1998)
MPQ	800	130	45p		Cu/Ni	2×10^{18}	200	K$_\alpha$	Pretzler (2000)

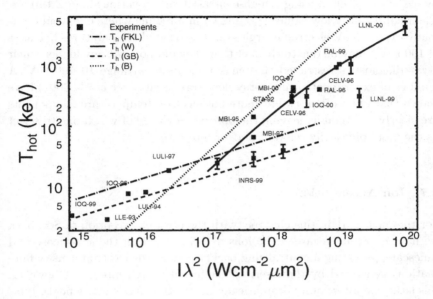

Fig. 5.19 Hot electron temperature measurements in femtosecond laser-solid experiments (squares) compared with various models: long pulse — Eq. (5.81); Brunel — Eq. (5.84); Gibbon & Bell — Eq. (5.85) and Wilks — Eq. (5.86). See the table on p. 178 for an explanation of the labels.

Closer inspection of the chart reveals two distinct low- and high-intensity regimes. From $I\lambda^2 = 10^{15}$–10^{17} Wcm$^{-2}\mu$m^2, the temperatures follow the scaling for vacuum heating predicted by Eq. (5.85). There follows an abrupt transition to the ponderomotive scaling described by Eq. (5.86). Whether this is really a change of absorption mechanism or an adjustment in, say, the energy/momentum balance as the intensity is increased, is probably impossible to say for sure: the adherence of recent measurements to the T_h^W curve in the relativistic regime ($I\lambda^2 > 10^{18}$ Wcm$^{-2}\mu$m^2) is impressive, nonetheless.

Notice that for non-relativistic intensities, T_h^W has the same scaling as the naîve Brunel model, $T_h^B \sim I\lambda^2$. Thus, it might be possible to observe a very strong scaling within a small range of intensities around 10^{17} Wcm$^{-2}\mu$m^2, as indeed claimed by Schnürer *et al.* (1995). Beyond this point, however, the asymptotic scaling is $\sim (I\lambda^2)^{1/2}$, which is also roughly what one obtains by trying to fit a curve to *all* the points displayed. One anomaly in the chart appears to be the series of values (INRS-99) obtained by Yu *et al.*, (1999a) using a high-contrast laser on various high-Z targets. Their consistently low values (≤ 50 keV) in a regime where one would normally expect MeV electron energies (and associated temperatures in excess of 100 keV), may be due to the fact that they did not *look* any higher: their x-ray diagnostics restricted the measurements to a range 10–100 keV. A number of experiments using γ-ray detectors (Schwoerer *et al.*, 2001) have shown that quite often, two or more hot electron temperature components are clearly identifiable, leaving the observer a choice of which one to select as the more physically relevant and/or interesting.

5.7 Ion Acceleration

For the most part in this chapter (with the obvious exception of Sec. 5.4), we have generally regarded the ions as being fixed on the sub-picosecond timescale, providing a neutralizing background to the electron density fluctuations generated by the laser. At sufficiently high intensities however, this situation alters quite dramatically due to the large electric fields, typically $\sim$ GV m^{-1}, induced when many electrons are rapidly displaced from their initial positions. As a result, a substantial fraction of ions may be accelerated to energies in the multi-MeV range. Before analyzing this process in detail, one may deduce a few general points concerning laser-ion acceleration from simple physical considerations. First, we notice that the ion quiver motion in a laser field is negligible compared to that of the electrons:

$$\frac{v_i}{c} = \frac{ZeE}{M\omega c} = \frac{Zm}{M}a_0, \qquad (5.87)$$

where $a_0 = v_{os}/c$ has its usual definition. Thus, to accelerate ions to relativistic velocities ($v_i \sim c$) directly by the laser field, one would require $a_0 \sim 2000$, or intensities in the region of $I\lambda^2 > 10^{24}$ Wcm$^{-2}\mu$m^2, still well beyond the reach of current CPA technology. In a plasma, however, the electrons *mediate* between the laser field and the ions via charge separation. In other words: the laser pushes the electrons around; the displaced electrons pull on the ions. Because of their higher inertia, the ion response is delayed by a factor $(M/Zm)^{1/2}$ relative to the electrons, which is just the ratio of electron and ion plasma frequencies ω_p/ω_{pi}, or the square-root of their quiver velocities. We will see how this factor actually arises shortly.

5.7.1 *Long pulse blow-off*

Just as for hot electron generation discussed in the previous section, there is a standard picture of energetic ion production associated with long pulse interactions, which again holds for short pulses only up to a point. Indeed, recent experiments on either side of the Atlantic have provoked some controversy over the origin and acceleration mechanism of energetic ions.

Before we confront these new mechanisms, however, we review the standard picture of ion blow-off, which is based on simple charge separation arguments: the hot electrons escaping the plasma in front of the target create an electrostatic Debye sheath which tugs on the ions. The mean ion energy will therefore be directly related to the potential generated by the hot electrons, whose kinetic energy is of course proportional to T_h. Thus, we expect

$$U_{\text{ion}} \propto k_B T_h,$$

for ions directed back towards the laser; a scaling which has been confirmed in an extensive series of experiments at the Los Alamos National Laboratory in the early 1980's (Gitomer *et al.*, 1986).

A quantitative estimate of this energy (more precisely, the maximum ion energy) can be obtained by dividing the plasma up into hot and cold components (Wickens *et al.*, 1978). For simplicity, let us assume that the energetic particles are significantly hotter than their cold counterparts, so that $T_h \gg T_e$, and on a short timescale (≤ 1 ps, say), the *bulk* of the ions will remain fixed. The hot electrons will form a cloud in front of this step-like target, resulting in an induced electric field

$$n_h e E = -\frac{\partial P_h}{\partial x} = -k_B T_h \frac{\partial n_h}{\partial x}, \qquad (5.88)$$

where $P_h = n_h k_B T_h$ is the hot electron pressure. Thus we immediately obtain for the field:

$$E \sim \frac{T_h}{L_h}, \qquad (5.89)$$

where $L_h \equiv n_h^{-1} \partial n_h / \partial x$ is the characteristic density scale-length of the hot electrons. This field will be felt by a number of (hot) ions in the vicinity of the target surface, which we will assume to obey the usual continuity and momentum equations (5.57, 5.58), reproduced here for convenience:

$$\frac{\partial n_i}{\partial t} + \frac{\partial}{\partial x}(n_i v_i) = 0, \tag{5.90}$$

$$\frac{\partial v_i}{\partial t} + v_i \frac{\partial v_i}{\partial x} = \frac{Ze}{M} E. \tag{5.91}$$

Substituting Eq. (5.88) into Eq. (5.91) and applying quasineutrality between hot electrons and ions ($n_h \simeq Z n_i$) to the pressure term, the momentum equation can be rewritten:

$$\frac{\partial v_i}{\partial t} + v_i \frac{\partial v_i}{\partial x} = -C_h^2 \frac{1}{n_i} \frac{\partial n_i}{\partial x}, \tag{5.92}$$

where $C_h = (ZT_h/M)^{1/2}$ can be identified as the *hot ion sound speed*.

This system is identical to the usual isothermal rarefaction wave in a freely expanding plasma (Kruer, 1988; Zel'dovich and Raizer, 1966), except that we have replaced the usual sound speed, $c_s = (ZT_e/M)^{1/2}$ by C_h. In our case, the expansion is driven by *hot* electrons rather than thermals. Following Kruer, it is a straightforward matter to write down the analogous self-similar solutions:

$$v_i = C_h + x/t$$
$$n_i = n_0 \exp(-x/C_h t). \tag{5.93}$$

A number of authors have pointed out that the final ion velocity, rather than increasing indefinitely as predicted by Eq. (5.93), will ultimately be limited by truncation of the electron velocity distribution (Pearlman and Morse, 1978; Denavit, 1979; Kishimoto *et al.*, 1983): the ions 'catch up' with the electrons and then continue to coast towards the detector with constant velocity. Traditionally, for low ion energies, Faraday cups are used to measure the ion current, which allows the velocity distribution dn_i/dv_i to be reconstructed. From Eq. (5.93) we see that $dn_i/dv_i \sim \exp(-v_i/C_h)$, so that in principle, this diagnostic yields the hot electron temperature through the hot ion sound speed C_h. A plasma containing two electron components will result in a characteristic dip in the ion distribution determined by this time-of-flight technique, thereby permitting both cold and hot temperatures to be deduced (Wickens *et al.*, 1978).

The position of the ion front, x_f, can be estimated by truncating the density profile at the point where its gradient equals the local Debye length of the hot electrons. That is:

$$L_n = \left| \frac{n_i}{\partial n_i / \partial x} \right|_{x_f} = C_h t_f = \lambda_D(x_f).$$ (5.94)

Noting that

$$\lambda_D(x_f) \equiv \left(\frac{k_B T_h}{4\pi Z n_f e^2} \right)^{1/2} = \frac{C_h}{\omega_{pi0}} \left(\frac{n_0}{n_f(x_f)} \right)^{1/2},$$

where $\omega_{pio} = (4\pi n_0 Z^2 e^2 / M)^{1/2}$ is the ion plasma frequency at the initial density n_0; we have from Eq. (5.93) and Eq. (5.94)

$$\begin{aligned}
v_{\max} &= C_h + \ln\left(\frac{n_0}{n_f} \right) \\
&= C_h + \ln\left(\omega_{pio} \frac{\lambda_D(x_f)}{C_h} \right)^2 \\
&= C_h \left[1 + 2\ln(\omega_{pio} t_f) \right].
\end{aligned}$$ (5.95)

Notice that even for highly relativistic electron temperatures the ion velocities remain largely *non*-relativistic: for example, for $T_h = 10$ MeV, we have $C_h \sim 0.1c$ for protons. Thus, noting that the logarithmic term in Eq. (5.95) is slowly varying, the maximum ion energy will scale as

$$U_{\max} \sim \frac{1}{2} M v_{\max}^2 \sim \frac{1}{2} M C_h^2 \sim Z T_h.$$ (5.96)

The first substantial study of fast ion generation with a short pulse laser was made by the Rochester group. Here, Faraday cups were used to determine the ion distribution and infer the hot electron temperature in the fashion described above. Meyerhofer *et al.*, (1993) showed a clear correlation between ion energies and the p-polarized component of the laser intensity; strong evidence that an oblique-incidence absorption mechanism such as resonance absorption was prevalent in the experiment. Shortly afterwards, a more sensitive diagnostic for MeV ions — a CR-39 plastic nuclear track detector — was successfully deployed in a short pulse experiment at the Rutherford laboratory. Fews *et al.*, (1994) found that the 2 ps Vulcan laser produced ions with a mean energy scaling consistent with the earlier data collected with CO_2 and Nd:glass nanosecond systems (Gitomer *et al.*, 1986), and a maximum ion energy conforming to Eq. (5.96), implying that $U_{max} \sim (I\lambda^2)^{1/3}$ if the long-pulse temperature scaling Eq. (5.81) is valid in this case (Beg *et al.*, 1997).

5.7.2   *Forward acceleration*

Over the last five years, a number of groups have reported multi-MeV ion measurements at intensities between 10^{18} Wcm^{-2} and 10^{20} Wcm^{-2} drawing markedly different conclusions as to the origin and acceleration mechanism of the ions. The argument was sparked off by the IC-RAL team, who presented measurements of energetic protons (≤ 18 MeV) again produced with the Vulcan CPA system but with a shortened pulse duration of ~ 1 ps (Clark *et al.*, 2000). In particular, ion detectors placed at the *back* of a 100 μm Al target reveal clear ring-patterns formed by a proton beam, see Fig. 5.20. Also a common feature of earlier nanosecond experiments, protons come from impurities (hydrocarbons or water vapour) adsorbed onto the metal target surface. The London team (Clark *et al.*,

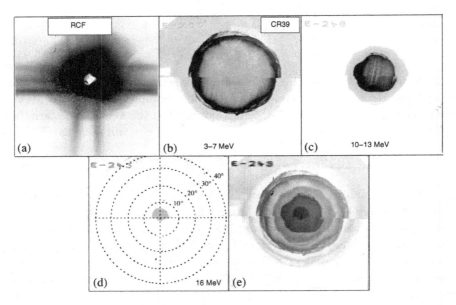

Fig. 5.20   Rear-side imaging of protons produced from a 100 μm Al target irradiated at 10^{19} Wcm^{-2}. a) Radio-chromic film (RCF) image sensitive to protons ≥ 3 MeV; b)– d) data from CR39 detector stack showing monoenergetic proton rings/discs in energy ranges 3–7 MeV, 10–13 MeV and 16 MeV respectively. *Courtesy:* E. Clark, Imperial College London, from (Clark, 2001).

2000; Zepf *et al.*, 2003) interpret these results by assuming that the protons were shock-accelerated near the critical surface (at the *front* of the target) by charge separation (see Sec. 5.8), and were subsequently deflected by ~ 30 MG magnetic fields as they travelled to the rear. Such large magnetic fields could be created by hot electrons 'leading' the ions through the target (Davies *et al.*, 1997), see also Sec. 5.9. A similar explanation was offered by the Michigan group (Maksimchuk *et al.*, 2000), who also claimed to observe 'frontside' acceleration of protons through to the rear of thin (2 μm) foil targets, albeit at much lower intensities and a higher contrast ratio by virtue of a frequency-doubled laser pulse.

Not to be outdone, the Livermore team responded with a new record (≤ 58 MeV) for protons produced by their Petawatt laser system focused onto plastic and gold targets (Snavely *et al.*, 2000). They ruled out proton acceleration from the front side on the grounds that preheating would have removed most of the adsorbed protons from the focal region. Moreover, observations made using wedge-shaped targets revealed collimated proton emission normal to the two rear surfaces; a fact which can more easily be explained by electrostatic forces created by hot electrons penetrating through to (and leaving) the rear surface. The accelerating field created this way is given roughly by

$$eE_{\max} = \frac{T_h}{\max(L_i, \lambda_{Dh})},$$

where L_i is the ion scale length at the rear and λ_{Dh} the hot electron Debye length. For hot electron temperatures of 6 MeV, this field is ~ 2 MeV μm^{-1} for an ion scale-length of 3 μm. A more rigorous analysis of rear-side (or target-normal) sheath acceleration (TNSA) has recently been given by Mora (2003). PIC simulations under somewhat idealized conditions appear to support this picture (Wilks *et al.*, 2001; Pukhov, 2001; Ruhl *et al.*, 2001), and also point to the possibility of *focusing* ion beams produced from a concave indentation at the rear. Numerical support for the front-side mechanism has been reported by Gibbon *et al.*, (2004) using a new particle simulation approach which takes into account finite target resistivity, see Sec. 5.10.1. The latter effect can prevent hot electrons from reaching the rear surface, trapping them instead in a region just in front of the ion shock front, thus enhancing the acceleration of front-side protons.

Needless to say, opinions on the origin of fast protons in this regime are highly polarized, while experimental evidence remains contradictory, so we can expect this topic to stay among the more controversial in short pulse interactions for some time yet.

5.8   Hole Boring

As we saw earlier in Sec. 5.6, bulk ion motion can alter the electron dynamics by changing the density profile near the critical surface. As long as this motion remains normal to the gradient, the absorption and hydrodynamics can still be modeled in 1D, even for obliquely incident light. This picture becomes inadequate if a hole is formed, or if the surface develops ripples. Both of these situations can occur for finite focal spot sizes (which are typically diffraction-limited to 2–10 μm) and at extreme irradiances ($I\lambda^2 > 10^{18}$ Wcm$^{-2}\mu$m^2). This regime was first studied in detail by Wilks *et al.*, (1992) using 2D PIC simulation. They found that tightly focused, normally incident light can bore a hole several wavelengths deep through moderately overdense plasma on the sub-ps timescale.

Hole boring results primarily from pressure imbalance. The inequality (5.76) describing the skin-layer regime is thus reversed: instead, $P_L/P_e \gg 1$, causing the plasma to be pushed inwards preferentially at the center of the focal spot. This results in the formation of an electrostatic bow-shock; a density discontinuity travelling into the target with constant velocity. To get a more quantitative handle on this process, we can make use of the simple one-dimensional shock model introduced by Estabrook *et al.*, (1975) to describe the asymptotic step-plateau density profiles found in particle-in-cell and hybrid fluid-kinetic simulations (Lee *et al.*, 1976) of resonance absorption and self-consistent profile steepening, see Fig. 5.21. As in the previous section, we start from the one-dimensional continuity and

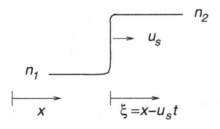

Fig. 5.21   Geometry of electrostatic shock.

momentum conservation equations for the ions (Eqs. 5.57, 5.58, this time with charge density ρ and velocity u), which including charge separation effects, are:

$$\frac{\partial \rho}{\partial t} + \frac{\partial}{\partial x}(\rho u) = 0 \qquad (5.97)$$

$$\frac{\partial \rho u}{\partial t} + \frac{\partial}{\partial x}(\rho u u) + \frac{\partial}{\partial x}(Z P_e + P_L) = 0. \qquad (5.98)$$

In a frame moving with the shock front, the fluid quantities will be stationary in time, so that we can set $\partial/\partial t = 0$ in Eqs. (5.97) and (5.98). Allowing for a fraction η_a of laser absorption at the surface, and assuming $P_L \gg Z P_e$, the fluid equations can then be integrated across the density step to give:

$$\rho u = \text{const.},$$

and

$$\rho u^2 = P_L = \frac{I_0}{c}(2 - \eta_a) \cos \theta.$$

Solving for u, we obtain:

$$\begin{aligned} \frac{u}{c} &= \left(\frac{(2 - \eta_a) I_0 \cos \theta}{2 \rho c} \right)^{1/2} \\ &= \left(\frac{Z m}{M} \frac{n_c}{n_e} \frac{(2 - \eta_a) \cos \theta}{4} \frac{I_{18} \lambda_\mu^2}{1.37} \right)^{1/2}, \end{aligned} \qquad (5.99)$$

which is essentially the result quoted by Zepf *et al.*, (1996). Setting $\eta_a = 0$ and $\cos \theta = 1$ (normal incidence with zero laser absorption), we recover the expressions of Wilks *et al.*, (1992) and Denavit (1992). To give a numerical example, a Ti:sapphire pulse focused to 10^{20} Wcm^{-2} onto a fully ionized Al plasma surface at solid density (10^{23} cm^{-3}) will result in a hole-boring velocity of $0.03c$. After one picosecond, the depth drilled by the laser will be 9 microns. These simple estimates from Eq. (5.99) are actually well reproduced in 2D PIC simulations by Wilks (1993).

Direct experimental evidence for hole boring has been given by Kalashnikov *et al.*, (1994), who observed a transition from blue-shift to red-shift in the wavelength of the reflected laser fundamental. The frequency shift is given by the usual Doppler formula:

$$\omega_r = \omega \frac{1 - \beta}{1 + \beta}, \qquad (5.100)$$

where $\beta = u/c$ is the hole-boring (or expansion) velocity of the critical surface. A systematic study of the intensity dependence of u using spectra of the 4th laser harmonic was made by Zepf *et al.*, (1996), who found excellent agreement with Eq. (5.99) assuming 50% laser absorption, see Fig. 5.22.

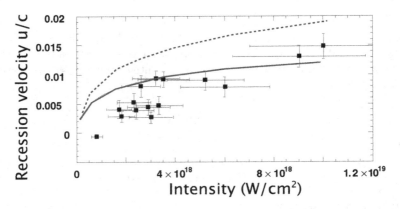

Fig. 5.22   Red shift from hole boring as function of laser intensity. *Courtesy:* M. Zepf, Queen's University Belfast; Fig. 4 from (Zepf *et al.*, 1996). ©1996, American Institute of Physics.

5.9   Magnetic Field Generation

Magnetic fields have been a source of fascination in the field of laser-plasma interactions for over 30 years. Of particular interest are the quasi-stationary, 'DC' fields arising from electron transport in the vicinity of the focal spot, which can reach 10^6–10^9 Gauss (G). A comprehensive review of the numerous effects which can lead to magnetic field generation in high-density plasmas, together with their saturation mechanisms, has been summarized recently by Haines (1997). Basically, a magnetic field will grow spontaneously in the absence of one-dimensional symmetry, where the electrons return along a *different* path, ultimately creating a current loop.

In short-pulse interactions, there are at least three mechanisms which can generate B-fields:

1)   *Radial thermal transport*, in which the electron temperature and

density gradients are not parallel (Stamper *et al.*, 1971), giving a *thermo-electric* source term:

$$\frac{\partial B}{\partial t} = \frac{\nabla T_e \times \nabla n_e}{en_e}. \tag{5.101}$$

Assuming the density gradient is primarily directed along the target normal, the magnitude of this field can be estimated as follows (Bell *et al.*, 1993):

$$B \sim 2 \left(\frac{\tau}{\text{ps}}\right) \left(\frac{k_B T_e}{\text{keV}}\right) \left(\frac{L_\perp}{\mu\text{m}}\right)^{-1} \left(\frac{L_\parallel}{\mu\text{m}}\right)^{-1} \text{ MGauss}, \tag{5.102}$$

where $L_\perp$ and $L_\parallel$ are the (transverse) temperature and (longitudinal) density gradients respectively, see Fig. 5.23. This field results in a magnetic pressure comparable to the plasma pressure — even for temperatures in excess of 100 keV — and ultimately leads to a *pinching* of the underdense plasma ablating from the target. Such elongated plumes have indeed been observed in experiments with picosecond pulses (Bell *et al.*, 1993; Beg *et al.*, 1997), and are also reproduced in fully self-consistent magneto-hydrodynamic (MHD) simulations (Bell, 1994).

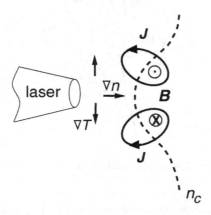

Fig. 5.23   Geometry of thermoelectric magnetic field generation ($\nabla T \times \nabla n_e$).

2) *DC currents in steep density gradients* driven by temporal variations in the ponderomotive force (Sudan, 1993; Wilks *et al.*, 1992; Tripathi and Liu, 1994; Mason and Tabak, 1998). As we have already seen in Sec. 4.6.3, a gradient in the time-averaged laser intensity causes electrons to be pushed

away from the focal spot. In the adiabatic limit (constant laser intensity), this drift is eventually halted by charge separation, leading to a local electron density depletion for roughly the duration of the laser pulse. A ponderomotive force $\boldsymbol{f}_p(t)$ increasing in time, however, will ensure that the 'push' is never completely balanced, so that a group of electrons will acquire a velocity

$$\boldsymbol{v}_p \propto m_e^{-1} \boldsymbol{f}_p \sim \nabla E_L^2,$$

yielding a net *DC ponderomotive current* $\boldsymbol{J} = e n_e \boldsymbol{v}_p \sim n_e \nabla I_0$. Naïve application of Ampère's law then gives us:

$$\nabla^2 \boldsymbol{B} \sim \nabla \times \boldsymbol{J}$$
$$\sim \nabla n_e \times \nabla I_0. \qquad (5.103)$$

Thus, a radial intensity profile incident on a steep density gradient will

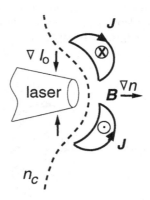

Fig. 5.24   Geometry of ponderomotively generated magnetic field $\nabla n_e \times \nabla I_0$.

drive a magnetic field in the sense depicted in Fig. 5.24. The theoretical treatments cited above differ on the magnitude of this ponderomotive magnetic field, but all agree on its polarity, which is exactly the *opposite* to the thermoelectric field in Fig. 5.23. In the highly relativistic limit ($a_0 \gg 1$), Sudan (1993) predicts a static field of the same order of the laser B-field, or 10^9 G for $a_0 = 10$, extended over a collisionless skin depth (c/ω_p).

3) *Fast electron currents* either along the target surface (Brunel, 1988; Gibbon, 1994; Ruhl and Mulser, 1995), or directed into the target (Pukhov

and Meyer-ter-Vehn, 1997; Bell and Kingham, 2003). These currents are a natural consequence of collective absorption mechanisms (Sec. 5.5), and develop independently of the ponderomotive effects considered above. A hot electron flux $n_h v_h$ can be significant at intensities above 10^{18} Wcm^{-2}, but will eventually be balanced by a cold return current, see Sec. 5.10. For this reason, the fields generated in this way are much lower than the ponderomotively driven ones near the surface: particle-in-cell simulations give values in the region of 100 MG for $I_0 = 10^{20}$ Wcm^{-2}(Pukhov and Meyer-ter-Vehn, 1997).

The first of the above mechanisms, thermoelectric magnetic field generation, occurs on the hydrodynamic timescale and thus persists long after the laser pulse is incident. By contrast, the other two mechanisms will act predominantly at early times ($t \leq \tau_p$) in the overdense (solid) region. For intensities of 10^{19} Wcm^{-2}, the magnitude of the B-field can be in excess of 10^9 G.

Direct measurements of MG B-fields in short pulse interactions have been made by Borghesi *et al.*, (1998) using a standard Faraday rotation technique. Although this method is restricted to the underdense part of the density profile, the authors found two distinct regions of magnetic field: the first around 50 μm from the laser spot, consistent with the thermoelectric $\nabla T \times \nabla n$ mechanism; the second closer to the focal spot with the field *reversed* in sign, indicating that either fast electrons or ponderomotively driven currents were responsible. Similar field reversals are also seen in 2D hydrodynamic simulations including both mechanisms (Mason and Tabak, 1998).

More recently, attempts have been made to measure the much higher and shorter-lived fields generated at near solid densities, i.e. in a region beyond the reach of optical probes. In particular, an experiment by Tatarakis *et al.*, (2002) at RAL using polarization measurements of the self-generated laser harmonics (see Sec. 5.12) suggest the existence of peak fields 340 MG $\leq B \leq$ 460 MG at intensities just above 10^{19} Wcm^{-2}.

5.10 Hot Electron Transport

In Sections 5.6–5.7 we have seen that energetic electrons and ions are an inevitable consequence of intense irradiation of solid surfaces. As far as electrons are concerned, the physical processes responsible for their acceleration are reasonably well understood, so that under controlled experimental conditions, one can predict the number and temperature of fast electrons produced — as encapsulated by Fig. 5.19. What happens next is less certain: the simplest and most frequently applied option is to assume that all these electrons, carrying, say, 20–30% of the laser energy, are absorbed somewhere inside the target, producing hard x-rays along the way, see Sec. 7.2.

This assumption appears to work well for modest laser intensities — below 10^{17} Wcm^{-2} say — independently of pulse length or absorption mechanism. On the other hand, a number of studies in the long-pulse regime already showed that induced electric and magnetic fields could complicate matters by inhibiting or deflecting fast electron flow (Bond *et al.*, 1980; Amiranoff *et al.*, 1982; Wallace, 1985; Luther-Davies *et al.*, 1987). Most of these effects can be attributed to the substantial underdense coronal layer in front of the target, a feature usually absent from femtosecond interactions. Geometrical and pre-plasma issues aside, it is worth asking how our naïve picture of hot electron transport — a beam-like flux of high-energy particles streaming freely into near-solid density plasma — holds as the laser intensity is increased.

5.10.1 *Inhibition by electric fields*

The flaw in this assumption was neatly exposed by Bell *et al.*, (1997) by way of the following paradox. Suppose we dump a portion of the laser energy into hot electrons with a characteristic energy T_h, and that this is carried away at a velocity $v_h = (k_B T_h/m)^{\frac{1}{2}}$ into the target. This is just the situation we had before in Sec. 5.6 to describe energy balance in resonance absorption, so dropping constants $O(1)$ arising from the shape of the hot electron distribution, we can rewrite Eq. (5.80) as:

$$\eta_a I \simeq n_h v_h T_h, \qquad (5.104)$$

where n_h is the hot electron density. The details of the absorption process are not important here, so we just replace $\eta_a I$ by I_a, the effective *absorbed* laser intensity. The hot electron flow implied by the RHS of Eq. (5.104) contains a particle flux

$$\phi_h = n_h v_h,$$

which has an associated current density

$$j_h = e\phi_h = en_h v_h = \frac{I_a}{T_h/\mathrm{eV}} \quad \mathrm{A\ cm}^{-2}. \tag{5.105}$$

Assuming the hot electrons are generated uniformly over the laser focal spot area, $\pi\sigma_L^2$, the total hot electron current carried into the target is

$$I_h = \frac{\pi\sigma_L^2 I_a}{T_h} = \frac{U_a}{T_h \tau_L}, \tag{5.106}$$

where $U_a = \eta_a U_L$ is the absorbed laser energy. As an example, consider a representative laser system with $U_L = 1$ J, $\tau_L = 100$ fs, focused to 10^{18} Wcm^{-2}. From the chart in Fig. 5.19, we expect a fast electron temperature $T_h \simeq 100$ keV at this intensity. If 50% of the laser energy is converted to hot electrons, Eq. (5.106) predicts a current $I_h = 50$ Megaamps, flowing in a cylinder with radius $\sigma \simeq 25$ μm. Elementary magnetostatics (Ampère's law) says that the magnetic field generated by this current flow is given by

$$\oint d\boldsymbol{l} \cdot \boldsymbol{B}_s = \frac{4\pi I_h}{c},$$

or

$$B_s \simeq 2000 \frac{I_{\mathrm{MA}}}{\sigma_\mu}, \tag{5.107}$$

with B_s in Megagauss, I_{MA} in Megaamps, σ_μ in microns.

For our example above, the magnetic field generated by the hot electrons would be a massive 4 Gigagauss (GG); an unimaginable value by terrestrial standards. (The Earth's magnetic field peaks at a mere 0.7 G near the poles.) Now suppose that this field is produced along the whole length of the hot electron trajectory. The penetration depth of an electron beam is normally determined by collisions in the target material, and can be estimated by the Bethe-Bloch formula, which for aluminium predicts (Harrach and Kidder, 1981):

$$R_e \simeq 2.5 \times 10^{-2} T_{\mathrm{keV}} \simeq 70\mu\mathrm{m}.$$

The magnetic energy contained in this cylinder can be estimated by integrating the energy density:

$$U_B = \int_0^\sigma 2\pi R_e \frac{B_s^2(r)}{8\pi} \cdot r dr$$

$$= \frac{B_s^2}{8\pi}\pi\sigma^2 R_e \text{ erg.} \tag{5.108}$$

Plugging in our numbers for the surface field $B_s = 4 \times 10^9$ G, $\sigma = 2.5 \times 10^{-3}$ cm, we find $U_B \simeq 9 \times 10^{10}$ erg, or 9 kJ. This is nearly 1000 times the amount of energy absorbed by the laser in the first place, implying that at least one of our assumptions was horribly wrong! Clearly a Megaamp current cannot exist in the 'naked' form we supposed: Bell *et al.*, (1997) reasoned that a *return* current must be supplied in addition so that the net current and its associated magnetic field inside the target are kept within physically reasonable bounds. In a hot plasma this is no problem, since it acts like a superconductor with almost zero resistivity ($\nu_{ei} \sim T_e^{-3/2} \rightarrow 0$), and one can easily satisfy $j_h + j_c \simeq 0$. In a cold solid, or cold, solid-density plasma, there *will* be a finite resistivity, so that $j_c \ll j_h$. In this case, an electric field will be rapidly induced, effectively inhibiting further hot electron penetration. The magnitude of this field is just given by Ohm's law:

$$j_c = -j_h = \sigma_e E, \tag{5.109}$$

where σ_e is the electrical conductivity.

Instead of hot electrons streaming freely into a passive, cold target, we have to picture an electrostatic equilibrium in which electrons are reflected back from the solid region out towards the surface by the potential barrier associated with this field. For a Maxwellian distribution, the density of hot electrons in the region of this barrier is given by (Chen, 1974):

$$n_h = A \exp(\phi/T_h),$$

from which we obtain

$$E = -T_h \frac{\nabla n_h}{n_h}. \tag{5.110}$$

Moreover, the continuity equation requires that the density obey

$$\frac{\partial n_h}{\partial t} + \nabla \cdot \boldsymbol{j}_h = 0.$$

Combining Ohm's law Eq. (5.109) with Eq. (5.110), we have finally:

$$\frac{\partial n_h}{\partial t} + \nabla \cdot \left[\frac{\sigma_e T_h}{e n_h} \nabla n_h \right], \tag{5.111}$$

which is just a nonlinear diffusion equation with diffusion coefficient $D = \sigma_e T_h / e n_h$.

This can be solved by separation of variables: $n_h(x,t) = g(x)h(t)$, eventually yielding a solution (Bell *et al.*, 1997):

$$n_h = n_{h0} \frac{t}{\tau_L} \left(\frac{x}{x + R_h} \right)^2, \tag{5.112}$$

where

$$n_{h0} = \frac{2 I_a^2 \tau_L}{9 e T_h^3 \sigma_e}, \quad R_h = \frac{3 T_h^2 \sigma_e}{I_a}. \quad (SI)$$

These two relations translate to the more convenient forms:

$$n_{h0} = I_{18}^2 \frac{\tau_L}{100 \text{ fs}} \left(\frac{T_h}{100 \text{ keV}} \right)^{-3} \left(\frac{\sigma_e}{10^6 \ \Omega^{-1} m^{-1}} \right)^{-1} 1.4 \times 10^{23} \text{ cm}^{-3}$$

$$R_h = \left(\frac{T_h}{100 \text{ keV}} \right)^2 \left(\frac{\sigma_e}{10^6 \ \Omega^{-1} m^{-1}} \right) I_{18}^{-1} \ 3 \ \mu\text{m}, \tag{5.113}$$

where $I_{18} = I_a / 10^{18}$ Wcm^{-2}. The constant R_h is a measure of the effective penetration depth of fast electrons while the laser is incident. For the parameters used in the earlier example, $I_{18} = 1$, $T_h = 100$ keV, and taking $\sigma_e = 10^6 \ \Omega^{-1} m^{-1}$ for a solid aluminium target (Milchberg *et al.*, 1988), we find $R_h \simeq 3 \ \mu$m, over 20 times smaller than the value for collisional stopping ($R_e \sim 70 \ \mu$m) predicted by the Bethe-Bloch formula. This is transport inhibition with a vengeance! Bell *et al.*, (1997) also derive a threshold condition for resistively inhibited transport by equating R_e and R_h, assuming classical Spitzer conductivity, $\sigma_e = n_e e^2 / m_e \nu_{ei}$, with ν_{ei} given by Eq. (5.10):

$$I_{\text{inhib}} > \left(\frac{T_h}{100 \text{ keV}} \right)^{3/2} \left(\frac{Z}{13} \right)^{1/2} \left(\frac{\rho}{2.7 \text{ gcm}^{-3}} \right) 10^{17} \text{ Wcm}^{-2}.$$

This modest intensity level implies that fast electron inhibition may already have been a significant feature of nearly all femtosecond-solid experiments performed since the mid-1990s.

One consequence of this effect is that values of T_h measured using standard K_α 'sandwich' diagnostics (Hares *et al.*, 1979) will tend to be lower than expected from the general scalings suggested by Fig. 5.19, or values predicted by PIC simulation (Wilks *et al.*, 1992; Gibbon, 1994; Pukhov and Meyer-ter-Vehn, 1997). In fact, resistive inhibition may explain apparent anomalies in experiments by, for example, Teubner *et al.*, (1996a), Jiang *et al.*, (1995) and Yu *et al.*, (1999a). The sensitivity of T_h and n_h to target material (via its electrical conductivity) could also have contributed to the wide variations in absorption fraction reported in the past, as witnessed in Fig. 5.16.

Direct experimental evidence for hot electron transport inhibition has emerged from extensive campaigns performed with the LULI TW laser. Using K-alpha emission from buried fluorescent layers to measure electron transport in conducting and insulating targets, Pisani *et al.*, (2000) and Batani *et al.*, (2002) found significantly reduced propagation in insulators, pointing to the presence of large electric fields created inside the target.

5.10.2 *Magnetic field effects*

From an energetic point of view, we have already seen that fast electron currents inside the target must be largely cancelled by a return flow. This does not preclude magnetic field generation entirely, however: a small local spatial imbalance between j_h and j_c can still lead to substantial magnetic field growth via Faraday's law,

$$\frac{\partial \boldsymbol{B}}{\partial t} = -\nabla \times \boldsymbol{E} = \nabla \times (\eta_e \boldsymbol{j}_h), \qquad (5.114)$$

where $\eta_e = \sigma_e^{-1}$ is the electrical resistivity. Here we assume that the displacement current $\partial E/\partial t$ will only be significant before the return current has time to establish itself (Glinsky, 1995; Davies *et al.*, 1997).

We can easily estimate these fields by combining Eq. (5.105) with Eq. (5.109) to get the electric field, given by

$$E = -\frac{e I_a}{\sigma_e k_B T_h},$$

or

$$E_{\max} = \left(\frac{\sigma_e}{10^6 \ \Omega^{-1} m^{-1}}\right)^{-1} I_{17}^{2/3} \eta_a \lambda_{\mu m}^{-2/3} \ 10^{10} \ \mathrm{Vm}^{-1}.$$

Applying Eq. (5.114) in cylindrical geometry, we have

$$\frac{\partial B_\theta}{\partial t} = \frac{1}{\sigma_e} \frac{\partial j_h}{\partial r},$$

which after inserting a Gaussian intensity profile of the form

$$I(r) = I_0 \exp(-r^2/\sigma_L^2) \exp(-t^2/\tau_L^2),$$

and integrating the electric field over time, leads to (Davies *et al.*, 1997):

$$B_{\text{max}} \simeq \left(\frac{\sigma_e}{10^6 \ \Omega^{-1}\text{m}^{-1}}\right)^{-1} \frac{\tau_L}{\text{ps}} \left(\frac{\sigma_L}{10 \ \mu\text{m}}\right)^{-1} \eta_a I_{17}^{2/3} \lambda_{\mu m}^{-2/3} \cdot 4 \text{ MG}, \quad (5.115)$$

values which are consistent with numerical simulations based on a stochastic Fokker-Planck model. These simulations explicitly illustrate the diffusive nature of fast electron transport in this regime; the maximum of the electric field being carried into the target on a picosecond timescale. Further complications to this picture arise from magnetic pinching of the fast electron beam (Tatarakis *et al.*, 1998; Bell and Kingham, 2003), and by local changes in resistivity either due to fast electron preheat, or in layered targets (Bell *et al.*, 1998).

5.10.3 *Relativistic currents*

A starkly contrasting picture of hot electron transport has emerged from PIC simulations geared to assessing the feasibility of fast ignition, a recently proposed method to 'kick start' thermonuclear burn in ICF using fast electrons, see Sec. 7.5. The scenario suggested by these simulations is that at very high intensities ($I \geq 10^{19}$ Wcm^{-2}), one or more relativistic beams of fast electrons are produced (Pukhov and Meyer-ter-Vehn, 1997), which initially break up into filaments via the Weibel instability (Weibel, 1967; Pegoraro *et al.*, 1996), but then subsequently coalesce due to mutual magnetic attraction (Honda *et al.*, 2000; Sentoku *et al.*, 2003).

The main difference between these PIC simulations and the transport models described previously is that in the former, collisions are either ignored or considered to be negligible in the high temperature (~ 10 keV), moderate-density ($\sim 10n_c$) plasmas modeled to date. Under these conditions, resistive flux inhibition effects are strongly (but artificially) suppressed, hence the wide discrepancies between the two types of model. Which scenario is correct for the fast ignition regime ($n_e \sim 1000n_c, T_e \sim$ keV), is currently a wide-open question, and one which will not be resolved easily. Either to scale PIC simulation up to realistic parameters or perform 3D Fokker-Planck simulations will require several orders of magnitude improvement in computing power.

5.11   Self-Induced Transparency

We have already seen in Sec. 4.6 how relativistic electron motion in a light wave can modify the beam propagation characteristics in underdense plasmas. At extreme intensities, this motion can even modify the refractive index in a nominally *overdense* plasma to permit propagation. This phenomenon, self-induced transparency (SIT), is easy enough to imagine on physical grounds: for quiver velocities $= a_0 \gg 1$, the relativistic factor reduces to $\gamma_0 = \left(1 + a_0^2/2\right)^{1/2} \simeq a_0/\sqrt{2}$. The effective plasma frequency appearing in the wave equation Eq. (4.72), $\omega_p' = \omega_p/\gamma_0^{1/2}$ is then proportional to $(n_e/a_0)^{1/2}$, so there will be a critical intensity above which the plasma loses its natural opacity, i.e. $\omega_p' \leq \omega$, of:

$$a_0^{\text{sit}} \geq 2^{\frac{1}{2}} \frac{n_e}{n_c}. \tag{5.116}$$

This simple observation was first pointed out by Kaw and Dawson (1970) in an analysis based on the general solutions derived by Akhiezer and Polovin (1956), see Sec. 4.1. Since this early work, a number of authors have returned to the subject intermittently, generally looking at the steady-state behavior of such propagation modes (Decoster, 1978; Shukla *et al.*, 1986). Motivated by the possibility of experimental verification of SIT with short-pulse lasers, Lefebvre and Bonnaud (1995) used relativistic PIC simulation to tackle the problem afresh from a *dynamic* point of view. Basically the question they asked was: given a thin slab of pre-ionized, overdense plasma, when and how quickly does it become transparent?

Actually, the situation is reminiscent of hole boring, where the excess laser pressure sets up a shock front moving into the target with a constant velocity. In this case, however, the ions are fixed and there is partial transmission of the wave, ultimately leading to a *penetration* front moving into the plasma with velocity v_f. Using Lorentz transformations to and from this moving reflection surface, one can show that the reflected light frequency takes the value:

$$\frac{\omega_r}{\omega} = \frac{1 + \beta_f^2 - 2\beta_f/\beta_\phi}{1 - \beta_f^2}, \tag{5.117}$$

where $\beta_f = v_f/c$ and $\beta_\phi c = \omega/k$ is the phase velocity of the wave:

$$\beta_\phi^2 = 1 + \frac{\omega_p^2}{\gamma\omega^2}.$$

An analogous expression is also found for the wave vector k_r.

The normalized reflection front velocity, β_f, is clearly less than unity because the plasma electrons need a finite time to get up to speed, thus delaying the modification to the refractive index. Guérin *et al.*, (1996) showed that this front velocity can be determined by balancing the Poynting flux with the rate of change of electromagnetic energy in the region of plasma occupied by the wave, yielding a rather complicated expression involving the incident, reflected and longitudinal wave fields. The reflectivity at the front is given by:

$$R = \frac{k_r\omega_r}{k\omega},$$

which can be evaluated if ω_r and k_r are known, which in turn depend on β_f. The end result is shown in Fig. 5.25, which compares estimates of the front velocity and reflectivity given by Lefebvre and Bonnaud (1995) and Guérin *et al.*, (1996) respectively.

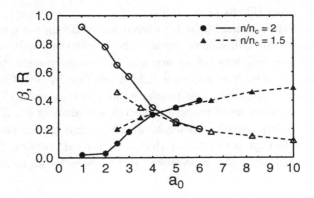

Fig. 5.25   Penetration front velocity β (filled symbols) and reflectivity R (open symbols) as a function of normalized laser amplitude a_0. The $n_e/n_c = 2$ points (triangles) are taken from Lefebvre and Bonnaud (1995); the $n_e/n_c = 1.5$ curves (circles) from Guérin *et al.*, (1996).

To date, only a few attempts at experimental verification of SIT have been reported. The first experiment, a collaboration between the Pisa and LOA groups (Giulietti *et al.*, 1997), stirred up some controversy. Their results suggested that a 30 fs pulse was transmitted through a thin plasma of near solid density at intensities way below those nominally predicted by Eq. (5.116) or Fig. 5.25. The main objection to this experiment was that there was no independent density diagnostic, so the authors were unable to tell precisely what density the main pulse actually 'saw'. Critics argued that the ASE pedestal (see Sec. 5.5.3) — in this case 10 ns at a power contrast ratio of 10^{-7} — was probably heating the foil long enough to burn a hole through it, leaving a region of *underdense* plasma which the main pulse could pass through unhindered.

Support for the findings of Giulietti's team nonetheless came soon afterwards from Limeil in a large experimental campaign involving four French and Canadian groups. Under somewhat different conditions (pulse lengths 10-20 times longer than the LOA laser; intensities up to 1.5×10^{19} Wcm^{-2}) Fuchs *et al.*, (1998b) concluded that they were also observing anomalously high transmittivity through both preformed and near-solid-density plasmas. Their estimated densities (10 n_c for preformed plasma; 50n_c for solid foils) were obtained from fits to x-ray line spectra with an atomic physics post-processing code. They interpreted their results as a combination of rapid heating — causing the foil to expand and the density to fall — followed by SIT and hole boring (Fuchs *et al.*, 1998a).

Reduction of the foil density by hot-electron driven expansion may work for longer pulses, but still cannot explain the Giulietti's results. Various theories have been put forward to account for the anomalous transmission, invoking effects such as magnetic field generation by ultrafast ionization (Teychenné *et al.*, 1998) and ponderomotive compression (Yu *et al.*, 1999c), resonant transport of transverse currents (Cairns *et al.*, 2000) and filamentation accompanied by strong longitudinal and transverse heating (Adam *et al.*, 1997). It is fair to say that even several years on, the SIT-phenomenon is still shrouded in mystery, and we have by no means heard the end of this story.

5.12   High Harmonic Generation from Solid Surfaces

From the previous Secs. 3.5 and 4.5 in which this topic is discussed, one might conclude that harmonic generation from free electrons is an inherently inefficient process compared to HHG from atoms (Sec. 2.5). We recall that in underdense plasmas, the efficiency of the 3rd harmonic scales according to Eq. (4.93) as $\eta_3 \sim (\omega_p/\omega)^4 a_0^4/(1 + a_0^2)$, which cannot easily be improved on in a tenuous medium where $\omega_p < \omega$.

However, laser interaction with *overdense* plasmas also offers a means of frequency conversion which is not handicapped by phase-matching requirements. It has been known for some time that low order harmonics can be generated in such plasmas via resonance absorption, parametric instabilities (Bobin, 1985), and transverse density gradients (Stamper *et al.*, 1985). Integer and non-integer harmonics generated in laser-ablated plasmas have been extensively studied because they can yield useful diagnostic information on the properties of the plasma (density gradient, expansion velocity, *etc.*) near the critical surface n_c (Veisz *et al.*, 2002, 2004).

Resonance absorption is of particular significance here because it provides an efficient means of converting an electromagnetic wave $(\omega, \mathbf{k})$ into a localized electrostatic mode $(\omega_p, \mathbf{k}_p)$ (Ginzburg, 1964). At the critical surface $\omega = \omega_p$, these two waves can mix to produce a 2nd harmonic via the current $\mathbf{J}_{2\omega} = n_\omega \mathbf{v}_\omega$. Thus $\omega_2 = \omega + \omega_p = 2\omega$. This wave is mainly reflected, but part of it can propagate up the density profile to $4n_c$, where it excites a plasma wave at 2ω. This in turn generates a 3rd harmonic, $\omega_3 = 3\omega$, which is resonant at $9n_c$, and so on, see Fig. 5.26. Note that unlike in the underdense, homogeneous case, both *odd and even* harmonics can be generated in an overdense profile. This simple mode-coupling picture

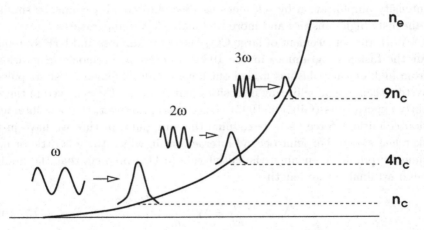

Fig. 5.26 Schematic picture of harmonic generation via linear mode-coupling in a plasma density gradient.

was supported by a number of early experiments using long-pulse lasers by Burnett *et al.*, (1977) and McLean *et al.*, (1978), who measured 11 and 5 harmonics respectively, with efficiencies consistent with an $I\lambda^2$-scaling.

Interest in this phenomenon intensified after Carman, Rhodes and Benjamin detected up to 46 harmonics in a series of experiments with the powerful CO_2 laser systems (300 J in 1 ns) at Los Alamos (Carman *et al.*, 1981a). To analyze their results, the Los Alamos team used 2D PIC simulations to model the laser-plasma interaction in a steepened density profile (Carman *et al.*, 1981b) — conditions consistent with steady-state model of resonance absorption outlined in Sec. 5.6.1. Assuming that the maximum density was determined by pressure balance between the laser and dense plasma (Eq. 5.79), they reasoned that according to the mode-coupling picture, the harmonic spectrum should exhibit a *cutoff* given simply by:

$$M = \sqrt{\frac{n_e}{n_c}}. \tag{5.118}$$

This scaling appeared to be supported both by the simulations and by fluid models developed shortly afterwards (Bezzerides *et al.*, 1982; Grebogi *et al.*, 1983), implying that the highest harmonic detected would correspond to the maximum electron density. Applying this rule to their later results however, the authors inferred densities of 500–2000 n_c, which, according to the pressure-balance relation Eq. (5.79), would have implied unusually low plasma temperatures of a few tens of eV. However, they argued that intensity amplification by self-focusing (Sec. 4.6) could account for simultaneously high densities and more reasonable keV-temperatures.

With the replacement of large CO_2 lasers by Nd:glass and KrF systems for the fusion programmes in the 1980s, interest in harmonic generation from high density plasmas fizzled out until the development of short pulse systems capable of delivering intensities matching the $I\lambda^2$ achieved in these early experiments (Gibbon, 1997). Once again, however, the new interaction conditions force us to reexamine the long pulse picture we have just sketched above. For femtosecond interactions in which there is little or no time for profile steepening, the cutoff rule (5.118) predicts that the minimum attainable wavelength,

$$\lambda_{\min} = \frac{\lambda_L}{M} = \lambda_L \left(\frac{n_c}{n_e}\right)^{1/2}, \tag{5.119}$$

which is *independent* of the laser wavelength and intensity. For example, a Ti:sapphire laser on a fully ionized carbon target (see Eq. (5.3)) would generate 25 harmonics with $\lambda_{min} \simeq 30$ nm. A KrF* laser, which has a $10\times$ higher critical density, would generate only 8 harmonics for the same λ_{min}. This is not too impressive compared to the hundreds of harmonics generated routinely from atoms, see Sec. 2.5. Taking our cue from the prediction of harmonics in underdense plasmas, we might suppose that the nonlinear electron motion at relativistic laser intensities might improve the prospects for HHG here, but it is not immediately clear how to replace the appealing mode-coupling description.

5.12.1 *The moving mirror model*

Alternative mechanisms were proposed by Wilks *et al.*, (1993) and Bulanov *et al.*, (1994), who analyzed PIC simulations of harmonics generated by normally and obliquely incident pulses in terms of an oscillating boundary layer. These ideas were further developed for arbitrary angles of incidence and polarization by Lichters *et al.*, (1996) and by von der Linde and Rzazewski (1996) — a picture now commonly referred to as the *moving-mirror* model. From this model, a rigorous set of selection rules can be derived. For example, whereas a *p*-polarized pump produces harmonics which are all *p*-polarized, for *s*-light the odd and even harmonics are *s* and *p*-polarized respectively.

To demonstrate this unambiguously, it is helpful to invoke the relativistic fluid model derived back in Sec. 4.3.3 in order to analyze nonlinear wakefield generation in underdense plasmas. In this case however, our geometry is somewhat different: obliquely incident light onto an overdense plasma slab. Normally this requires two spatial dimensions, but the problem can be reduced to a single one by means of a specially chosen relativistic boost parallel to the surface of the target (Bourdier, 1983; Gibbon and Bell, 1992; Gibbon *et al.*, 1999), see Fig. 5.27. It turns out that this is a particularly convenient way to analyze harmonic generation because both the incident and reflected light appear *normal* to the surface in this frame of reference.

Denoting the boost (S) frame quantities by primes, the inverse Lorentz transformations for the wave frequency and k-vector are:

$$\omega' = \gamma_0(\omega - v_0 k_y)$$
$$k'_y = \gamma_0(k_y - \frac{v_0}{c^2}\omega)$$
$$k'_{x,z} = k_{x,z}.$$

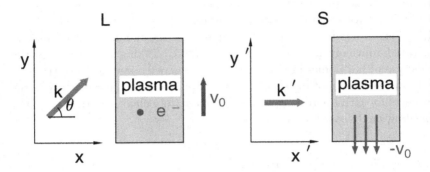

Fig. 5.27   The boost technique for oblique-incidence interactions: variables are transformed from the lab frame (L) to the boost frame (S) using a relativistic velocity translation v_0 along the y-axis.

Since $k_y = k \sin\theta = \omega/c \sin\theta$, we must choose $v_o = c \sin\theta$ in order for k'_y to vanish. The above transformations then reduce to:

$$\omega' = \omega/\gamma_0 = \omega \cos\theta$$
$$k'_y = 0 \tag{5.120}$$
$$k'_x = k' = |\,\boldsymbol{k'}\,| = k/\gamma_0 = k\cos\theta,$$

where

$$\gamma_0 = \left(1 - \frac{v_0^2}{c^2}\right)^{-1/2} = \sec\theta.$$

In the boosted frame of reference, the EM wave interacts with a *streaming* plasma. The transverse motion of the electrons thus obeys:

$$\boldsymbol{p}_\perp = \boldsymbol{p}_0 + \frac{e\boldsymbol{A}}{c},$$

where the drift momentum is given by:

$$\boldsymbol{p}_0 = m\gamma_0\boldsymbol{v}_0 = -\hat{y}mc\tan\theta.$$

Applying our usual system of normalization, $u = v_x/c$, $\tilde{p} = p/mc$, $a = eA/mc^2$, we see that the longitudinal equation of motion (4.49) acquires an additional driving term arising from the cross-product of the transverse drift with the laser magnetic field $B_z = \partial a / \partial x$:

$$
\begin{aligned}
\frac{dp_x}{dt} = \frac{d}{dt}(\gamma u) &= c\frac{\partial \phi}{\partial x} - \frac{c}{\gamma}(\boldsymbol{a} - \boldsymbol{p}_0) \cdot \frac{\partial \boldsymbol{a}}{\partial x} \\
&= c\frac{\partial \phi}{\partial x} - \frac{c}{2\gamma}\frac{\partial a^2}{\partial x} + \frac{cp_0}{\gamma}\frac{\partial a_y}{\partial x},
\end{aligned}
\tag{5.121}
$$

where

$$
\gamma = \left(\frac{1+|\boldsymbol{p}_\perp|^2}{1+u^2}\right)^{1/2} \ ; \quad \boldsymbol{p}_\perp = \boldsymbol{a} - p_0\hat{y}.
$$

The continuity equation (4.51) is unchanged:

$$
\frac{\partial n_e}{\partial t} + \nabla\cdot(n_e\boldsymbol{v}) = 0,
\tag{5.122}
$$

but anticipating a fixed ion background charge, the Poisson equation is more conveniently written as follows:

$$
\frac{\partial^2 \phi}{\partial x^2} = k_p^2\gamma_0\left[n_e(x,t) - Zn_i(x)\right].
\tag{5.123}
$$

The additional γ_0-factor in Eq. (5.123) arises because we have normalized both electron and ion densities to the upper shelf density of the unperturbed plasma, n_0', which transforms according to:

$$
n_0' = \gamma_0 n_0,
$$

implying that

$$
\omega_p'^2 = \gamma_0\omega_p^2 = \gamma_0 c^2 k_p^2.
$$

With the exception of ω_p and k_p, all fluid and wave variables $(x, t, u, p, n_{e,i}, \phi)$ henceforth refer to the boost frame.

The EM wave equation also acquires additional current source terms:

$$
\begin{aligned}
\boldsymbol{J} &= -en_e\boldsymbol{v}_e + Zen_i\boldsymbol{v}_i \\
&= -\frac{en_e}{\gamma}(\boldsymbol{a} - \boldsymbol{p}_0) - ZEn_i\boldsymbol{v}_0.
\end{aligned}
\tag{5.124}
$$

Initially, before the laser reaches the plasma surface, we have $\boldsymbol{a} = 0$, $n_e = Zn_i$ and $u = 0$, so that $\gamma = (1 + p_0^2)^{1/2} = \gamma_0$ and $\boldsymbol{J} = 0$, as it should. Substituting Eq. (5.124) into the RHS of the wave equation Eq. (4.53) and normalizing as before, we have:

$$
\left(\frac{\partial^2}{\partial t^2} - c^2\frac{\partial^2}{\partial x^2}\right)\boldsymbol{a} = -\omega_p^2\frac{\gamma_0}{\gamma}n_e(\boldsymbol{a} - \boldsymbol{p}_0) - \omega_p^2 Zn_i\boldsymbol{p}_0.
\tag{5.125}
$$

As shown by Lichters *et al.*, (1996), the angular dependence of the fluid equations in the boost frame can be expressed explicitly by making the substitutions $p_0 = \tan\theta$, $\gamma_0 = \sec\theta$. Analogous equations for the electric and magnetic fields have been derived by Bulanov *et al.*, (1994). As we hinted previously, the advantage of this formalism is that the current sources of the harmonics are transparent. For *s*-polarized light, $\boldsymbol{a} = (0, 0, a_z)$, we have:

$$J_z^s = -\frac{\gamma_0 n_e a_z}{\gamma(u^2, a_z^2)}$$
$$J_y^s = \frac{\gamma_0 p_0 n_e a_z}{\gamma(u^2, a_z^2)} - p_0 Z n_i. \tag{5.126}$$

For *p*-polarized light, $\boldsymbol{a} = (0, a_y, 0)$, we have:

$$J_y^p = \frac{\gamma_0 n_e}{\gamma(u^2, a_y, a_y^2)}(p_0 - a_y) - p_0 Z n_i$$
$$J_z^p = 0. \tag{5.127}$$

The relativistic factor for arbitrary polarization can be written:

$$\gamma^{-1} = \frac{(1 - u^2)^{1/2}}{(1 + |\boldsymbol{a}|^2 \cos^2\theta - a_y \sin 2\theta)^{1/2}}.$$

The electron density, n_e, is determined by the longitudinal fluid motion, which in turn is driven by the laser field: the second and third terms on the RHS of Eq. (5.121). The continuity and Poisson equations (5.122, 5.123) immediately tell us that u, n_e and ϕ will share the same harmonic content, whether odd or even. Inspection of Eq. (5.121) and the relativistic factor shows that

$$\frac{\partial u}{\partial t} \sim \frac{a_y^2 + a_z^2}{\gamma} + \frac{a_y}{\gamma},$$

which contains purely even harmonics, $f(a_z^2)$, for *s*-light; but both even and odd harmonics, $f(a_y^2, a_y)$ for *p*-light. *S*-polarized light will therefore generate odd *s*-polarized harmonics and even *p*-polarized harmonics due to the density oscillations originating from the $\boldsymbol{v} \times \boldsymbol{B}$ force at 2ω. *P*-polarized light, on the other hand, directly drives the plasma surface at the laser frequency, and will therefore generate both even and odd *p*-polarized harmonics. At normal incidence the even harmonics in both cases will vanish. These selection rules, first pointed out by Lichters *et al.*, (1996) and von der Linde and Rzazewski (1996), are summarized in Table 5.3.

Table 5.3 Selection rules for harmonics specularly reflected from a plasma surface. Possible polarizations of the incident light are P, S, linear ($L = P$ or S) and circular (C).

Laser polarization	Odd harmonics ω^{2n+1}	Even harmonics ω^{2n}
S	S	P
P	P	P
normal, L	L	—
normal, C	—	—

The *amplitude* of the harmonics depends primarily on the product of the electron density and the transverse fluid velocity $v_\perp \sim a_\perp/\gamma$ at the surface ($x = 0$). For a highly overdense step-profile, the laser driving field inside the plasma is evanescent, and from Sec. 5.2.2 we recall the following (strictly speaking, non-relativistic) expression:

$$a_s(x) = 2a_0 \frac{\omega}{\omega_p} \exp(-x/l_s), \qquad (5.128)$$

where $l_s \simeq c/\omega_p$, the collisionless skin depth. The moving mirror model approximates the electron density by a rigid Heaviside step function:

$$n_e(x,t) = \Theta[x - \xi(t)],$$

where $\xi(t)$ represents the position of the (oscillating) plasma mirror surface. Calculation of the harmonic spectrum thus reduces to solving for $\xi(t)$, giving us the nonlinear currents, which we then insert into the wave equation (5.125) to obtain the reflected light amplitudes. This rather complex task is usually side-stepped by initially assuming that the surface is driven harmonically by the laser. For example, at normal incidence, we would have

$$\xi(t) \simeq \xi_s \sin 2\omega t.$$

The oscillation amplitude, ξ_s, may be estimated from Eq. (5.121), substituting $u = \dot{\xi}_s/c$, which after linearizing and retaining only the leading driving term, reduces to

$$\frac{1}{c} \frac{\partial^2 \xi_s}{\partial t^2} \sim 12\gamma^2 \omega_p a_s^2.$$

Inside the plasma, the timescale is determined by the overdense plasma frequency. Thus, we have the simple result

$$\xi_s \simeq l_s a_s^2,$$

or substituting for the surface amplitude $a_s(x = 0)$ from Eq. (5.128):

$$k\xi_s = 2 \left(\frac{\omega}{\omega_p} \right)^3 a_0^2. \qquad (5.129)$$

This result tells us straight away that harmonics will be driven more efficiently at higher intensities and, in contrast to underdense harmonics, at *lower* densities. However, the precise scaling of the harmonic amplitudes requires a little more work. The reflected light can be calculated by solving the wave equation with the help of Green's function techniques (Arfken and Weber, 1995). For example, the normalized electric field outside the plasma ($x \leq 0$), $\tilde{E} = eE/m\omega c = a(x,t)/\omega$ can be expressed approximately as:

$$\tilde{\boldsymbol{E}}(x,t) = \frac{\omega_p}{2\omega} \boldsymbol{J}(\xi(t_{\rm ret}), t_{\rm ret}),$$

where the current source is now taken at the 'mirror' surface position $\xi(t_{\rm ret})$ at a *retarded* time (see Chap. 3, Fig. 3.7):

$$t_{\rm ret} = t - \frac{\xi(t_{\rm ret})}{c} + \frac{x}{c}.$$

For small pump amplitudes, $a_0 \ll 1$, and a mirror oscillation of the form $\xi(t) \simeq \xi_s \sin \omega t$ (which would apply for oblique incidence), the current can be expanded to give:

$$\begin{aligned} \tilde{E}(0,t) &\simeq \frac{\omega_p}{2\omega} a_0 \sin \left(\omega t - \frac{\omega \xi(t_{\rm ret})}{c} \right) \\ &= \frac{\omega_p}{2\omega} a_0 \sin(\omega t + k\xi_s \sin \omega t). \end{aligned}$$

$$(5.130)$$

This can be expressed in terms of Bessel functions, but the harmonic content is already apparent, and results in a saw-tooth reflected electric field, see Fig. 5.28.

5.12.2   *Numerical simulations*

Early simulations of harmonic generation were hampered by lack of spatial and temporal resolution available with 2D PIC codes; a fact which was readily acknowledged by Carman *et al.*, (1981b), who were restricted to < 10 harmonics at that time. However, application of the boost technique illustrated (Fig. 5.27) to a PIC simulation (Gibbon and Bell, 1992)

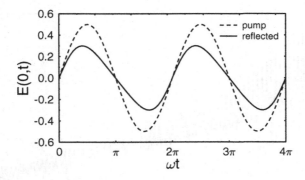

Fig. 5.28   Nonlinear reflection of laser electric field on a plasma mirror: the saw-tooth pattern is a typical signature of harmonic content.

allows order-of-magnitude improvements in temporal and spatial resolution over traditional 2D codes, and thus enables the high harmonics to be retained. In particular, Gibbon (1996) exploited this spectral advantage in demonstrating that large numbers of harmonics could be generated this way, with no cutoff at the upper plasma density as implied by Eq. (5.119), see Fig. 5.29. This simulation also shows that the harmonic efficiencies for p-polarized light are 1–2 orders of magnitude higher than for s-polarized or normally incident light. In the latter case, only the relativistic $\boldsymbol{v} \times \boldsymbol{B}$ mechanism is present, whereas *density-bunching*, via the $n_e a_y$ product in Eq. (5.127), makes an important contribution to the nonlinear current for p-polarized light.

The harmonic efficiencies obtained by particle simulations for p-polarized light at 45 degrees incidence can be summarized by an empirical relation valid for high orders ($m \gg 1$):

$$\eta_m \simeq 9 \times 10^{-5} \left(\frac{I\lambda^2}{10^{18} \ \text{Wcm}^{-2}\mu\text{m}^2} \right)^2 \left(\frac{m}{10} \right)^{-\alpha}, \qquad (5.131)$$

where α also depends weakly on intensity and is $\simeq 5$ at 10^{19} Wcm^{-2}. This scaling means that the highest harmonic order (down to the detection threshold) is simply determined by $I\lambda^2$, although we also expect some dependence on upper shelf density and incidence angle. Thus, contrary to Eq. (5.119), a short wavelength pump beam will generate shorter wavelength harmonics for the same irradiance. Indeed, the simulations by Gib-

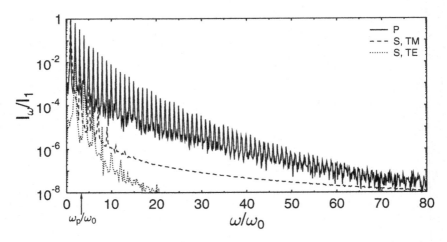

Fig. 5.29   Simulated power spectra for a $10^{19}Wcm^{-2}\mu m^2$ pulse incident on a plasma with density $n_e/n_c = 10$ for oblique, p-polarized light.

bon (1996) predict that a KrF pump with intensity of 1.6×10^{20} Wcm^{-2} will generate harmonics with efficiencies $O(10^{-6})$ into the 'water-window' of 2.3–4.4nm (which is reached by the 56th).

5.12.3   *Experiments on harmonic generation*

As mentioned earlier, the technological strides in short pulse lasers over the last decade have rekindled experimental interest in the phenomenon unveiled by Carman et al., (1981a). In the first of a series of experiments in the mid-1990s, Kohlweyer et al., (1995) successfully demonstrated 'high' harmonic generation with a 100fs Ti:sapphire (794nm) system in Lund, Sweden . Using focused laser intensities of 10^{17} Wcm^{-2}, up to 7 harmonics were unequivocally observed on various targets. In a very similar experiment, von der Linde et al., (1995) observed up to 15 harmonics from the Ti:sapphire system at LOA, France. They estimated the efficiency of the highest harmonic to be $\geq 10^{-9}$, which is about 50 times lower than the value given by Eq. (5.131).

In both of these experiments, the highest harmonic observed was still consistent with the linear cutoff model, assuming densities of $\sim 2 - 4 \times 10^{23}$ cm^{-3} for aluminium targets. However, this model had to be ditched after a spectacular experiment with the Vulcan Nd:glass CPA system at

RAL, UK. In this experiment, the laser delivered moderate contrast-ratio pulses of 2.5 ps duration, focused to 10^{19} Wcm^{-2} creating plasma conditions probably not too different to those in the Los Alamos experiments, albeit at 100× higher densities thanks to the 1 μm laser wavelength. In the event, Norreys *et al.*, (1996) found up to 68 harmonics in first order diffracted signal (75 in 2nd order), with efficiencies exceeding 10^{-6}, implying power conversions into spectral lines from 14–15 nm in the Megawatt range, see Fig. 5.30. According to the cutoff theory Eq. (5.118), the highest harmonic observed would imply a barely credible electron density of 5600 n_c, or 17× solid density for the CH-coated glass targets used!

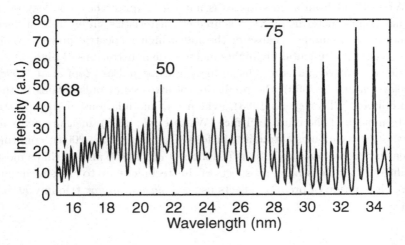

Fig. 5.30 High-order harmonic spectrum measured from interaction of a 2.5ps Nd:glass laser with a plastic target at 10^{19} Wcm$^{-2}\mu$m^2. *Courtesy:* P. Norreys, Rutherford-Appleton Laboratory, and M. Zepf, Queen's University Belfast.

A number of features in these spectra were in good agreement with the (1D) PIC simulations of Gibbon (1996), particularly the efficiency scaling with intensity and harmonic order. In particular, the power conversion into the 38th harmonic at 27.7 nm is found to be about 24 MW, corresponding to an absolute efficiency of 3×10^{-6}, which is about a factor of 3 lower than the simulation result for the same intensity. This is somewhat surprising given that the interaction was probably colored by 2D effects such as surface rippling and hole boring (Wilks *et al.*, 1992). A dimpled surface will blur any differences between spectra measured with *p*- and *s*-polarized light

(Wilks and Kruer, 1997), and lead to a more isotropic harmonic emission into 2π steradian.

The coherence properties of harmonics generated from laser-solid interaction were studied in a follow-up paper on the same experiment by Zhang *et al.*, (1996). They estimate a source size of 12 μm and an intensity-dependent bandwidth of $0.02 - 0.4$ ps, both of which are competitive with other coherent XUV sources such as x-ray lasers. Broadening of harmonic lines at high intensity can arise from relativistic self-phase modulation in the underdense plasma in front of the target, suggesting that the bandwidth could be improved by using shorter pulses with sufficiently high contrast to avoid pre-plasma formation.

While HHG from solid surfaces is not yet in quite the same league as gas harmonics in terms of XUV source characteristics, interest in surface harmonics is booming because of the much higher spectral powers possible. Recent experimental highlights include: demonstrations of high-order, specularly emitted harmonics using high-contrast pulses (Zepf *et al.*, 1998; Tarasevitch *et al.*, 2000); the prediction of attosecond pulse trains in the time domain (Playa *et al.*, 1998); evidence of modulations in the spectra due to motion of the critical surface (Watts *et al.*, 2002); measurements of *transmitted* harmonics through thin foils (Teubner *et al.*, 2004), confirming earlier theoretical predictions (Gibbon *et al.*, 1997); and, perhaps most significantly for the scaling prospects of this technique up to extreme intensities, elimination of prepulse effects using a *plasma mirror* (Doumy *et al.*, 2004).

Exercises

(1) Show that the collision frequency in Eq. (5.10) reduces to the convenient normalized form:

$$\frac{\nu_{ei}}{\omega_p} \simeq \frac{Z \ln(9N_D/Z)}{10N_D}.$$

(2) a) Why can't we solve the Helmholtz equation for B in the limit $L/\lambda=0$?

b) Using the Helmholtz solver provided in Chapter 6 to compute the energy deposition rate

$$\frac{\partial U}{\partial t} = < \boldsymbol{J}(x) \cdot \boldsymbol{E}(x) >$$

as a function of density for different scale-lengths, see the article by Kieffer *et al.*, (1989a) for guidance. This example emphasizes the fact that the overdense part of the plasma is directly exposed to the laser field during femtosecond interactions.

(3) Derive the heat-front penetration velocity for constant absorption on a plasma slab, using a) heuristic arguments based on classical Spitzer heat-flow and b) directly from the self-similar solutions derived in Sec. 5.3.

(4) Verify the self-similar behavior of resonance absorption of p-polarized light on the parameter $(kL)^{1/3} \sin \theta$ by mapping the curves shown in Fig. 5.2 onto Fig. 5.10. (kL is the normalized density scale-length and θ angle of incidence)

(5) a) Using the momentum and energy conservation arguments from Sec 3.3 to describe electron motion in a laser focus, find that the transverse momentum acquired by an electron along the surface of a density step profile irradiated by a p-polarized laser incident at an angle θ to the target normal.

b) Show that the emission/penetration angle θ' of the electron from the focal region is given by:

$$\tan \theta' = \frac{p_y}{p_x} = \frac{(\gamma - 1) \sin \theta}{\sqrt{2(\gamma - 1) + (\gamma - 1)^2 \cos^2 \theta}}.$$

Compare this to Eq. (3.23) and comment on the difference.

(6) For newcomers to the field of laser-plasma interaction, the parameter combination $I\lambda^2$ invariably provokes a degree of scepticism, even outright disbelief among laser-builders, who often have to watch their hard-fought peak intensities get downgraded by factors ranging from 4 to 16. Using the data in Table 5.2, construct a chart of T_h vs. laser intensity I instead of $I\lambda^2$ and compare the result with Fig. 5.19.

(7) Starting from the Lorentz equation 4.1, derive a relativistic version of the $j \times B$ force term for arbitrarily polarized light. Show that the heating component vanishes for circular light. (Hint: insert a polarization vector such as the one in Eq. (4.77).)

(8) Show that the hot electron temperature scales as $(I\lambda^2)^{1/3}$ in the 'long-pulse' regime where the density profile is steepened by pressure balance. What scaling would result if the density remained constant? (See Sec. 5.6.)

Chapter 6

Numerical Simulation of Short Pulse Laser Interactions

By now, the reader will have noticed the high proportion of illustrations in the preceding chapters lacking the usual accreditation. These have not been conjured up out of thin air, but for the most part have been created with the help of original numerical models written by the author. The purpose of this chapter is to make these models accessible to those who wish to try their own hand at simulation — either at the 'black box' level, or as an intermediate step for numerically inclined theoreticians who ultimately wish to develop their own codes.

For each of the numerical algorithms outlined here, the reader will find references to do-it-yourself projects using codes available from a software archive specially set up to accompany the material in this book. At the time of going to press, this is located at:

http://www.fz-juelich.de/zam/files/splim

Some of these projects are already structured, having been created for a computer-lab teaching module. The idea is to start from a skeletal implementation of the algorithm, gradually introducing more features until the model resembles a fully-fledged simulation code.

Where possible, the source code is provided in two languages: C and Fortran-90, a choice which which should also satisfy adherents of C++ and Fortran-77. Self-contained compilation instructions can be found in the relevant source directories, which assume that the programs will be run under a Unix-based operating system, such as a Linux-PC. Most of the programs should run under other systems too (Windows and MAC-OS), but certain features, such as file handling, may need a bit of adaptation. In the author's experience, computer graphics come second only to the subject of programming language itself in its ability to cause verbal warfare during the coffee break, so to maximize portability and minimize complaints, graphical display is left entirely to the reader to worry about. Generally, the output data from the codes can be read by a simple x-y or surface $z(x, y)$ plotting program such as GLE, XMGRACE or GNUPLOT.

6.1   Plasma Modeling

With the exception of parts of Chapter 2, most of the phenomena discussed in this book involve the interaction with — or creation of — plasmas by lasers. Thus, it makes sense to consider some general issues in plasma simulation before we tackle the specific problems considered earlier. The first issue to address is to decide on the *detail* needed to describe the plasma for the application at hand. This depends on various factors: the initial state we expect the plasma to be in; the typical time- and length- scales of interest; and, most importantly, what kind of effects we are looking for, such as changes in density or temperature, waves, or production of energetic particles. As the 'fourth state' of matter, plasmas invariably pose

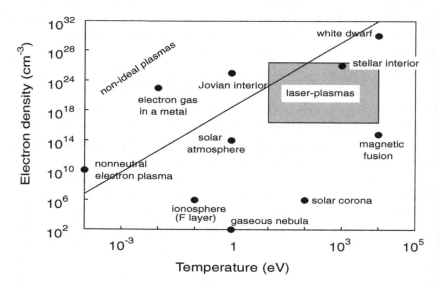

Fig. 6.1   Plasma classification in the density-temperature plane.

a something of a dilemma because they span a huge range of densities and temperatures, see Fig. 6.1. The analogy with gases or liquids is often helpful for classical (ideal), low-density plasmas in thermal equilibrium, where they can be described by the Navier-Stokes fluid equations. High-density plasmas have more in common with solid-state systems, however, and require special treatment (see p. 142). Laser interaction with solid targets is particularly irksome because it inevitably involves transitions through various solid and plasma states. In the high-intensity, short pulse context, most plasma models can be placed into one of 3 categories: static, fluid or kinetic. The practical consequences of this over-simplified grouping will become clear as we proceed, but for now it is sufficient to note that there is a formal relationship between these pictures which is applicable in a much broader plasma theory context, see Fig. 6.2.

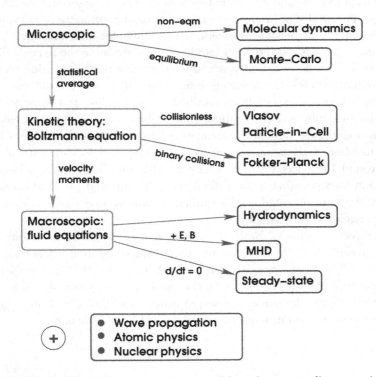

Fig. 6.2 Physical basis of common plasma models and corresponding numerical approaches.

The first of these approaches treats the plasma more as a passive medium which is created (by ionization) or altered as the laser propagates through it. This description is particularly appropriate for very low density systems where the plasma period $2\pi/\omega_p$ is long compared to the interaction time. Thus, the plasma makes itself felt essentially via local changes in the refractive index, which, in turn, may influence the laser pulse propagation, as seen frequently in Chapter 4.

Fluid, or hydrodynamic modeling goes a step beyond this static picture, but usually assumes that the plasma is at least locally in thermal equilibrium, that is: its particle distribution functions are Maxwellian with well-defined temperatures T_e and T_i (often equal). Under these conditions, the plasma state can be described by its macroscopic variables: density, pressure, fluid velocity *etc.* The main use of hydro-models is to follow the large-scale dynamic behavior of the plasma, perhaps under the influence of external electric and magnetic fields, or heating by laser and/or particle beams. Timescales of interest are governed by the ion motion — typically from around a picosecond to several nanoseconds for laser-plasmas.

Kinetic models, on the other hand, seek to determine the particle distributions $f_\alpha(v)$ self-consistently, and do not assume that everything remains Maxwellian. Typical examples are laser propagation and plasma wave generation, where we expect large oscillation amplitudes, and some form of wave-particle interaction (trapping or wave-breaking). In its most elementary form, kinetic simulation involves solving the N-body problem for a huge number of electrons and ions. For example, a solid-density plasma has around 10^{23} electrons plus ions per cubic cm. To describe a laser interaction incorporating a focal diameter of 10 μm and penetration depth of, say, 0.2 μm, we would need a simulation volume in excess of 15 μm^3, or 10^{18} particles at this density.

A direct solution of Newton's equation needs $O(N^2)$ operations, and is not feasible with 'brute-force' techniques for more than a few thousand 'bodies'. Statistical averaging can be used to reduce the discrete N-body problem to a continuum $f(\boldsymbol{x}, \boldsymbol{v}, t)$, the mathematical proof of which I will not discuss here: the reader is referred instead to Elliot (1993) for a gentle introduction to the formalities and nuances of kinetic theory.

6.2 Finite-Difference Equations

Just as we take some basic plasma physics for granted, it would also not be in the spirit of this chapter to provide a rigorous theory of numerical analysis to justify the models used here. The only 'proof' of mathematical correctness offered for these models is that they *work*. For the purist, there are enough texts covering the fields of numerical methods and computational physics generally, see, for example, Richtmyer and Morton (1967) or Potter (1972); as well as a number dealing specifically with plasma simulation, such as Birdsall and Langdon (1985), Hockney and Eastwood (1981) and Tajima (1988). Some excellent introductory articles can be found in the 5-volume series of the Scottish Universities Summer School Proceedings on Laser-Plasma Interactions (1979–1994); in particular by Evans (1985), Nicholas (1982) and Pert (1979, 1988). The remainder of this section is loosely based on this material, which will serve to introduce some terminology and provide a basic toolkit for getting started in numerical modeling.

Most physical systems can be described by an equation of the form:

$$\frac{\partial f}{\partial t} = G(f, x, \nabla.f, \nabla^2 f),$$

which may not necessarily be linear. Take, for example, the diffusion equation describing heat conduction in one dimension:

$$\frac{\partial T}{\partial t} = \sigma \frac{\partial^2 T}{\partial x^2} + S(x, t), \tag{6.1}$$

where σ is the material-dependent diffusion coefficient and $S(x, t)$ the heat source. Or consider the Lorentz equation (4.1) for a charged particle in a laser field:

$$\frac{d}{dt}(\gamma v) = E + v \times B \; ; \; \frac{dx}{dt} = v. \tag{6.2}$$

Example (6.2) is in effect doubly nonlinear: first, because of the relativistic γ−factor and second, because $v = v\{x(t), t\}$. When we sought analytical solutions to Eq. (6.1) and Eq. (6.2) in Secs. 5.3 and 3.1 respectively, we implicitly assumed that these solutions were mathematically well-behaved; that is, the functions and their derivatives were continuous, non-singular and so on. In order to represent our function $f(x, t)$ on a digital computing machine, we must part with our hard-earned training in differential calculus and consider instead its discrete counterpart f_i^n, defined over a finite number of grid points $x_i = \{i\Delta x, i = 0, 1, .., n\}$, at set points in time

$t^n = n\Delta t$, see Fig. 6.3. In the numericists jargon, Δx is referred to as the *grid spacing* or *mesh size*; Δt as the *timestep*. The important design

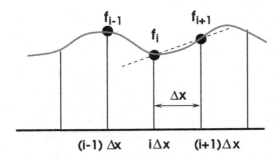

Fig. 6.3   Representing a continuous function on a finite, evenly spaced mesh.

part of computational physics involves finding a suitable numerical *scheme*, that is: approximations to $\partial f/\partial x, \partial^2 f/\partial x^2$ etc., providing a solvable set of algebraic equations. Because f is no longer exact, its derivatives must be approximated too, and it turns out that there are several ways of choosing them. A crude approximation to $\partial f/\partial x$ at x_i can be found by performing a Taylor expansion of f about $x_i + \Delta x$:

$$f(x_i + \Delta x) = f(x_i) + \Delta x \frac{\partial f}{\partial x}\bigg|_i + \frac{1}{2}\Delta x^2 \frac{\partial^2 f}{\partial x^2}\bigg|_i + ..O(\Delta x^3).$$

Thus, rearranging, we can write:

$$\frac{\partial f}{\partial x}\bigg|_i = \frac{f(x_i + \Delta x) - f(x_i)}{\Delta x} - \frac{1}{2}\Delta x \frac{\partial^2 f}{\partial x^2}\bigg|_i - O(\Delta x^2)$$

$$= \frac{f_{i+1} - f_i}{\Delta x} - \frac{1}{2}\Delta x \frac{\partial^2 f}{\partial x^2}\bigg|_i - O(\Delta x^2). \tag{6.3}$$

The largest error in the derivative is given by the 2nd term on the RHS, which is in effect $O(\Delta x)$. This means that by doubling the number of grid points, we would only halve the errors in $\partial f/\partial x$ at the cost of twice the computational effort.

The accuracy convergence rate can be improved upon by *centering*. Supposing we instead expand our function f about $x_i - \Delta x$:

$$f(x_i - \Delta x) = f(x_i) - \Delta x \frac{\partial f}{\partial x}\bigg|_i + \frac{1}{2}\Delta x^2 \frac{\partial^2 f}{\partial x^2}\bigg|_i - ..O(\Delta x^3).$$

Then, by analogy with Eq. (6.3), we obtain an alternative expression for $\frac{\partial f}{\partial x}\big|_i$:

$$\frac{\partial f}{\partial x}\bigg|_i = \frac{f_i - f_{i-1}}{\Delta x} + \frac{1}{2}\Delta x \frac{\partial^2 f}{\partial x^2}\bigg|_i - O(\Delta x^2). \tag{6.4}$$

Adding Eq. (6.3) to Eq. (6.4) — or taking the average of the two, we end up with:

$$\frac{\partial f}{\partial x}\bigg|_i = \frac{f_{i+1} - f_{i-1}}{2\Delta x} + O(\Delta x^2). \tag{6.5}$$

This expression is *second-order accurate*, and as such, converges faster than either of the $O(\Delta x)$ differences. Its disadvantage is that it does not actually depend on f_i at all, which can lead to a potentially dangerous decoupling of adjacent grid points. A commonly used technique which neatly combines the best of both worlds is simply to define gradients at the *mid-points* of the mesh:

$$\frac{\partial f}{\partial x}\bigg|_{i+\frac{1}{2}} = \frac{f_{i+1} - f_i}{\Delta x} + O(\Delta x^2). \tag{6.6}$$

Higher derivatives can generally be obtained by 1st-order differencing of the next lowest order gradient. For example, the 2nd derivative can be found by differencing Eq. (6.3), thus:

$$\begin{aligned}
\frac{\partial^2 f}{\partial x^2}\bigg|_i &= \frac{d}{dx}\left[\frac{f_{i+1} - f_i}{\Delta x}\right] \\
&= \frac{f_{i+1} - 2f_i + f_{i-1}}{\Delta x^2} + O(\Delta x^2),
\end{aligned} \tag{6.7}$$

which, we observe, is naturally centered.

Centering is even more important in the time domain, where we aim to keep errors down to $O(\Delta t^2)$ or smaller. An error $O(\Delta t)$ will almost inevitably result in numerical nonsense, because decreasing the timestep will leave the *cumulative* (time-integrated) error unchanged: $\epsilon = \sum_i^N O(\Delta t) \propto N\Delta t \propto t_{\text{run}}$. The motion of a particle in an electric field serves as a good example of how to construct a second-order-accurate integrator. The *non*relativistic version of Eq. (6.2) is simply:

$$\frac{dv}{dt} = E; \quad \frac{dx}{dt} = v.$$

Straightforward differencing gives:

$$v^{n+1} = v^n + \Delta t E^{n+\frac{1}{2}},$$
$$x^{n+\frac{3}{2}} = x^{n+\frac{1}{2}} + \Delta t v^{n+1}, \qquad (6.8)$$

which is naturally centered if we let the particle position and field be shifted by $\Delta t/2$ with respect to its velocity. For obvious reasons, this scheme is known as a *leap-frog* integrator, see Fig. 6.4.

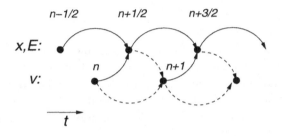

Fig. 6.4   Leap-frog integrator.

Higher order schemes are possible using intermediate steps, but these are only necessary where minimising cumulative errors is of paramount importance (for example in computing planetary orbits). For our purposes, the Golden Rule is: Keep it simple! Numerical robustness and simplicity are the keys to successful plasma simulation. Quite often it is necessary to 'add more physics' to a difference equation; a task which may take a few minutes with a leap-frog scheme, but days to implement within a 6th-order Runge-Kutta algorithm.

6.3   Laser Propagation in Gases

The tour of numerical models begins with a more detailed look at optical field ionization. The spatio-temporal model introduced here improves on the simplified picture used in Sec. 2.4 to describe ionization-induced refraction. It will be extended further when we consider relativistic propagation in plasmas. For the moment, the first problem to tackle is the field ionization itself, the theory of which was covered in Sec. 2.2. However, we stopped short of actually computing ion populations. These are given by rate equations similar to those for collisional ionization, see Eq. (5.8) in Sec. 5.1. If we ignore all forms of recombination, the appropriate equation for OFI will be:

$$\frac{dn_Z}{dt} = \alpha_Z n_{Z-1} - \alpha_{Z+1} n_Z.$$ (6.9)

The first term represents *creation* of ions with charge state Z from ions with charge $Z - 1$; the second term *depletion* of the same ions due to further ionization. The rates α_Z for each ion are given by the Keldysh formula Eq. (2.11) or some extension thereof.

The numerical solution of this coupled set of rate equations is best illustrated by way of a concrete example, for which we will take helium. In this case we have just two coupled equations:

$$\begin{aligned}
\frac{dn_1}{dt} &= \alpha_1 n_0 - \alpha_2 n_1, \\
\frac{dn_2}{dt} &= \alpha_2 n_1.
\end{aligned}$$ (6.10)

The number densities n_0, n_1 and n_2 represent the concentrations of neutral He gas, He$^+$ ions, and He^{2+} ions respectively. Obviously these must add up to the total amount of gas present before ionization, ρ_0, therefore:

$$n_0 + n_1 + n_2 = \rho_0.$$ (6.11)

Also, for a charge-neutral system, the total number of electrons must satisfy:

$$n_e = n_1 + 2n_2.$$ (6.12)

Helium is a particularly convenient example because both bound electrons are in the $1s$ level, with quantum numbers $n = 1$, $l = m = 0$. This means that we can apply the Keldysh model as is, giving ionization rates:

$$\alpha_i = 4\omega_a \left(\frac{E_i}{E_h}\right)^{\frac{5}{2}} \frac{E_a}{E_L(t)} \exp\left[-\frac{2}{3}\left(\frac{E_i}{E_h}\right)^{\frac{3}{2}} \frac{E_a}{E_L(t)}\right],$$ (6.13)

where ω_a and E_a are the atomic frequency and field as defined by Eq. (2.12) and Eq. (2.1) respectively. The ionization potentials are: $E_1 = 24.59$ eV, $E_2 = 54.418$ eV, $E_h = 13.61$ eV.

Using the conservation relation Eq. (6.11) to eliminate n_0, we can write down a set of 2nd order difference equations corresponding to Eq. (6.10):

$$n_1^{m+1} = n_1^m + \Delta t \alpha_1^{m+\frac{1}{2}} \left(\rho_0 - n_1^{m+\frac{1}{2}} - n_1^{m+\frac{1}{2}} \right) - \alpha_2^m n_1^{m+\frac{1}{2}}$$

$$n_2^{m+1} = n_2^m + \alpha_2^{m+\frac{1}{2}} n_1^{m+\frac{1}{2}} \tag{6.14}$$

$$n_i^{m+\frac{1}{2}} = \frac{1}{2} \left(n_i^{m+1} + n_i^m \right), \quad i = 1, 2.$$

These are implicit, but can be iterated fairly rapidly to get n_1 and n_2 at timestep $m + 1$. In fact, Eq. (6.15) can be rearranged in this case to solve for n_1, but for more complex atoms this procedure requires inversion of a large matrix. Given the local ion densities and corresponding electron density through Eq. (6.12), we can proceed to determine the influence of variations in the refractive index on the laser pulse via the wave equation:

$$\left(\frac{\partial}{\partial t^2} - \nabla^2 \right) A = -n_e A,$$

where t and x and n_e normalized to ω, c/ω and n_c respectively.

Applying the slowly varying envelope approximation as on p. 99, but this time using the *vacuum* dispersion relation $\omega^2 = c^2 k^2$ to cancel the low-order terms (initially we have no plasma, so ω_p is not defined), we end up with:

$$\frac{\partial a}{\partial t} + v_g \frac{\partial a}{\partial z} - \frac{i}{2} \nabla_\perp^2 a = -\frac{1}{2} n_e a. \tag{6.15}$$

As it stands, this electromagnetic (EM) envelope equation is a 2-dimensional (2D) (r,z,t) hybrid advection-diffusion equation that needs some care in the choice of difference scheme. In Sec. 4.6.2 we derived moment or 'beam' equations for the simpler steady-state system ($r, z = v_g t$), but here we need a method for handling propagation with an arbitrary (dynamic) plasma response. An explicit Lax-Wendroff algorithm for the advective part tends to be dispersive in z, so to avoid this we instead integrate along the characteristics, fixing the mesh size to $\Delta z = v_g \Delta t$. This is equivalent to setting $\partial/\partial z = 0$ and integrating Eq. (6.15) in a computational window moving with the pulse. The gas or plasma (represented here by n_e) is then advected backwards through the grid with velocity v_g (Gibbon, 1992). The required difference equation on a uniform grid with

$r_i = i\Delta r; z_k = k\Delta z$, is then simply:

$$a_{i,k}^{m+1} = a_{i,k}^m - \frac{i\Delta t}{2} \left[\frac{a_{i+1} - 2a_i + a_{i-1}}{(\Delta r)^2} + \frac{a_{i+\frac{1}{2}} - a_{i-\frac{1}{2}}}{r_i \Delta r} \right]_k^{m+\frac{1}{2}}$$
$$- \frac{i\Delta t}{2} (n_e)_{i,k}^{m+\frac{1}{2}} a_{i,k}^{m+\frac{1}{2}}. \tag{6.16}$$

This scheme is fully implicit, and is solved iteratively: 2–3 iterations are typically required for convergence. This extra work pays for itself because the scheme conserves power, satisfying Eq. (4.111) for each z-slice:

$$P^{m+1} - P^m = 2\pi \sum_i |a_i^{m+1}|^2 r_i \Delta r - 2\pi \sum_i |a_i^m|^2 r_i \Delta r = 0.$$

It also turns out to be marginally stable, i.e. the transverse modes of interest, $k_\perp > \pi/\Delta x$ are neither damped nor amplified. The EM boundary conditions are reflective at $r = 0$, i.e. $a_{-1} = a_1$, and an absorbing layer is placed at $r = R$ to avoid unphysical reflection. The system size R is chosen according to the initial (unfocused) beam width.

The electron density is computed from the solution to the rate equations at each new timestep:

$$(n_e)_{ik}^{m+\frac{1}{2}} = (n_1)_{ik}^{m+\frac{1}{2}} + 2(n_2)_{ik}^{m+\frac{1}{2}}. \tag{6.17}$$

Depending on the ionization rates, Eq. (6.15) may need to be sub-cycled with a smaller timestep inside each completed EM step. The electric field entering the ionization rates α_i is just taken from the vector potential $a(r, z, t)$. This procedure is simplified if we renormalize the fields and frequencies in Eq. (6.13) to the corresponding laser quantities ω and $m\omega c/e$, giving:

$$\tilde{\omega}_a = 22.13\lambda_\mu,$$
$$\tilde{E}_a = 0.16\lambda_\mu.$$

Within this unit system, $\tilde{E_L}(t) = a(t)$.

Source Code:

Directory: `propagation`
f77-version: `src/drelp.f, src/*.f src/makefile`
Documentation: `README, README.code`
Compilation: `make drelp`

Projects:

1) Propagation of focused, Gaussian pulse in vacuum
2) Ionization defocusing in hydrogen (see Fig. 2.3)
3) Refractive effects in helium

6.4 Single Particle Motion in an Electromagnetic Field

Here an algorithm is described for integrating the fully relativistic equation of motion for a particle in an electromagnetic field. This so-called electromagnetic 'particle-pusher', originally developed by Buneman (1973), actually forms the kernel of many of the particle-in-cell codes currently in use. Indeed, we will meet this scheme again when we discuss PIC in its various guises in Sec. 6.9. Despite its apparent simplicity, the accompanying model can be used to explore most of the material presented already in Chapter 3.

To keep things self-contained, we restate our equation of motion for an electron in a linearly polarized EM wave $A = a_0 \cos(t - x)\hat{y}$ (equivalent to setting $\delta = 1$ in Eq. (3.3). The z-component being ignorable, we have from Eq. (4.1):

$$\frac{dp_x}{dt} = -v_y B_z$$

$$\frac{dp_y}{dt} = -E_y + v_x B_z$$

$$\gamma = (1 + p_x^2 + p_y^2)^{1/2}; \quad \boldsymbol{v} = \frac{\boldsymbol{p}}{\gamma}; \quad \frac{dx}{dt} = \boldsymbol{v}.$$

Unlikely though it may seem at first, it is possible to find a centered, explicit difference scheme for these equations. Buneman's trick was to split the forces acting on the particle into linear acceleration (electric field) and rotation (magnetic field) components. In our simplified geometry — $E = (0, E_y, 0), \boldsymbol{B} = (0, 0, B_z)$ — the scheme reduces to a particularly simple and elegant form. Normalizations (already applied above) are chosen as before in Sec. 3.1, that is: $t \rightarrow \omega t; x \rightarrow kx; v \rightarrow v/c; p \rightarrow p/mc; E \rightarrow eE/m\omega c; B \rightarrow eB/m\omega$.

Assuming $p_x^{n-\frac{1}{2}}, p_x^{n-\frac{1}{2}}, \gamma^{n-\frac{1}{2}}, E_y^n$ and B_z^n are known, the scheme then proceeds as follows:

1/2-acceleration:

$$p_x^- = p_x^{n-\frac{1}{2}}$$
$$p_y^- = p_y^{n-\frac{1}{2}} + \frac{\Delta t}{2} E_y^n$$

rotation:

$$\gamma^n = (1 + (p_x^-)^2 + (p_y^-)^2)^{1/2} \; ; \quad t = \frac{\Delta t}{2} \frac{B_z^n}{\gamma^n} \; ; \quad s = \frac{2t}{1+t^2}$$
$$p_x' = p_x^- + p_y^- t$$
$$p_y^+ = p_y^- - p_x' s \qquad\qquad (6.18)$$
$$p_x^+ = p_x' + p_x^+ t$$

1/2-acceleration:

$$p_x^{n+\frac{1}{2}} = p_x^+$$
$$p_y^{n+\frac{1}{2}} = p_y^+ + \frac{\Delta t}{2} E_y^n.$$

Note that the velocity v of the particle is actually redundant, although the γ-factor is still needed for the rotation part. Of course, we need to recover $v^{n+\frac{1}{2}}$ from $p^{n+\frac{1}{2}}$ to get the new position x^{n+1}, but it is numerically safer to compute everything in terms of the momentum rather than velocity. If we only stored v, and used this to compute γ, we would get severe rounding errors when the motion became strongly relativistic. There is an exhaustive discussion of the Buneman pusher — including a geometrical proof of its second-order accuracy — in Birdsall and Langdon (1985), Chapter 4.

Source Code:

Directory: `FIGURE8`
f90-version: `singlee.f, makefile`
C-Version: `singlee.c, makefile`
Compilation: `make orbit`

Projects:

1) Electron orbits in an EM plane wave
2) Finite pulse shape $a_0(t)$
3) Laser focus

6.5 Nonlinear Propagation in Underdense Plasmas (NLSE)

The envelope model presented here can be thought of as an extension of the one already introduced earlier for simulating ionization-induced refraction (Sec. 6.3). Two important additions are made: first, relativistic drag effects in the transverse electron current; second, ponderomotive expulsion of electrons from the beam axis. For the time-being, we leave aside ionization effects and consider a pre-ionized, initially homogeneous plasma. Adopting the same EM unit system as before, our starting equation (4.108) can be rewritten:

$$\frac{\partial a}{\partial t} + v_g \frac{\partial a}{\partial z} + \frac{i}{2}\nabla^2 a + \frac{i}{2}\left(n_0 - \frac{n_e}{\gamma}\right)a = 0, \qquad (6.19)$$

where the electron density is determined from Poisson's equation (see Sec. 4.6.3):

$$n_e(r,t) = n_0 + \nabla_\perp^2 \gamma(r,t). \qquad (6.20)$$

For self-consistency, the envelope approximation demands that $\gamma(r,t)$ varies slowly compared to ω and ω_p, so we neglect harmonic components (as, for example in Sec. 4.5) and take $\gamma = [1 + a^2(r,t)/2]^{1/2}$.

As before, we consider a computational window moving with the pulse group velocity v_g, fix $\Delta z = v_g \Delta t$, and advect the plasma back through the grid. The difference equation in this case is:

$$a_{i,k}^{m+1} = a_{i,k}^m - \frac{i\Delta t}{2}\left[\frac{a_{i+1} - 2a_i + a_{i-1}}{(\Delta r)^2} + \frac{a_{i+\frac{1}{2}} - a_{i-\frac{1}{2}}}{r_i \Delta r}\right]_k^{m+\frac{1}{2}}$$
$$- \frac{i\Delta t}{2}\left[(n_0)_{i,k} - \frac{(n_e)_{i,k}^{m+\frac{1}{2}}}{\gamma_{i,k}^{m+\frac{1}{2}}}\right]a_{i,k}^{m+\frac{1}{2}}. \qquad (6.21)$$

The density is specified according to:

$$(n_e)_{i,k}^{m+\frac{1}{2}} = (n_0)_{i,k}^{m+\frac{1}{2}} + \left[\frac{\gamma_{i+1} - 2\gamma_i + \gamma_{i-1}}{(\Delta r)^2} + \frac{\gamma_{i+\frac{1}{2}} - \gamma_{i-\frac{1}{2}}}{r_i \Delta r}\right]_k^{m+\frac{1}{2}}. \qquad (6.22)$$

This scheme is again fully implicit, this time necessarily so because of the nonlinear n_e/γ term, and must therefore be solved iteratively. Although 3 iterations may be sufficient for weakly relativistic conditions, more may be needed for convergence if strong self-focusing occurs when $P/P_c > 1$. To recover the steady-state model, we let the pulse length $\tau \to \infty$, and follow the evolution of a single cell in z, setting $N_z = k = 1$.

The background density can n_0 can also be varied to include finite longitudinal and radial gradients, such as those typical for a gas jet or a plasma waveguide. Any imposed gradient length-scale should be much larger than a plasma wavelength to avoid violating the adiabatic approximation. As in previous works (Sun *et al.*, 1987; Borisov *et al.*, 1992b; Bonnaud *et al.*, 1994), we handle complete cavitation by setting $n_e = \text{max}(0, n_e)$. The effect of this is that the electron charge is no longer conserved, and the walls of the density channel created by self-focusing tend to be higher than they should be. This non-physical aspect of the adiabatic model will be most severe for powers greatly exceeding P_c, and can exaggerate the tendency for self-channeled propagation (Borisov *et al.*, 1992b). This breakdown of the fluid model has been demonstrated in 2D particle-in-cell (PIC) simulations (Mori *et al.*, 1988), albeit in slab geometry, where wave-breaking and filamentation are observed at the channel walls.

Source Code:

Directory: `propagation`
f77-version: `src/drelp.f`, `src/*.f src/makefile`
Documentation: `README, README.code`
Compilation: `make drelp`

Projects:

1) Numerical verification of relativistic power threshold
2) Relativistic self-channeling in pre-ionized plasma

6.6 Wave Propagation in Overdense Plasmas: Helmholtz Solver

Just as the wave-envelope model in Sec. 6.5 provides insight into nonlinear propagation behavior in tenuous plasmas, so the Helmholtz-solver here offers a fast and intuitive introduction to laser-solid interactions. In the following section a numerical solution to the one-dimensional steady-state Helmholtz equations Eq. (5.20) and Eq. (5.24) is described for both s- and p-polarized light. The model can be used to calculate collisional absorption in arbitrary, (1D) density profiles.

S-polarized light

Normalizing the x-coordinate to the wave-vector, we rewrite our equation for the electric field $E = E_z$ (see Fig. 5.1) as:

$$\frac{\partial^2 E_z}{\partial x^2} + (\varepsilon - \sin^2 \theta)E_z = 0,$$

where

$$\varepsilon(x) = 1 - \frac{n_e}{1 + i\nu_{ei}(x)}.$$

Straightforward differencing of the 2nd derivative gives:

$$E_{i+1} - 2E_i + E_{i-1} + \Delta x^2(\varepsilon_i - \sin^2 \theta)E_i = 0,$$

which can be rewritten in the standard tridiagonal form:

$$\alpha_i E_{i-1} + \beta_i E_i + \gamma_i E_{i+1} = y_i, \tag{6.23}$$

with coefficients:

$$\begin{aligned} \alpha_i &= \gamma_i = 1 \\ \beta_i &= (\varepsilon_i - \sin^2 \theta)\Delta x^2 - 2 \\ y_i &= 0. \end{aligned} \tag{6.24}$$

This is a linear equation system which can be easily solved by standard matrix-inversion techniques, such as LU decomposition (Press *et al.*, 1989). Before we can apply this technique, however, we must provide some missing information at the boundaries. Assuming the light is incident from the left and is reflected back out again, we have at this boundary:

$$\alpha_1 E_0 + \beta_1 E_1 + \gamma_1 E_2 = 0. \tag{6.25}$$

We suppose that $x_1 = \Delta x$ lies in a region of vacuum, so that we can split the wave into forward- and backward-travelling plane waves $\sim \exp\{i\boldsymbol{k}.\boldsymbol{x}\}$ with amplitudes $A = a_0$ and B respectively:

$$E(\Delta x) = E_1 = A + B.$$

If the waves are moving at an angle θ to the x-axis, then the neighbouring grid point will see a phase either advanced or receded by an amount

$\Delta\phi = k\Delta x \cos\theta$. Therefore,

$$E_0 = Ae^{-i\Delta x \cos\theta} + Be^{i\Delta x \cos\theta}$$
$$= E_1 e^{i\Delta x \cos\theta} - 2ia_0 \sin(\Delta x \cos\theta). \tag{6.26}$$

Eliminating E_0 in Eq. (6.26) then gives:

$$\beta_1' E_1 + E_2 = y_1, \tag{6.27}$$

where

$$y_1 = 2ia_0 \sin(\Delta x \cos\theta)$$
$$\beta_1' = \beta_1 + e^{i\Delta x \cos\theta}. \tag{6.28}$$

In other words, boundary conditions are implemented here by modifying the $\beta-$ and $y-$ coefficients of the first row of the matrix prior to the inversion.

P-polarized light

We recall from Sec. 5.2.1 that for p-light the electric field components E_x and E_y are coupled, leading, among other things, to resonant driving of plasma waves. For this reason, it is simpler to solve for the magnetic field B_z first, and derive the other fields from its solution. the normalized wave equation for B_z is:

$$\frac{\partial^2 B_z}{\partial x^2} - \frac{1}{\varepsilon}\frac{\partial \varepsilon}{\partial x}\frac{\partial B_z}{\partial x} + (\varepsilon - \sin^2\theta)B_z = 0.$$

The difference form of this equation is:

$$\frac{B_{i+1} - 2B_i + B_{i-1}}{\Delta x^2} - \frac{\varepsilon_{i+1} - \varepsilon_{i-1}}{2\Delta x \varepsilon_i}(B_{i+1} - B_{i-1}) + (\varepsilon_i - \sin^2\theta)B_i = 0.$$

Once again, this can be cast into tridiagonal form by grouping together the terms at common grid points, giving:

$$\alpha_i B_{i-1} + \beta_i B_i + \gamma_i B_{i+1} = y_i, \tag{6.29}$$

where:

$$\alpha_i = 1 + \frac{\Delta\varepsilon_i}{\varepsilon_i},$$

$$\gamma_i = 1 - \frac{\Delta\varepsilon_i}{\varepsilon_i},$$

$$\beta_i = (\varepsilon_i - \sin^2\theta)\Delta x^2 - 2, \qquad (6.30)$$

$$y_i = 0,$$

$$\Delta\varepsilon_i = \frac{\varepsilon_{i+1} - \varepsilon_{i-1}}{4}.$$

For a dielectric function $\varepsilon(x)$ of the form given by Eq. (5.19), we observe that ε'/ε has a singularity due to the plasma resonance at $n_e/n_c = 1$, so some care needs to be taken with the differencing near this point. As before, the boundary conditions are also determined by considering the relative phase between mesh points at the vacuum boundary. Because the dielectric function vanishes here ($\varepsilon(0) = 0$), it turns out that the required coefficients y_1 and β'_1 are identical with those for s-light in Eq. (6.28).

Having solved for the fields E_z or B_z, the remaining fields can be easily obtained from the following relations:

$$B_y = iE'_z,$$

$$E_y = -i\frac{c}{\varepsilon\omega}B'_z, \qquad (6.31)$$

$$E_x = -\frac{\sin\theta}{\varepsilon}B_z.$$

An example of this solution for both $s-$ and $p-$light is shown in Fig. 6.5. Here, the collision frequency was set at $\nu_{ei}/\omega_p = 0.01$ at the maximum density ($n/n_c = 1.5$), giving net absorption rates of 0.5 and 0.7 for the two polarizations respectively. Note the sharp resonance in E_x at $n = n_c$, demonstrating the classical tunneling behavior for p-light. It is also instructive to compare the fluid solution with the fully kinetic simulation shown earlier in Fig. 5.11. In the latter, the resonance is less well-defined because of wave-breaking, and there is significant harmonic content in the electrostatic field below n_c.

The absorption is straightforward to compute directly from the fields. At the vacuum boundary, we have:

$$E_1 = B_1 = a_0(1 - r),$$

where r is the relative amplitude of the reflected light compared to the

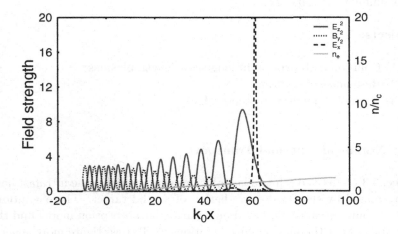

Fig. 6.5 Stationary electric and magnetic fields computed from the Helmholtz equations for a linear density profile with $L/\lambda \simeq 8, \theta = 5^o$.

incident wave. The reflectivity is then just:

$$
R = |r|^2 = \begin{cases} \left|1 - \dfrac{E_1}{a_0}\right|^2, & s\text{-pol.} \\[3ex] \left|1 - \dfrac{B_1}{a_0}\right|^2, & p\text{-pol.} \end{cases} \tag{6.32}
$$

Alternatively, we can use the Poynting flux to get the absorption:

$$
A = \frac{|\boldsymbol{E} \times \boldsymbol{B}|_x}{E^2 + B^2} = \begin{cases} \dfrac{E_z(0)B_y(0)}{2a_0^2}, & s\text{-pol.} \\[3ex] \dfrac{B_z(0)E_y(0)}{2a_0^2}, & p\text{-pol.} \end{cases} \tag{6.33}
$$

Since we should normally have $A = 1 - R$, evaluating both expressions serves as a convenient check on the numerics.

Source Code:

Directory: `helmholtz`
f77-version: `helmholtz.f makefile`
Documentation: `README`

Compilation: `make helm`

Projects:

1) Collisional absorption in long scale-length plasmas
2) Resonance absorption
3) Normal skin effect in step-profile

6.7 Nonlinear Thermal Transport

In Sec. 5.3 we derived self-similar solutions characterizing the ultrafast heat front created by short-pulse irradiation of a solid target. These solutions were of course ideal cases, based on a particular absorption model and the assumption that the ions remained stationary. To investigate more general behavior, in which the absorption is perhaps unknown *a priori*, or where hydrodynamics sets in early, the transport equation (5.38) must be solved numerically. We recall that for an ideal gas, this has the form:

$$\frac{\partial T_e}{\partial t} = \frac{2}{3n_e} \frac{\partial}{\partial x} \left(\kappa_e \frac{\partial T_e}{\partial x} \right) + \frac{2}{3n_e} \frac{\partial \Phi_L}{\partial x}, \qquad (6.34)$$

where the Boltzmann constant k_B has been absorbed into T_e, and the thermal conductivity, κ_e, is assumed to take the usual Spitzer expression:

$$\begin{aligned} \kappa_e &= \frac{16\sqrt{2\pi}}{3} \frac{n_e v_{te}^2}{\nu_{ei}} \\ &= \frac{16\sqrt{2}}{\pi^{3/2}} \frac{T_e^{5/2}}{Ze^4 m^{1/2} \ln \Lambda}. \end{aligned} \qquad (6.35)$$

Note that the two expressions in Eq. (6.35) are consistent if we take $v_{te} = (k_B T_e/m)^{1/2}$ and the collision frequency ν_{ei} is defined according to Eq. (5.10).

To reduce the number of constants floating around in the model, we now introduce a somewhat unorthodox system of units, the point of which will soon become apparent. First, we note that Eq. (6.34) is actually expressed in Watts or Js^{-1} (SI), whereas it is more usual here to speak of temperature changes in eV. Also, the natural time- and length-scales for this problem are femtoseconds and nanometres respectively. These considerations lead us to the following renormalizations:

$$\frac{k_B}{T_e} = e\left(\frac{k_B T_e}{\text{eV}}\right) = e\tilde{T},$$

$$t = 10^{15}\, \tilde{t}_{\text{fs}},$$

$$x = 10^{-9}\, \tilde{x}_{\text{nm}}, \tag{6.36}$$

$$\frac{n_e}{\text{m}^{-3}} = 10^{27}\left(\frac{\tilde{n}}{\text{nm}^{-3}}\right),$$

$$\frac{\Phi_L}{\text{Wcm}^{-2}} = 10^{33}\left(\frac{\tilde{\Phi}}{\text{eV nm}^{-2}\text{fs}^{-1}}\right).$$

Within this system, the effective diffusion coefficient becomes:

$$\frac{D}{\text{ms}^{-1}} = \frac{2\kappa_e}{3n_e} = 10^{-3}\frac{\tilde{D}}{\text{nm}^2\text{fs}^{-1}} = \frac{2}{3}\cdot 10^{-3}\frac{\tilde{\kappa}}{\tilde{n}},$$

where

$$\tilde{\kappa} = 0.81\frac{T_{\text{eV}}^{5/2}}{Z\ln\Lambda}\ \text{nm}^{-1}\text{fs}^{-1}.$$

Substituting these rescaled quantities into Eq. (6.36) gives a numerically much cleaner form:

$$\frac{\partial\tilde{T}}{\partial\tilde{t}} = \frac{\partial}{\partial\tilde{x}}\left(\tilde{D}\frac{\partial\tilde{T}}{\partial\tilde{x}}\right) + \frac{2}{3\tilde{n}}\frac{\partial\tilde{\Phi}}{\partial\tilde{x}}. \tag{6.37}$$

This can be differenced according to the well-known Crank-Nicholson scheme (Potter, 1972):

$$\frac{T_i^{m+1} - T_i^m}{\Delta t} = \Delta D_i^{m+\frac{1}{2}}\Delta T_i^{m+\frac{1}{2}} + \frac{D_i^{m+\frac{1}{2}}}{\Delta x^2}(T_{i+1} - 2T_i + T_{i-1})^{m+\frac{1}{2}} + S_i^{m+\frac{1}{2}}. \tag{6.38}$$

Here, all quantities at half-timesteps are averaged over the new and old values; the difference operator Δ takes the 2nd order difference:

$$f^{m+\frac{1}{2}} = \frac{1}{2}(f^m + f^{m+1}); \quad \Delta f \equiv \frac{1}{2}(f_{i+1} - f_{i-1}).$$

The problem with Eq. (6.37) is that unlike for classical diffusion problems, the 'coefficients' D_i are actually nonlinear functions of T_i, ruling out any explicit approach. What we do instead, is assume that the D_i are small in some sense, and concentrate on the linear components of T_i. Letting

$\overline{D}_i = D_i^{m+\frac{1}{2}}$, we rearrange the difference equation according to 'knowns' (RHS) and 'unknowns' (LHS), giving:

$$\alpha_i T_{i-1}^{m+1} + \beta_i T_i^{m+1} + \gamma_i T_{i+1}^{m+1} = W_i, \tag{6.39}$$

where

$$\alpha_i = \frac{\delta}{8}(\overline{D}_{i+1} - \overline{D}_{i-1} - 4\overline{D}_i),$$

$$\beta_i = 1 + \delta\overline{D}_i,$$

$$\omega_i = 1 - \delta\overline{D}_i \tag{6.40}$$

$$\gamma_i = \frac{\delta}{8}(\overline{D}_{i-1} - \overline{D}_{i+1} - 4\overline{D}_i),$$

$$W_i = -\alpha_i T_{i-1}^m + \omega_i T_i^m - \gamma_i T_{i+1}^m + S_i^m \Delta t,$$

$$\delta = \frac{\Delta t}{\Delta x^2}. \tag{6.41}$$

This set of equations can be solved iteratively, using our tridiagonal solver from Sec. 6.6, to obtain the new temperatures at time $(m+1)$ from the old set. At each iteration p, the coefficients are updated:

$$\overline{D}_i = \frac{1}{2}(D_i^p + D_i^m).$$

The boundary conditions in this case include the absorbed laser flux (via the source term S), which can be handled either by modifying the RHS W_1 accordingly, or by holding the LH boundary ($x = 0$) at the required temperature. A further subtlety in this heat flow model is that if the irradiation is intense enough, the temperature gradients become so large that the heat flow Q_e becomes (unphysically) larger than the 'free-streaming limit', $Q_{fs} = mv_{te}^3/2$ (Bell *et al.*, 1981). The workaround for this is usually to clamp Q_e at some fraction of Q_{fs} by means of a flux limiter f (Pert, 1988):

$$\frac{1}{Q'} = \frac{1}{Q} + \frac{1}{fQ_{fs}},$$

where f is typically given a value in the range 0.1–0.3.

Source Code:

Directory: `heatflow/`
Fortran version: `spitzer.f`, `ssode.f`, `makefile`

Projects:

(1) Constant absorption (Rosen)

(2) Normal skin effect (Rosmus & Tikhonchuk)

6.8 Hydrodynamic Simulation

Hydrodynamic and magnetohydrodynamic (MHD) models are now pretty standard tools in plasma research, so it is not really necessary to spell out all the details of these codes here. Many of these use adaptive meshes to follow zones of expansion or compression, and include such sophisticated atomic and nuclear physics packages that it is rarely worthwhile to attempt building a code from scratch these days.

Nevertheless, our treatment of plasma simulation would not be complete without some description of how these models are actually constructed, so we consider a one-dimensional example similar to the approach used in MEDUSA. We begin by restoring our 'divergence form' fluid equations (5.57–5.60) back into explicit form, taking $T_i = 0$ for simplicity:

$$\frac{\partial \rho}{\partial t} + u\frac{\partial \rho}{\partial x} + \rho\frac{\partial u}{\partial x} = 0,$$

$$\frac{\partial u}{\partial t} + u\frac{\partial u}{\partial x} = -\frac{1}{\rho}\frac{\partial}{\partial x}(P + P_L), \tag{6.42}$$

$$\frac{\partial P}{\partial t} + u\frac{\partial P}{\partial x} = -\gamma P\frac{\partial u}{\partial x} + (\gamma - 1)\frac{\partial}{\partial x}(Q + \Phi_L).$$

$$\tag{6.43}$$

We now define a specific volume $V \equiv 1/\rho$ and suppose that the target has some initial, uniform mass density ρ_0. This allows us to define a Lagrangian grid such that $\rho_0 dx_0 = \rho dx = dm$, so that the *mass in each cell is conserved*. Applying this change of variable $(\partial/\partial x = \rho\partial/\partial m)$, Eq. (6.42) becomes:

$$\frac{dV}{dt} = \frac{\partial u}{\partial m},$$

$$\frac{du}{dt} = -\frac{\partial}{\partial m}(P + P_L), \tag{6.44}$$

$$\frac{d\varepsilon}{dt} + P\frac{dV}{dt} = -\frac{\partial}{\partial m}(Q + \Phi_L),$$

$$\frac{dr}{dt} = u,$$

where

$$\varepsilon = \frac{P}{\gamma - 1} + \frac{1}{2}\rho u^2.$$

These equations are then solved cyclically, starting from a cold, solid target and introducing a heat flux via Φ_L close to the critical surface. Alternatively, a wave solver such as that in Sec. 6.6 can be used to compute the laser fields and thus a more self-consistent absorption rate (via resonance absorption, skin effect, or some other model).

Source Code:

Directory: MEDUSA/
Fortran version: med101.f makefile

Projects:

(1) Short prepulse with Al target
(2) Long pulse interaction

6.9   Particle-in-Cell Codes

In the simplest kinetic description of a plasma, one starts from the single-particle velocity distribution function $f(\boldsymbol{r}, \boldsymbol{v})$, which evolves according to the Vlasov equation (Elliot, 1993):

$$\frac{\partial f}{\partial t} + \boldsymbol{v}\cdot\frac{\partial f}{\partial \boldsymbol{x}} + q(\boldsymbol{E} + \frac{\boldsymbol{v}}{c} \times \boldsymbol{B})\cdot\frac{\partial f}{\partial \boldsymbol{p}} = 0. \qquad (6.45)$$

The distribution function $f(\boldsymbol{r}, \boldsymbol{v})$ is 6-dimensional, so the general solution of Eq. (6.45) is intractable for most practical purposes. Even for problems reducible to a 1D geometry, one typically still needs to retain 2 or 3 velocity components in order to incorporate the appropriate electron motion and its coupling to Maxwell's equations, which effectively results in a 3- or 4-dimensional code. A much more economical method developed in the 1960s is the so-called Particle-in-Cell (PIC) technique. In this case the distribution function is represented instead by a large number of discrete *macro-particles*, each carrying a fixed charge q_i and mass m_i, see Fig. 6.6.

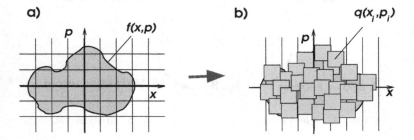

Fig. 6.6   Relationship between a) Vlasov and b) PIC approaches.

The particles are moved individually in Lagrangian fashion according to the Lorentz equation (4.1):

$$\frac{d}{dt}(\gamma_i \boldsymbol{v}_i) = \frac{q_i}{m_i}\left(\boldsymbol{E} + \frac{\boldsymbol{v}_i}{c}\times\boldsymbol{B}\right)$$

$$\gamma_i = (1 - \frac{v_i^2}{c^2}), \qquad\qquad i = 1...N. \qquad (6.46)$$

The density and current sources needed to integrate Maxwell's equations are obtained by mapping the local particle positions and velocities onto a grid as follows:

$$\rho(\boldsymbol{r}) = \sum_j q_j S(\boldsymbol{r}_j - \boldsymbol{r}),$$

$$\boldsymbol{J}(\boldsymbol{r}) = \sum_j q_j \boldsymbol{v}_j S(\boldsymbol{r}_j - \boldsymbol{r}), \qquad j = 1...N_{\text{cell}} \qquad (6.47)$$

where $S(\boldsymbol{r}_j - \boldsymbol{r})$ is a function describing the effective shape of the particles. Usually it is sufficient to use a linear weighting for S — the 'Cloud-in-Cell' scheme — although other more accurate (higher order) methods are also possible. Once $\rho(\boldsymbol{r})$ and $\boldsymbol{J}(\boldsymbol{r})$ are defined at the grid points, we can proceed to solve Maxwell's equations to obtain the new electric and magnetic fields. These are then interpolated back to the particle positions so that we can go back to the particle push step Eq. (6.46) and complete the cycle, see Fig. 6.7.

Because of its simplicity and ease of implementation, the PIC-scheme sketched above is currently one of the most important plasma simula-

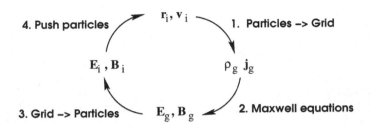

4. Push particles r_i, v_i **1. Particles -> Grid**

E_i, B_i ρ_g, j_g

3. Grid -> Particles E_g, B_g **2. Maxwell equations**

Fig. 6.7   Schematic illustration of the particle-in-cell algorithm.

tion methods. It is particularly suitable for the study of *kinetic* or *non-Maxwellian* effects. The simplest variation of this technique is a '1D1V'-configuration: 1 space coordinate plus 1 velocity, the numerical behavior of which was first considered by John M. Dawson, one of the pioneers of kinetic plasma simulation, some forty years ago (Dawson, 1962). Today, helped by advances in computing power, the PIC method has been developed to the point where fully electromagnetic 3D3V simulations can be routinely carried out (Esirkepov, 2001; Fonesca *et al.*, 2002; Pukhov, 2003).

In keeping with the spirit of this book, however, I will stick to the former reduced geometry for the sample PIC codes offered here, which offer both an apprenticeship in 'professional' plasma simulation, as well as plenty of insight into the behavior of laser-produced plasmas. The heart of the first (electrostatic) code is based on the following difference equations:

Particle pusher: $v_i^{n+\frac{1}{2}} = v_i^{n-\frac{1}{2}} + \dfrac{q_i}{m_i} E_i^n \Delta t,$

$$x_i^{n+1} = x_i^n + v_i^{n+\frac{1}{2}} \Delta t. \qquad (6.48)$$

Density gather: $\rho_j^{n+1} = \displaystyle\sum_i q_i S(x_i - x_j),$

$$S = 1 - \frac{|x_i - x_j|}{\Delta x}. \qquad (6.49)$$

Field integration: $E_{j+\frac{1}{2}}^{n+1} = E_{j-\frac{1}{2}}^{n+1} + \rho_j^{n+1} \Delta x. \qquad (6.50)$

The routines for the above steps — PUSH, DENSITY and FIELD — can be found in the code provided, PICES. The control routine at the top of the code, MAIN, calls each of these routines in turn, as well as some initialization routines (INIT, LOADX, LOADV) and diagnostic routines (DIAGNOSTICS,

PLOTS, HISTORIES). All main variables and storage arrays are defined in the header-file es.h. The normalizations used in the code are the natural ones for electrostatic particle simulation, where the timescale is governed by the plasma frequency:

$$t \to \omega_p t,$$
$$x \to \omega_p x/c$$
$$v \to v/c$$
$$E \to eE/m\omega_p c \tag{6.51}$$
$$\phi \to e\phi/mc^2$$
$$n_{e,i} \to n_{e,i}/n_0.$$

Source Code:

Directory: ESPIC/
Files for Fortran version: f90/ *.f, es.h, makefile
Files for C version: C/ *.c, es.h, makefile

Projects:

(1) Getting to know an electrostatic PIC code

   (a) Thermal equilibrium and numerical heating
   (b) Plasma oscillations: nonlinear Langmuir-waves

(2) Collisionless laser absorption in an overdense plasma

   (a) Modeling a plasma-vacuum interface
   (b) Brunel-effect

(3) Laser wakefield accelerator

   (a) Simulating a finite laser pulse via the ponderomotive force
   (b) Wakefield excitation and particle acceleration

The next level of sophistication in PIC simulation is to introduce transverse velocity components (v_y, v_z), which allows one-dimensional electromagnetic wave propagation, with $\boldsymbol{B} = (0, B_y, B_z)$; $\boldsymbol{E} = (E_x, E_y, E_z)$. This is the system needed for modeling oblique-incidence irradiation of solid targets (Secs. 5.5, 5.6, 5.11, 5.12) using the boost technique (Fig. 5.27). This technique was first implemented in the BOPS code (Boosted Oblique-incidence Particle-in-cell Simulation) by the present author (Gibbon and Bell, 1990) and has been widely adopted in PIC codes since.

Source Code:

Directory: BOPS/
Files for Fortran version: src/*.f, *.h, makefile

(1) Resonance absorption in long scale-length plasmas
(2) Vacuum heating in steep-gradient plasmas
(3) High-harmonic generation from overdense plasma surface

Chapter 7

Applications of Short-Pulse Laser-Matter Interactions

This book has so far concentrated primarily on basic physics issues of femtosecond laser-plasma interactions. While it is true that much current research is curiosity-driven, an equally important motivating factor is that laser-plasmas have a tremendous potential as primary sources of photons, electrons and ions for other purposes. These sources are interesting not only for their relative compactness, but also for the extraordinary beam qualities made possible by the high energy density created: high brightness, low emittance, short duration and so on. In this chapter, some of these applications are considered in a little more detail, where appropriate comparing laser-plasma sources to more conventional ones.

7.1 Bench-Top Particle Accelerators

The demise of the Superconducting Super Collider (SSC) (Drell, 1994) has remotivated the search for alternatives to conventional particle accelerator technology, currently approaching its limit with the Large Hadron Collider (LHC) project at CERN. It has been realized for some time that plasma could form the basis of a new generation of compact accelerators thanks to their ability to support much larger electric fields. Conventional synchrotrons and LINACS operate with field gradients limited to around 100 MVm^{-1}. A plasma, on the other hand, which is already ionized and therefore immune to electrical breakdown, can theoretically sustain a field 10^4 times larger, given by:

$$E_p = \frac{m_e c \omega_p}{e} \varepsilon$$
$$\simeq n_{18}^{1/2} \varepsilon \ \mathrm{GV} \ \mathrm{cm}^{-1}, \tag{7.1}$$

where n_{18} is the electron density in units of 10^{18} cm^{-3}.

7.1.1 Electron acceleration

To accelerate light particles (electrons or positrons, say), these fields must propagate with velocities approaching the speed of light. In a seminal paper, Tajima and Dawson (1979) proposed two methods of coupling the transverse electromagnetic energy of a high power laser into longitudinal plasma waves with high phase velocity. The first requires a pulse length matched to the plasma period such that $\tau_p \simeq \pi/\omega_p$ (Sec. 4.3), which translates into a technical requirement:

$$t_{\text{fwhm}} \simeq 50 \, n_{18}^{-1/2} \text{ fs.} \tag{7.2}$$

This condition could not be met with the available technology at that time, so they proposed the beat-wave scheme, in which two lasers are used with frequencies chosen so that $\omega_1 - \omega_0 = \omega_p$. In contrast to the wakefield scheme, where a plasma wave is forcibly driven up by the pulse, the beat-wave method relies on a more gentle build-up over several plasma periods, see Sec. 4.3.4. In either case, the plasma wave has a phase velocity:

$$v_p = c \left(1 - \frac{\omega_p^2}{\omega_o^2}\right)^{\frac{1}{2}} \simeq c \left(1 - \frac{1}{2\gamma_p^2}\right), \tag{7.3}$$

where $\gamma_p = \omega_0^2/\omega_p^2$. An electron trapped in such a wave will be accelerated over at most half a wavelength (after which it starts to be decelerated), giving an effective acceleration length

$$L_a = \frac{\lambda_p c}{2(c - v_p)} \simeq \lambda_p \gamma_p^2$$

$$\simeq 3.2 \, n_{18}^{-3/2} \lambda_{\mu m}^{-2} \text{ cm.} \tag{7.4}$$

Combining Eq. (7.1) and Eq. (7.4), we obtain the maximum energy gain of an electron accelerated by the plasma wave:

$$\Delta U = eE_p.L_a = 2\pi\gamma_p^2 \varepsilon \, mc^2$$

$$\simeq 3.2 \, n_{18}^{-1} \lambda_{\mu m}^{-2} \text{ GeV.} \tag{7.5}$$

Thus, a multi-Terawatt Ti:sapphire laser is in principle capable of accelerating an electron to 5 GeV in a distance of 5 cm through a plasma with density 10^{18} cm^{-3}. In practice, acceleration will be limited by other factors, such as laser diffraction or instabilities. For example, comparing the

Rayleigh length Eq. (4.100) with Eq. (7.4), we typically have $Z_R \ll L_a$, so some means of guiding the laser beam over the dephasing length must be found to optimize the energy coupling Katsouleas *et al.* (see 1996). Experimentally, both relativistic self-guiding (Wagner *et al.*, 1997) and preformed channel-guiding (Geddes *et al.*, 2004) have been successfully combined with wakefield acceleration, raising the hope that this limitation can be overcome as laser-plasma schemes mature.

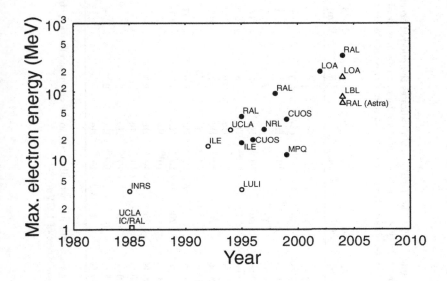

Fig. 7.1   A 'Livingstone' chart for laser-plasma electron accelerators. Open circles represent early beat-wave experiments; filled circles single pulse wakefield experiments. The newest results (triangles) show that *monoenergetic* electron beams with low emittance can also be produced with short pulse lasers.

Over the last 20 years there have been some landmark experiments demonstrating various regimes and aspects of particle acceleration, the progress of which is charted in Fig. 7.1. The early experiments exploited the beat-wave scheme (Sec. 4.3.4) to generate large-amplitude plasma wakes with long pulses, achieving electron acceleration up to several 10s of MeV (Kitagawa *et al.*, 1992; Clayton *et al.*, 1993; Everett *et al.*, 1994; Amiranoff *et al.*, 1995). Since the mid-1990s, efforts in this field have naturally concentrated on short-pulse wakefield schemes (Nakajima *et al.*, 1995; Modena *et al.*, 1995), regularly producing new record electron energies with each

Table 7.1 Experiments on laser-electron acceleration since 1985. The 'field' column indicates the inferred accelerating field in the plasma of effective length L_p, typically created with a gas jet. The 'wake type' refers to the acceleration mechanism with which the authors interpret their results. Further details on the experimental conditions can be obtained from the original source cited in the first column.

Authors (year)	Laser	λ (μm)	τ_L (fs)	Energy (J)	I_L (W/cm^2)	Target	L_p (mm)	Field (GV/m)	Wake type	Max energy (MeV)
Clayton (1985)	CO$_2$	9.56/10.59	2 ns	16	2×10^{13}	H$_2$	1	1	beat-wave	1.0[a]
Ebrahim (1985)	CO$_2$	9.55/10.55	1.2 ns	50	3×10^{13}	C-foil	0.5	6	beat-wave	3.5
Dangor (1990)	Nd:glass	1.064/1.053	1 ns	30	1×10^{14}	H$_2$	–	1	beat-wave	1[a]
Kitagawa (1992)	CO$_2$	9.57/10.6	1 ns	150	2×10^{13}	H$_2$	7	1.5	beat-wave	16
Everett (1995)	CO$_2$	10.28/10.59	150 ps	80	5×10^{14}	H$_2$	10	2.8	beat-wave	28 (inj)
Amiranoff (1995)	Nd:glass	1.053/1.064	90 ps	8.1	3.1×10^{14}	D$_2$	4.8	0.6	beat-wave	3.7 (inj)
Nakajima (1995)	Nd:glass	1	1 ps	30	10^{17}	He	0.6	30	SMWF	18
Modena (1995)	Nd:glass	1	0.8 ps	20	6×10^{18}	He	0.3	100	WB/RFS	44
Umstadter (1996)	Ti:sapphire	0.8	400 fs	3	4×10^{18}	He	0.75	–	guided WF	2.5
Moore (1997)	Nd:glass	1	400 fs	1	5×10^{18}	He	0.64	47	SMWF	30
Gordon (1998)	Nd:glass	1	1 ps	20	6×10^{18}	He	1	240	WB	94
Chen (1999)	Nd:glass	1.053	400 fs	2	10^{19}	He	–	–	SMWF	40
Gahn (1999)	Ti:sapphire	0.8	200 fs	0.25	4×10^{18}	He	0.4	–	DLA	12
Malka (2002)	Ti:sapphire	0.8	30 fs	1	3×10^{18}	He	1	–	forced WF	200
Mangles (2004)	Ti:sapphire	0.8	40 fs	0.5	2.5×10^{18}	He	0.6	–	forced WF	70 (mono)
Geddes (2004)	Ti:sapphire	0.8	55 fs	0.5	10^{19}	H$_2$	2	–	guided WF	86 (mono)
Faure (2004)	Ti:sapphire	0.8	30 fs	1	3.2×10^{18}	He	–	–	bubble WF	170 (mono)

[a] The first experiments by the UCLA and IC/RAL groups did not actually measure electron acceleration, but determined the beat-wave amplitude instead, thereby inferring accelerating fields in excess of 1 GVm^{-1}.

major upgrade in laser power, see Table 7.1.

A striking feature of these experiments is the sheer variety of explanations given for the acceleration mechanism: wakefields are no longer the simple ponderomotively driven plasma waves familiar from Sec. 4.3, but are self-modulated, forced, broken, guided, bubble-like, or even dispensed with entirely. Despite these differing interpretations, most long-pulse ($\tau_L \gg \omega_p^{-1}$) experiments reveal a Maxwellian or even flat electron energy spectrum — not an ideal property for an accelerator. The most recent results using matched or very short, strongly relativistic pulses reveal much more promising quasi-monoenergetic spectra (Mangles *et al.*, 2004; Geddes *et al.*, 2004; Faure *et al.*, 2004). Regardless of the preferred approach, the relentless progress evident in the chart is compeling, and even a conservative projection suggests that the GeV milestone will be reached before the decade is out, bringing the dream of the tabletop particle accelerator a step closer (Joshi and Katsouleas, 2003; Bingham *et al.*, 2004).

7.1.2 *Fast ion sources*

Owing to their higher mass, ions cannot be accelerated via fast plasma waves (with $v_p \sim c$) because they are simply too slow. A 10 MeV proton has a decidedly *non*relativistic velocity $v_i \simeq 0.1c$, which rules out any hope of trapping it in a fast-phase-velocity (Eq. 7.3) wakefield. As we have seen in Secs. 4.6.3 and 5.7, however, significant ion acceleration *can* be achieved by using the laser to create large, static electric fields via charge displacement. Both protons and heavier ions can be accelerated to many MeV by this means, presenting a new, compact ion source for existing and novel applications.

Although less developed than fast electron or x-ray sources, these already include exciting new possibilities such as proton imaging (Borghesi *et al.*, 2004; Cowan *et al.*, 2004), rare isotope production (Nemoto *et al.*, 2001), isochoric (constant-volume) heating (Patel *et al.*, 2004), hadron tumour therapy (Bulanov *et al.*, 2002), and proton-based fast ignition (Sec. 7.5). A favourable characteristic of the ion beams produced by lasers is that they are virtually charge neutral, resulting in a very low emittance, but which obviously poses a technical problem of stripping the electrons off prior to use.

7.2   Soft X-Ray Sources

One of the early motivating factors behind the development of femtosecond lasers was the possibility of generating short, high-brightness x-ray pulses (Kuhlke *et al.*, 1987; Harris and Kmetec, 1988; Stearns *et al.*, 1988; Murnane *et al.*, 1989; Zigler *et al.*, 1991; Teubner *et al.*, 1995). Such pulses can in principle be used as an x-ray microscope to study transient phenomena in physical, chemical or biological systems, or as backlighters for probing hot, dense matter on a sub-picosecond timescale (Murnane *et al.*, 1991). Ideally, one would like to be able to predict the yield and duration of a laser-plasma x-ray source, quantities which depend sensitively on the interplay of a number of effects: absorption, heating, ionization, recombination and transport, see Fig. 5.9. Because of these transport and hydrodynamic effects, x-ray pulses have durations several times longer than the laser pulse itself.

By treating the plasma as a high-density, blackbody radiator, Rosen (1990) and Milchberg *et al.*, (1991) demonstrated the possibility of generating sub-ps x-rays via rapid cooling of the target. More sophisticated treatments including hydrodynamics and detailed atomic modeling can be used determine and help control the pulse duration of particular lines (Audebert *et al.*, 1994; Workman *et al.*, 1995). Experimental measurements of x-ray pulse duration are currently instrument-limited to just under 1ps (Kieffer *et al.*, 1993; Shepherd *et al.*, 1995; Belzile *et al.*, 2003), and new techniques are being explored — for example, using atomic excitation and ionization (Barty *et al.*, 1995) — to measure pulses in the sub-100fs regime.

7.2.1 *Microscopy*

State-of-the-art electron microscopes (see, for example, Rochow and Tucker, 1994), can deliver both resolution and contrast right down to atomic (sub-Å) levels. The main advantage of photons over electrons is their ability to pass less destructively through aqueous solutions. Moreover, scanning electron microscopy normally requires carefully prepared, static biological samples, either freeze-dried or treated with hydrophobic agents — processes which can alter the cellular structure. Soft x-ray pulses from laser-produced plasmas, on the other hand, offer the chance to study *living* cells with a time resolution sufficient to capture dynamical processes on a sub-nanosecond or even sub-picosecond timescale.

A prerequisite for biological microscopy is a wavelength within the so-called *water-window* between the absorption K-edges of oxygen (23.2 Å) and carbon (43.7 Å) (Solem and Chapline, 1984). This choice allows transmission of the probe beam through the sample while providing natural contrast between proteins (i.e. carbon) and water (oxygen), which should yield information on protein structures in their natural (aqueous) environment.

A major disadvantage of laser-solid x-ray sources is that they produce a lot of debris — mainly ion blow-off, but also hot electrons — during the interaction, which tends to find its way onto the sample or the delicate and expensive x-ray optics. This can be avoided to a large extent by using liquid droplet targets, a technique pioneered by the Lund group (Rymell *et al.*, 1995), and which has since been widely adopted as a standard for compact x-ray microscopes (Hertz *et al.*, 2003)

7.2.2 EUV Lithography

While lithography is often cited as a potential application for laser-plasma x-ray sources, it is not immediately obvious that ultra-short pulse lengths bring any real advantage. In the semiconductor industry, excimer lasers (KrF 248 nm and ArF 193 nm) are currently used for imaging masks onto wafers, giving a minimum feature size of around 100 nm. To improve the etching resolution further, x-ray wavelength (or extreme ultraviolet) light will be needed. The industrial requirements for such a source are extremely demanding (small bandwidth, high average power, debris-free), but there is a consensus that EUV light with a wavelength of 13 nm should be able to meet the specifications. Nanosecond lasers currently appear to represent the most promising option for producing this radiation (from oxygen or fluorine lines, for example) (Chaker *et al.*, 1990; Maldonado, 1995; Abel *et al.*, 2004). On the other hand, shorter (ps) pulses combined with liquid drop targets (Rymell *et al.*, 1995; Düsterer *et al.*, 2001) might still produce 'cleaner' x-rays and reduce problems with debris.

7.3 Hard X-Ray Flash Lamps

In contrast to the soft, thermal x-ray emission considered in the previous section, hard x-rays are primarily a *direct result* of hot electrons produced in the vicinity of the focal spot (Sec. 5.6.2). Thanks to its long mean-free-path, a fast electron can penetrate into the cold region of the target beyond the heat front. Here it gets slowed down via elastic collisions with ions, emitting bremsstrahlung radiation in the process, or can undergo *in*elastic collisions with bound, inner-shell electrons, thereby prompting K_α line emission from the recombination process, see Fig. 7.2.

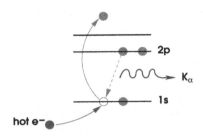

Fig. 7.2 K_α radiation produced by an energetic electron striking and removing a bound inner-shell (1s) electron from an atom, subsequently provoking a 2p → 1s transition.

Bremsstrahlung appears as a continuum anywhere in the 0.1 keV–MeV range depending on the laser intensity and plasma parameters (Kmetec *et al.*, 1992; Schnürer *et al.*, 1995; Schwoerer *et al.*, 2001), whereas inner shell line emission can be 1–100keV depending on the atomic number of the target material (Soom *et al.*, 1993; Rousse *et al.*, 1994; Jiang *et al.*, 1995). Cold line radiation is preferred for pump-probe structural investigations because of its typically narrow bandwidth, short pulse duration, small spot-size (Eder *et al.*, 2000; Reich *et al.*, 2003) and high potential brightness (Yu *et al.*, 1999b).

K_α radiation is also an important diagnostic tool. The spectral intensity of photons emitted is a direct function of the electron energy. This can be exploited by observing that electrons eventually stop in cold material, so that the total line emission will in general depend on the target thickness. The hot electron temperature can therefore be inferred from a sandwich target (e.g. Al/Si) in which the thickness of the front layer is varied (Hares *et al.*, 1979; Rousse *et al.*, 1994; Feurer *et al.*, 1997). The change in the K_α line ratios with thickness can then be fitted to a characteristic temperature or distribution function.

Much research effort has recently gone into optimizing hard x-ray sources by various means. Yield from solid targets can be enhanced by adapting the laser-target configuration (intensity, absorption, target thickness) depending on the atomic number of the material (Reich *et al.*, 2000);

or by exploiting the relativistic increase in the K-shell ionization cross-section for relativistic electrons (Ewald *et al.*, 2002). Extensive investigations at LULI/LOA have shown that the laser absorption and x-ray yield is highly sensitive to the density scale-length (Bastiani *et al.*, 1997). Other groups have used mass-limited (efficiently heated) targets such as atomic clusters (McPherson *et al.*, 1994; Ditmire *et al.*, 1995), or exotic 'nanoplasma' coated targets to create a 'lightning rod' effect on the surface (Rajeev *et al.*, 2003)

7.3.1 *Biomedical imaging*

One far-reaching application of ultrafast x-ray flash-lamps presently under serious evaluation by several groups is medical and biological imaging. Since their discovery by Roentgen over a century ago, x-rays have been exploited for this purpose with increasing sophistication, so it is natural to ask what new advances could be achieved by laser-plasma sources. The requirements of medical imaging are essentially threefold: first, x-ray photon energies need to be in the range 20–100 keV to allow transmission through the body; second, the bandwidth should be as narrow as possible to minimize the dose from unwanted soft x-rays (collateral damage); third, a high degree of tunability is needed to distinguish between different types of tissue. While the traditional x-ray tube does of course meet these general specifications, femtosecond laser-plasma sources do offer some additional advantages (Herrlin *et al.*, 1993):

 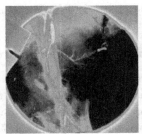 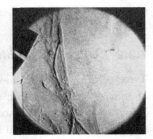

Fig. 7.3 Differential imaging of a rat heart-section: exposures with Nd K_α (left) and Gd K_α (middle) are subtracted from each other to yield a high-definition image of the region of interest (right). Courtesy, E. Andersson (2002).

A short pulse x-ray source delivers a high quantity of radiation within a picosecond or so, in contrast to the microsecond–millisecond exposure

Table 7.2   Time- and length-scales of protein motions.

	time (s)	deflection (Å)
Atomic vibration	10^{-15}–10^{-11}	0.01–1
Collective	10^{-12}–10^{-3}	0.01–5
Bond-breaking/-joining	10^{-9}–10^{-3}	0.5–10

times typical of conventional sources. This offers the possibility of substantial dose reduction by using a *time-gating* technique to cut out scattered x-rays which degrade the image contrast (Gordon III *et al.*, 1995). Also, *differential imaging* with rapid, simultaneous exposure is possible (Tillman *et al.*, 1996). This technique, also done using conventional synchrotron sources, requires rapid exposure by two x-ray lines with photon energies above and below the K-absorption edge of a contrast agent. Subtraction of the two resulting images enhances the parts of the sample containing the contrast agent and suppresses unwanted information — see Fig. 7.3. Finally, and perhaps most intriguingly, the small target size makes it possible to conceive novel image projections where the x-ray source is placed *inside* the object of investigation (Tillman *et al.*, 1995).

While laser-based x-ray sources are not quite ready for serious deployment in medicine (unlike femtosecond lasers themselves, which are now routinely used in surgery and dentistry), progress in resolving issues such as dosage control, high-intensity x-ray tissue interaction and increasing power throughput (e.g. by going to kHz laser systems) has been steady, so that imaging facilities may become viable within the next few years — see the recent reviews by Svanberg (2001) and Kieffer *et al.*, (2002).

7.3.2   Ultrafast probing of atomic structure

Two techniques which have been used for some time to probe the structure of matter at the atomic and molecular level are *in-situ* x-ray diffractometry and x-ray spectroscopy. These methods have quite different source requirements and optical arrangements, see Figs. 7.4 and 7.5, but share the exciting new possibilities offered by femtosecond time-resolution. A further technical challenge is to focus the x-rays onto the sample efficiently enough to allow single shot exposures (Attwood, 1992; Förster *et al.*, 1994).

Consider, for example, the typical time- and length-scales of protein motions shown in Table 7.2 (Petsko and Ringe, 1984) . Diffractometry exploits the interference effect created by adjacent atomic planes (Bragg

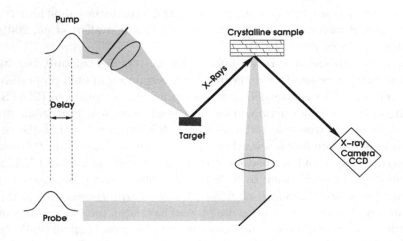

Fig. 7.4   X-ray optical arrangement for ultrafast pump-probe diffractometry.

scattering) to obtain global structural information about fluid or crystal samples. Since x-ray diffraction measurements can be directly inverted to atomic positions or bond lengths, it soon became clear that ultrafast exposures on the femtosecond timescale might eventually allow 'filming' of dynamic processes such as phase changes or chemical reactions (Barty *et al.*, 1995; Tomov *et al.*, 1995).

Over the last decade, this vision has indeed become reality, and femto-crystallography has already established itself as a new scientific field in its own right Rousse *et al.* (2001a). Most experiments to date have concentrated on detecting changes in the lattice structure of crystalline samples subjected to ultrafast heating (Sec. 5.3) by a second femtosecond pulse, see Fig. 7.4. The hard x-ray pulse generated by the pump via hot electrons (Sec. 5.6.2) is normally reflected from the unperturbed sample according to the usual Bragg diffraction condition. When the heating pulse arrives, however, the lattice is either perturbed or destroyed by a nonthermal heat-wave, leading to rapid changes in the x-ray reflectivity recorded on the camera.

From this, one can deduce information about the structural changes which occur under these conditions, such as nonthermal melting (Rischel *et al.*, 1997; Siders *et al.*, 1999; Feurer *et al.*, 2001; Rousse *et al.*, 2001b). The same technique has also been used to observe lattice phonon dynamics

(Rose-Petruck *et al.*, 1999; Cavalleri *et al.*, 2000), thereby revealing hitherto inaccessible details on solid-solid phase transitions (Cavalleri *et al.*, 2001; Sokolowski-Tinten *et al.*, 2003).

Spectroscopy can also reveal information on atomic structure, but its interpretation is generally complicated by uncertainties in bulk properties. An exception to this is extended x-ray absorption fine-structure, or EXAFS, which yields direct information on the near-neighbours of a given atom. An advantage of spectroscopic techniques over diffractometry is that the required x-ray photon flux is several orders of magnitude lower. Soft thermal x-rays (Sec. 7.2) from laser-plasmas have been successfully used as EXAFS sources for some time owing to their high brightness and sub-nanosecond recording capability (Eason *et al.*, 1984). Using an arrangement such as the one displayed in Fig. 7.5, short pulse sources have been used as a means of extending the time-resolution down to the sub-picosecond regime (Tallents *et al.*, 1990) and to shorter wavelengths (Forget *et al.*, 2004).

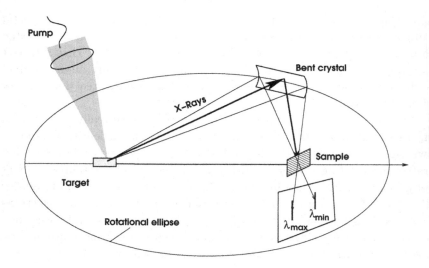

Fig. 7.5   Optical arrangement for ultrafast x-ray spectroscopy.

7.4 Coherent XUV Radiation

The realization of a tunable, coherent light source in the 1–100 nm wavelength range promises to open up as many new possibilities as the development of the laser itself in the 1960s. In a tutorial review of XUV sources, Attwood (1992) gives three basic definitions which characterize coherent radiation: brightness, coherence length and transverse resolving power. The brightness B is defined as:

$$B = \Phi/(\Delta A \cdot \Delta \Omega \cdot BW) \tag{7.6}$$

where Φ is the number of photons per second, ΔA is the source area, $\Delta \Omega$ is the solid angle of the emitted radiation, and BW is the bandwidth $\delta\lambda/\lambda$. The temporal coherence length,

$$l_{\text{coh}} = \lambda^2/2\Delta\lambda \tag{7.7}$$

determines the detail in which information can be resolved along the propagation path. The resolving power is set by the diffraction limit:

$$d \cdot \theta \geq \lambda/2\pi \tag{7.8}$$

where d is the minimum resolvable distance on the object and θ is the observation half-angle.

Prior to the advent of femtosecond lasers there were just two major high-brightness light sources with this wavelength range: synchrotron-undulators (Winick, 1994) and kilojoule-class x-ray lasers (XRL) (Elton, 1990; Maxon *et al.*, 1993). Short-pulse lasers now offer a number of cheaper, much more compact alternatives: hybrid schemes comprising a nanosecond pulse to create the large volume of plasma with suitable ion populations, plus a femtosecond pulse to pump the inversion (Nickles *et al.*, 1997); optical-field-ionization (OFI) schemes (Burnett and Enright, 1990; Amendt *et al.*, 1991; Eder *et al.*, 1994); inner-shell-photoionization (IPSI) schemes (Kapteyn, 1992; Eder *et al.*, 1994; Pretzler *et al.*, 2000), harmonic generation in gases (Salieres and Lewenstein, 2001) or plasmas (Gibbon, 1997); engineered 'hollow atom' configurations (Borisov *et al.*, 2003) and lately even hybrid harmonic-seeded XRL schemes (Zeitoun *et al.*, 2004). The compactness, good scaling properties, high efficiencies and superior time-resolution of these concepts compared to long-pulse XRL schemes combine to give these sources huge potential for opening up new areas of femto-science, such as biological (water-window) microholography (London *et al.*, 1989) and high-density plasma interferometry (Ress *et al.*, 1994).

7.5   Advanced Fusion Concepts

One of the most exciting developments in the field of inertial confinement fusion (ICF) to emerge as a result of short pulse laser technology is the *fast ignitor* concept (Tabak *et al.*, 1994). In the standard ICF scenario (Meyer-ter-Vehn, 1982), ignition is achieved by compressing a deuterium-tritium pellet to high density in such a way that a hot spot of around 5 keV is created at the center (see Sec. 1.5). This hot spot ignites first, providing a spark which then propagates outwards, burning the surrounding higher density fuel. The simultaneous requirements of high compression *and* central ignition translate into a tough set of constraints for the driver and target (Lindl, 1997). For example, laser energies of more than 1 MJ are needed before significant gain is reached (energy out > energy in), and implosion symmetry has to be controlled to better than 1% to prevent impurities mixing with the fuel at the stagnation point (where maximum compression is reached).

In the fast ignitor (FI) scheme, the hot spot is replaced by an *external* energy source, thus relaxing the driver, compression and uniformity requirements theoretically by up to an order of magnitude. Tabak *et al.* (1994) proposed to use a short pulse, high intensity laser to deliver energy to the compressed fuel via fast electrons. This scheme works in three stages:

First, the fuel is compressed to a density of 300 gcm^{-3} with a radius r_h of around 20 μm,[1] the equivalent of an α-particle range at an areal density of 0.4 gcm^{-2}. Next, a 'prepulse' several hundred picoseconds long with a tailored intensity profile is used to dig an optical channel through the underdense corona and also push the critical surface closer to the center of the target. Finally, a short pulse with intensity of around 10^{20} Wcm^{-2} is sent through the channel to the critical surface, where it heats electrons to energies up to 1 MeV. These fast electrons penetrate the fuel core, where they are stopped via collisions, heating up the ions to fusion temperatures (5–10 keV) in the process.

The original scheme has since been subjected to much closer exami-

---

[1]This DT 'droplet' has a density of three hundred times that of water — sometimes expressed as 300 XLD — conditions which have already been achieved in earlier implosion experiments at ILE, Osaka

nation, particularly concerning the coupling efficiency between the electron ignitor beam and the core. A detailed series of 2-dimensional hydrodynamic simulations by Atzeni (1999) set a minimum deposited energy of

$$E_{\text{ig}} \simeq 140 \left(\frac{\rho}{100 \ \text{gcm}^{-3}} \right)^{-1.85} \ \text{kJ} \qquad (7.9)$$

required to heat the DT fuel to an ignition temperature of 10 keV.

For the above target density of 300 gcm^{-3}, this means that the total kinetic energy U_h carried by hot electrons to the core must be around 20 kJ, so that the ignitor pulse energy has to be 30–50 kJ, depending on how efficiently the laser energy can be converted into MeV electrons (Sec. 5.5). Just how hot the electrons have to be can be deduced from the fact that their stopping distance should be matched to the core radius, equivalent to an areal density of 0.6 gcm^{-2}, implying energies of ~ 1 MeV.

A further constraint is dictated by the fuel disassembly time, given by

$$t_c = \frac{r_h}{c_s}. \qquad (7.10)$$

Inserting $r_h = 20 \ \mu$m and our expression for the sound speed Eq. (5.4), this gives $t_c \simeq 20$ ps at a plasma temperature of 10 keV: the hot electron energy must therefore be delivered within this time, *before* the fuel blows apart.

Putting these three constraints together leads to a requirement for the fast ignitor *current*:

$$J_{\text{FI}} = \frac{U_h}{T_h t_c} \simeq 1 \ \text{GA}. \qquad (7.11)$$

This would be fine, except that there is a well-known limit for relativistic beam propagation, the Alfvén current (Alfvén, 1939), beyond which the self-induced magnetic field becomes so strong that the particles actually get turned *back*, given by:

$$J_A = \beta\gamma \frac{mc^3}{e}$$
$$\simeq 17\beta\gamma \ \text{kA}, \qquad (7.12)$$

where $\beta = v/c$ and $(\gamma - 1)mc^2$ is the electron kinetic energy. For electrons with a mean energy $T_h \sim 1$ MeV, the Alfvén current is 47 kA, over 20000 times smaller than the required ignition current! Clearly, the hot electron flow has to be balanced by a nearly equal return current if fast ignition is to stand any chance of succeeding. On the other hand, this is precisely the scenario where Weibel instabilities are expected (Sec. 5.10.3), which would tend to cause a single relativistic electron beam to break up into smaller filaments.

Worse still, classical or anomalous resistive effects may even prevent a large enough return current from forming at all (Davies, 2004). The main hope for the fast electron scheme is that the stopping power can be enhanced well beyond the collisional rate (Deutsch *et al.*, 1996), either by strong Langmuir turbulence (Malkin and Fisch, 2002) or magnetic instabilities (Honda *et al.*, 2000). This would allow the mean beam energy (T_h) to be increased — by a factor of 10, say — thus reducing the gap between J_{FI} and J_A by 100.

Alternative approaches suggest depositing the electron energy outside the core — coronal ignition (Hain and Mulser, 2001) — or advocate using protons to ignite the fuel instead (Roth *et al.*, 2001). Protons (see Sec. 5.7) have the advantage that they propagate with very low divergence and can be integrated into both direct and indirect drive implosion geometries. An important point in favour of these schemes is that they avoid the need for hole boring/channel formation (Young *et al.*, 1995). In fact, this is regarded as one of the biggest potential obstacles to the FI concept, being prone to laser propagation instabilities and uncertainty over how far the critical surface can be pushed towards the core.

For this reason the first major experimental FI studies have side-stepped these issues by using a novel cone-pellet target configuration to provide a clean path for the ignitor pulse, see Fig. 7.6. This arrangement has already led to the first proof-of-principle demonstrations of fast electron heating of a compressed DT pellet (Kodama *et al.*, 2001), in which the neutron yield was enhanced by 1000 using a 0.5 PW heating pulse (Kodama, 2002; Tanaka *et al.*, 2003). Less spectacular results have emerged from similar campaigns at the RAL facility however (Norreys *et al.*, 2004), where the hot electron heating was much lower than expected, possibly due to significant preplasma formation inside the gold cone. These initial studies show that there are still many basic physics issues to resolve before fast ignition becomes a viable option in ICF.

7.6   Laser Nuclear Physics

The last five years have seen a flurry of novel, collaborative experiments at large and small laser facilities involving laser, plasma and nuclear physicists. The continual firsts chalked up by these campaigns — laser-induced

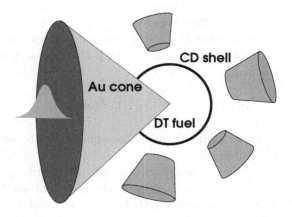

Fig. 7.6 Hollow cone-shell geometry for integrated implosion/heating studies in the advanced fast ignition scheme.

photofission of uranium; copious quantities of fusion neutrons produced during Coulomb explosions of atomic and molecular clusters; even laser-transmutation of unpleasant radionuclides into relatively harmless, short-half-life isotopes — have ushered in a new, very accessible field of TW laser nuclear physics.

7.6.1 *Fusion*

Controlled nuclear fusion is of course the main driving force behind the ICF programmes (Sec. 1.5), now entering the Megajoule era, in which it is hoped to finally demonstrate controlled ignition of DT-filled microballoons. Neutron diagnostics have therefore been a standard component of high-power implosion experiments for some time. Despite the head start on the diagnostic side, it has taken short pulse lasers a good 15 years to reach the intensity levels necessary to generate the ion energies and numbers needed to produce a detectable yield of fusion reactions.

The first unambiguous observations of D-D reactions were made in Garching by Pretzler *et al.*, (1998) using deuterated plastic targets. they measured a modest 140 neutrons per shot at intensities up to 10^{18} Wcm^{-2}, first using a prepulse to create a zone of underdense plasma in which relativistic channeling could take place (Sec. 4.7.1). Expulsion of electrons from the channel center causes a Coulomb explosion, sending D^+ ions radially outwards (just as in previous experiments using H_2 gas — Sec. 4.6.3). If sufficient quantities of these ions reach energies of a few hundred keV, they will start to induce fusion reactions with their stationary neighbours surrounding the channel:

$$D + D \rightarrow {}^3He + n(2.45 \text{ MeV}).$$

The characteristic energy carried away by the neutron can be identified by a time-of-flight (TOF) detector.

A very similar experiment was reported shortly afterwards by the Limeil team (Disdier *et al.*, 1999), this time using the high-contrast P102 laser at 10^{19} Wcm^{-2} and 529 nm wavelength focused onto (solid) deuterated ethylene targets. They measured 10^7 neutrons mainly in the forward direction, suggesting that the D^+ ions were shock-accelerated in this case (see Sec. 5.8). Neutrons from D_2 gas targets have also been measured with the VULCAN laser at RAL by Fritzler *et al.*, (2002), who also showed that the yield could be optimized by placing a CD_2 target adjacent to the focal region in the gas jet.

A very efficient and spectacular way of inducing fusion reactions was demonstrated by Ditmire *et al.*, (1999) using gas *cluster* targets. Atomic clusters can be considered as intermediate forms of matter in between the molecular and solid state, having low average density but local clumps of high density material. As a result, a laser pulse can pass through the gas jet, but gets strongly absorbed by the clusters, leading to multi-keV electron heating and subsequent localized ion Coulomb explosions even at relatively modest laser intensities $\sim 10^{16}$ Wcm^{-2}.

In $(D_2)_n$ clusters, the same $D(d, n)^3He$ fusion reactions may take place between freed deuterium ions and neighbouring clusters. Variations on this theme include molecular clusters such as $(CD_4)_n$ (Grillon *et al.*, 2002) and $(D_2O)_n$ (Last and Jortner, 2001), which tend to produce more monoenergetic deuteron energies — a feature which is still not satisfactorily explained by theoretical modeling.

Although none of these short pulse experiments can hope to compete with ICF facilities in terms of neutron numbers (if they could, the fusion nut would have been well and truly cracked), they still provide a economic and convenient high repetition-rate source for all kinds of nuclear physics studies (McKenna *et al.*, 2003) and could deliver accurate data on reaction rates of keen interest to the nuclear astrophysics community.

7.6.2 *Fission*

Demonstration of optically induced nuclear fission came hot on the heels of fusion neutrons. Experiments on the PW laser at LLNL and on VULCAN at RAL simultaneously reported photonuclear (γ-ray) fission of uranium-238 (Ledingham *et al.*, 2000; Cowan *et al.*, 2000). This process, $^{238}U(\gamma, f)$ for short, has many reaction channels, in effect leading to a distribution of decay products (nuclides) which can be identified via their own γ-ray emission lines. From the intensity of these lines, one can deduce the number of fusion events, which in these two experiments was in the range 10^6–10^7.

The physical conditions needed for laser fission were in fact worked out with remarkable prescience by Boyer *et al.*, (1988) some 12 years prior to these first PW-class experiments. They showed that by producing hot electrons with MeV-temperatures (Sec. 5.6), the subsequent bremsstrahlung emitted by them passing through a high-Z solid target would generate sufficient numbers of γ rays to induce a significant number (several thousand) fission events in heavy elements such as ^{238}U and ^{232}Th. These estimates actually agree quite well with the experimentally measured numbers deduced by Ledingham *et al.*, (2000) and Cowan *et al.*, (2000).

A series of landmark fission experiments have been performed by the Jena group using a high-repetition-rate (university-scale) TW Ti:sapphire system: first, using MeV photons from laser-irradiated Ta targets to induce $^9Be(\gamma, n)2\alpha$ reactions (Schwoerer *et al.*, 2001); subsequently demonstrating fission of heavy elements (Schwoerer *et al.*, 2003) and most recently, the first laser-induced transmutation of ^{129}I (Magill *et al.*, 2003). The latter is one of the most troublesome nuclides in the nuclear fuel cycle, with a half-life of 15.7 million years. Via a (γ, n) reaction it can be transformed into the comparatively benign ^{128}I, with a half-life of just 25 minutes.

Such proof-of-principle experiments are something akin to the Philosopher's Stone, raising the hope that all-optical transmutation technology might be scaled up to provide a long-term solution to the nuclear waste problem, or to manufacture isotopes for medical use.

7.7 A Taste of Things to Come ...

Throughout this chapter I have stuck to tangible applications of high-intensity laser interactions which have already passed the proof-of-principle stage. One might think that given the fantastic advances already made in these areas, that a period of consolidation might be in order to allow these new fields to bear fruit. In fact, CPA laser technology does appear to have a ceiling of around 10^{23} Wcm^{-2}, which will be reexamined below. Needless to say, progress in relativistic laser-plasma interaction and its associated spin-offs has only made researchers thirsty for more. To conclude this brief survey therefore, I indulge in some crystal-ball gazing and consider some of the topics which might shape the frontiers of this field in 10-20 years' time.

To appreciate the rationale behind this irrepressible dynamism, it is worth recalling that the present CPA technology is likely to be superceded by optical parametric CPA (OPCPA), which could bring pulse lengths down by at least a factor of 10 for Petawatt-class laser systems (Sec. 1.6). Assuming comparable improvements in adaptive optics also occur over the next few years, then we might well see intensities of several 10^{23} Wcm^{-2} by the end of the decade.

Can we go beyond this, or are we in for another 'plateau' in laser development such as that experienced in the 1970s-1980s? (See Fig. 1.1.) Of course as long as pulses can be made shorter, they can in principle be made more powerful, but only as far as the natural, single wave-cycle limit close to being reached by state-of-the-art technology. Whether single-cycle pulses can be amplified to PW levels, and then focused to the diffraction limit, still remains to be seen.

For the newly founded attophysics community this will be of minor concern: XUV harmonics generated with 5 fs lasers can already be harnessed as attosecond soft x-ray probes, capable of exploring and perhaps controlling the structural dynamics of atoms and molecules (Hentschel *et al.*, 2001). Also at the low-intensity end is the fascinating prospect of self-guiding of femtosecond pulses in the atmosphere, using the Kerr effect to balance diffraction and ionization-induced refraction (Sec. 2.4). In principle, *kilometre*-length channels can be created this way, opening up new means of detecting atmospheric pollutants, or even creating 'safe' routes for lightning (Kasparian *et al.*, 2003).

For those simply addicted to astronomical energy densities, however, some way of increasing the laser intensity will have to be found which does not sacrifice too many Joules. To this end, a number of speculative ideas have already been suggested for achieving intensities well beyond the optical limit of 10^{23} Wcm^{-2}. Like plasma-based accelerators, these amplification schemes exploit the fact that fully-ionized plasmas cannot be easily overheated or 'damaged', and so permit much higher gains than would be possible in more conventional media. The best-developed of these is probably *superradiant amplification* (Malkin *et al.*, 2000) — very similar in spirit to the free-electron laser — which uses Raman backscatter to pump a counter-propagating seed pulse. More fanciful schemes propose using a double-foil configuration to trap and feed and electromagnetic soliton-like structure, resulting in gains of 100 or more (Shen and Yu, 2002), or flying plasma mirrors (Bulanov *et al.*, 2003) to compress and focus a counterpropagating pulse to intensities approaching the Schwinger limit $\sim 10^{29}$ Wcm^{-2}.

In the eyes of many theorists, the latter represents the next big threshold in nonlinear optics: something akin to how the relativistic regime ($a_0 \sim 1$) must have appeared in the 1960s when the first lasers were boasting intensities of 10^{10} Wcm^{-2}. The QED critical field for spontaneous e$^-$e$^+$ pair-creation by an electromagnetic field in vacuum was famously predicted over half a century ago by Heisenberg and Euler (1936) and Schwinger (1951) to be:

$$E_{\mathrm{cr}} = \frac{m^2 c^3}{e\hbar} = 1.32 \times 10^{16} \ \mathrm{Vcm}^{-1}, \qquad (7.13)$$

which corresponds to an intensity:

$$I_{\mathrm{cr}} = 2.3 \times 10^{29} \ \mathrm{Wcm}^{-2}. \qquad (7.14)$$

Of course this milestone will probably remain beyond reach for the foreseeable future, but plenty of other genuinely quantum electrodynamical effects should be accessible well before this, including pair production itself (via third-party 'spectator' particles), vacuum polarization and photon-photon scattering (Becker, 1991; Popov, 2002).

There are also an almost limitless number of extreme relativistic (laser astrophysics) phenomena which already start to come into play at a mere 10^{22} Wcm^{-2}, such as radiation damping (Zhidkov *et al.*, 2002) and Larmor-radiating solitons (Esirkepov *et al.*, 2004). Exotic states of matter will also be on offer: plasma temperatures of 400 MeV would emulate conditions around 1 ms after the Big Bang; at GeV temperatures one can expect relativistic matter flows and the formation of quark-gluon plasmas. Finally, proposals have even been made for detecting Unruh radiation from laser-accelerated charges (Chen and Tajima, 1999): a vision for laboratory black-hole research which could perhaps bring sorely needed experimental impetus to the quest for a theory of quantum gravity.

Appendix A

List of Frequently Used Symbols

Symbol	Description
β	dimensionless fluid velocity v/c
ε	plasma dielectric constant
γ	relativistic factor
η_a	fractional laser absorption
η_r	refractive index
κ_e	thermal conductivity
λ_D	electron Debye length
λ_L, λ_0	vacuum laser wavelength
λ_p	plasma wavelength
ν_{ei}	electron-ion collision frequency
σ_e	electrical conductivity
σ_L, σ_0	laser spot size
ω, ω_0	laser frequency
ω_p	plasma frequency
Φ_L, Φ_p	ponderomotive potential
B_L, B_0	laser magnetic field strength
E_L, E_0	laser electric field strength
I_p, E_{ion}	ionization potential
I_L, I_0	laser intensity
L	density scale-length
M	ion mass
P, P_L	laser power
P_c	critical power for relativistic self-focusing

Symbol	Description
R_L	Rayleigh length
T_e	electron temperature
T_i	ion temperature
T_h	hot electron temperature
U_L	laser energy
a_0	dimensionless laser field amplitude or quiver veloc
c_s	ion sound speed
$\boldsymbol{f}_p$	ponderomotive force
$\boldsymbol{k}, \boldsymbol{k}_0$	laser wave vector
$\boldsymbol{k}_p$	plasma wave vector
j_h	hot electron current
l_s	collisionless skin depth
m_e, m	electron mass
n_c	critical density
n_e	electron density
n_h	hot electron fraction
t_L	laser pulse duration
v_h	electron quiver velocity
v_i	ion velocity
v_{os}	hot electron velocity
v_p	plasma wave phase velocity
v_{te}	electron thermal velocity

Appendix B

Slowly Varying Envelope Approximation

A quantity containing a periodic phase ψ may be represented in the complex plane by the product of a slowly varying complex amplitude, ε, and a factor $e^{i\psi}$. Consider a typical plane-wave-like variable such as the electric field of an electromagnetic (or electrostatic) wave, which can be written:

$$E(\psi) = \frac{1}{2} \left(\varepsilon e^{i\psi} + \varepsilon^* e^{-i\psi} \right).$$

This can be seen graphically in Fig. B.1 by inspecting the projection of E along the real axis. Note that the amplitude ε is *also* complex, and can thus be written as the product of a purely real quantity, r (the radius of the circle), and a slowly varying phase, θ. The relation between fast and

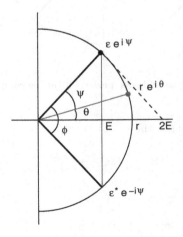

Fig. B.1   Field variable in the complex plane.

slow quantities is simply:

$$E = \Re\{\varepsilon e^{i\psi}\} = \Re\{re^{i(\theta+\phi)}\}$$
$$= |\varepsilon|\cos\phi$$
$$= r\cos\phi$$
$$= r\cos(\phi + \theta). \tag{B.1}$$

For another quantity, say v, which is 90° out of phase with E, we need to consider its projection along the imaginary axis, see Fig. B.2.

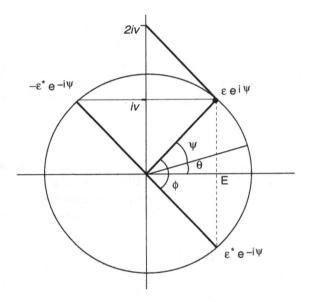

Fig. B.2   Ninety degree phase shift in the complex plane.

From this diagram, we see that

$$iv = \frac{1}{2}\left(\varepsilon e^{i\psi} - \varepsilon^* e^{-i\psi}\right).$$

Therefore,

$$v = \frac{1}{2}\left(-i\varepsilon e^{i\psi} + i\varepsilon^* e^{-i\psi}\right)$$
$$= \frac{1}{2}\left(\varepsilon e^{i(\psi-\pi/2)} + c.c.\right), \tag{B.2}$$

where 'c.c.' is short for 'complex conjugate'. This corresponds to

$$v = r\cos(\phi - \frac{\pi}{2})$$
$$= r\sin\phi, \tag{B.3}$$

where $\phi = \psi + \theta$, the sum of fast and slow phases, as before. Alternatively, we can write

$$v = \Im\{\varepsilon e^{i\psi}\}$$
$$= \Im\{ir\sin\phi\}$$
$$= r\sin\phi. \tag{B.4}$$

Bibliography

Abel, B., Assmann, J., Faubel, M., Gäbel, K., Kranzusch, S., Lugovoj, E., Mann, K., Missalla, T., and Peth, Ch. (2004). Characterization of extreme ultraviolet light-emitting plasmas from a laser-excited fluorine-containing liquid polymer jet target. *J. Appl. Phys.*, **95**, 7619–7623.

Abramyan, L. A., Litvak, A. G., Mironov, V. A., and Sergeev, A. M. (1992). Self-focusing and relativistic waveguiding of an ultrashort laser pulse in a plasma. *Sov. Phys. JETP*, **75**, 978–982.

Adam, J. C., and Héron, A. (1988). Parametric instabilities in resonant absorption. *Phys. Fluids*, **31**, 2602–2614.

Adam, J. C., Héron, A., Guérin, S., Laval, G., Mora, P., and Quesnel, B. (1997). Anomalous absorption of very high-intensity laser pulses propagating through moderately dense plasma. *Phys. Rev. Lett.*, **78**, 4765–4768.

Agostini, P., Barjot, G., Bonnal, J. F., Mainfray, G., and Manus, C. (1968). Multiphoton ionization of hydrogen and rare gases. *IEEE J. Quant. Elec.*, **QE-4**, 667–669.

Agostini, P., Fabre, F., Mainfray, G., Petite, G., and Rahman, N. K. (1979). Free-free transitions following 6-photon ionization of xenon atoms. *Phys. Rev. Lett.*, **42**, 1127–1130.

Akhiezer, A. I., and Polovin, R. V. (1956). Theory of wave motion of an electron plasma. *Sov. Phys. JETP*, **3**, 696–705.

Alfvén, H. (1939). On the motion of cosmic rays in interstellar space. *Phys. Rev.*, **55**, 425–429.

Amendt, P., Eder, D. C., and Wilks, S. C. (1991). X-ray lasing by optical-field-induced ionization. *Phys. Rev. Lett.*, **66**, 2589–2592.

Amiranoff, F., Fabbro, R., Fabre, E., Garban-Labaune, C., and Weinfeld, M. (1982). Experimental studies of fast electron generation in laser-produced plasmas at 1.06, 0.53 and 0.26 μm laser wavelength. *J. de Physique*, **43**, 1037–1042.

Amiranoff, F., Bernard, D., Cros, B., Jacquet, F., Matthieussent, G., Mine, P., Mora, P., Morillo, J., Moulin, F., Specka, A. E., and Stenz, C. (1995). Electron acceleration in Nd-laser plasma beat-wave experiments. *Phys. Rev. Lett.*, **74**, 5220–5223.

Ammosov, M. V., Delone, N. B., and Krainov, V. P. (1986). Tunnel ionization of

271

complex atoms and of atomic ions in an alternating electromagnetic field. *Sov. Phys. JETP*, **64**, 1191.

Andersson, E. 2002. *Entwicklung eines röntgenoptischen Systems für Angiographie des menschlichen Herzens unter Ausnutzung laserproduzierter Plasmen*. Ph.D. thesis, University of Jena, Dept. Physics.

Andreev, A. A., Gamaliy, E. G., Novikov, V.N., Semakhin, A. N., and Tikhonchuk, V. T. (1992). Heating of a dense plasma by an ultrashort laser pulse in the anomalous skin-effect regime. *Sov. Phys. JETP*, **74**, 963–973.

Andreev, A. A., Limpouch, J., and Semakhin, A. N. (1994). Absorption of the energy of a short laser pulse obliquely incident on a highly inhomogeneous plasma. *Bull. Russ. Acad. Sci.*, **58**, 1056–1063.

Andreev, A. A., Litvinenko, I. A., and Platonov, K. Y. (1999). Escape into vacuum of fast electrons generated by oblique incidence of an ultrashort, high-power laser pulse on a solid target. *Sov. Phys. JETP*, **89**, 632–639.

Antonsen Jr., T. M., and Mora, P. (1992). Self-focusing and Raman scattering of laser pulses in tenuous plasmas. *Phys. Rev. Lett.*, **69**, 2204–2207.

Antonsen Jr., T. M., and Mora, P. (1993). Self-focusing and Raman scattering of laser pulses in tenuous plasmas. *Phys. Fluids B.*, **5**, 1440–1452.

Arfken, G. B., and Weber, H. J. (1995). *Mathematical Methods for Physicists*. 4th edn. Academic Press, San Diego.

Ashcroft, N. W., and Mermin, N. D. (1976). *Solid State Physics*. Holt, Rinehart & Winston, New York.

Attwood, D. (1992). New opportunities at soft-x-ray wavelengths. *Phys. Today*, **45**, 24–31.

Atzeni, S. (1999). Inertial fusion fast ignitor: Igniting pulse parameter window *vs.* the penetration depth of the heating particles and the density of the pre-compressed fuel. *Phys. Plasmas*, **6**(8), 3316–3326.

Audebert, P., Geindre, J. P., Rousse, A., Gauthier, J. C., Mysyrowicz, A., Grillon, G., and Antonetti, A. (1994). K-shell emission dynamics of Be-like to He-like ions from a 100 fs laser-produced aluminium plasma. *J. Phys. B: At. Mol. Opt. Phys.*, **27**, 3303–3314.

Augst, S., Strickland, D., Meyerhofer, D., Cin, S. L., and Eberly, J. (1989). Tunneling ionization of noble gases in a high-intensity laser field. *Phys. Rev. Lett.*, **63**, 2212–2215.

Auguste, T., Monot, P., Lompré, L.-A., Mainfray, G., and Manus, C. (1992a). Defocusing effects of a picosecond terawatt laser pulse in an underdense plasma. *Optics Commun.*, **89**, 145–148.

Auguste, T., Monot, P., Lompré, L. A., Mainfray, G., and Manus, C. (1992b). Multiply charged ions produced in noble gases by a 1ps laser pulse at $\lambda = 1053$ nm. *J. Phys. B: At. Mol. Phys.*, **25**, 4181–4194.

Auguste, T., Monot, P., Mainfray, G., Manus, C., Gary, S., and Louis-Jacquet, M. (1994). Focusing behavior of a multi-terawatt laser pulse in a H_2 gas jet. *Optics Commun.*, **105**, 292–296.

Balcou, P., and L'Huillier, A. (1993). Phase-matching effects in strong-field harmonic generation. *Phys. Rev. A*, **47**(2), 1447–1459.

Bardsley, J. N., Penetrante, B. M., and Mittleman, M. H. (1989). Relativistic dynamics of electrons in intense laser fields. *Phys. Rev. A*, **40**, 3823–3835.

Barr, H. C., and Hill, L. J. (2003). Stimulated scattering and harmonic generation from relativistically intense laser light in plasma. *Phys. Plasmas*, **10**, 1135–1145.

Barr, H. C., Mason, P., and Parr, D. M. (1999). Electron parametric instabilities driven by relativistically intense laser light in plasma. *Phys. Rev. Lett.*, **83**, 1606–1609.

Barty, C. P. J., Ráksi, F., Rose-Petruck, C., Schafer, K. J., Wilson, K. R., Yakovlev, V. V., Yamakawa, K., Jiang, Z., Ikhlef, A., Côté, C. Y., and Kieffer, J.-C. (1995). Ultrafast x-ray absorption and diffraction. *Pages 246–257 of: Time resolved electron and X-ray diffraction*, vol. 2521. SPIE, Bellingham, USA.

Bastiani, S., Rousse, A., Geindre, J.P., Audebert, P., Quoix, C., Hamoniaux, G., Antonetti, A., and Gauthier, J.-C. (1997). Experimental study of the interaction of subpicosecond laser pulses with solid targets of varying initial scale lengths. *Phys. Rev. E*, **56**, 7179–7185.

Batani, D., Antonicci, A., Pisani, F., Hall, T. A., Scott, D., Amiranoff, F., Koenig, M., Gremillet, L., Baton, S., Martinolli, E., Rousseaux, C., and Nazarov, W. (2002). Inihibition in the propagation of fast electrons in plasma foams by resistive electric fields. *Phys. Rev. E*, **65**.

Bauer, D., Mulser, P., and Steeb, W.-H. (1995). Relativistic ponderomotive force, uphill acceleration and transition to chaos. *Phys. Rev. Lett.*, **75**, 4622–4625.

Becker, W. (1991). Quantum electrodynamics in intense laser fields. *Laser Part. Beams*, **9**, 603–618.

Beg, F. N., Bell, A.R., Dangor, A. E., Danson, C. N., Fews, A. P., Glinsky, M. E., Hammel, B. A., Lee, P., Norreys, P. A., and Tatarakis, M. (1997). A study of picosecond laser-solid interactions up to $10^{19}\mathrm{Wcm}^{-2}$. *Phys. Plasmas*, **4**, 447–457.

Bell, A. R. (1994). Magnetohydrodynamic jets. *Phys. Plasmas*, **1**, 1643–1652.

Bell, A. R., and Gibbon, P. (1988). Electron non-linearities in Langmuir waves with application to beat-wave experiments. *Plas. Phys. Contr. Fusion*, **30**, 1319.

Bell, A. R., and Kingham, R. J. (2003). Resistive collimation of electron beams in laser-produced plasmas. *Phys. Rev. Lett.*, **91**, 035003.

Bell, A. R., Beg, F. N., Chang, Z., Dangor, A. E., Danson, C. N., Edwards, C. B., Fews, A. P., Hutchinson, M. H. R., Luan, S., Lee, P., Norreys, P. A., Smith, R. A., Taday, P. F., and Zhou, F. (1993). Observation of plasma confinement in picosecond laser-plasma interaction. *Phys. Rev. E*, **48**, 2087–2093.

Bell, A. R., Davies, J. R., Guérin, S., and Ruhl, H. (1997). Fast-electron transport in high-intensity short-pulse laser-solid experiments. *Plasma Phys. Control. Fusion*, **39**, 653–659.

Bell, A. R., Davies, J. R., and Guérin, S. R. (1998). Magnetic field in short-pulse high-intensity laser-solid experiments. *Phys. Rev. E*, **58**, 2471–2473.

Bell, A.R., Evans, R.G., and Nicholas, D.J. (1981). Electron energy-transport in steep temperature gradients in laser-produced plasmas. *Phys. Rev. Lett.*, **46**, 243.

Belzile, C., Kieffer, J. C., Cote, C. Y., Oksenhendler, T., and Kaplan, D. (2003). Jitter-free subpicosecond streak cameras. *Rev. Sci. Instr.*, **73**, 1617–1620.

Berezhiani, V. I., and Murusidze, I. G. (1990). Relativistic wake-field generation by an intense laser pulse in a plasma. *Phys. Lett. A*, **148**, 338–340.

Bers, A. (1983). Space-time evolution of plasma instabilities – absolute and convective. Pages *451–517 of:* Rosenbluth, M. N., and Sagdeev, R. Z. (eds), *Handbook of Plasma Physics*, vol. 1. North Holland, Amsterdam.

Bethe, H. A., and Saltpeter, E. E. (1977). *Quantum Mechanics of One- and Two-Electron Atoms*. Plenum, New York.

Bezzerides, B., Gitomer, S. J., and Forslund, D. W. (1980). Randomness, Maxwellian distributions and resonance absorption. *Phys. Rev. Lett.*, **44**, 651–654.

Bezzerides, B., Jones, R. D., and Forslund, D. W. (1982). Plasma mechanism for ultraviolet harmonic radiation due to intense CO_2 light. *Phys. Rev. Lett.*, **49**, 202–205.

Bingham, R., Mendonca, J. T., and Shukla, P. K. (2004). Plasma based charged-particle accelerators. *Plas. Phys. Contr. Fus.*, **46**, R1–R23.

Birdsall, C.K., and Langdon, A.B. (1985). *Plasma Physics via Computer Simulation*. McGraw-Hill, New York.

Blyth, W. J., Preston, S. G., Offenberger, A. A., Key, M. H., Wark, J. S., Najmudin, Z., Modena, A., Djaoui, A., and Dangor, A.E. (1995). Plasma temperature in optical field ionization of gases by intense ultrashort pulses of ultraviolet radiation. *Phys. Rev. Lett.*, **74**(4), 554–557.

Bobin, J.-L. (1985). High-intensity laser plasma interaction. *Phys. Reports*, **122**, 173–274.

Bond, D. J., Hares, J. D., and Kilkenny, J. D. (1980). Demonstration of resistive inhibition of fast electrons from laser-produced plasmas in low-density gold targets. *Phys. Rev. Lett.*, **45**, 252–255.

Bonnaud, G., Gibbon, P., Kindel, J., and Williams, E. (1991). Laser interaction with a sharp-edged, overdense plasma. *Laser and Particle Beams*, **9**, 339–354.

Bonnaud, G., Brandi, H. S., Manus, C., Mainfray, G., and Lehner, T. (1994). Relativistic and ponderomotive self-focusing of a laser beam in a radially inhomogeneous plasma. II. *Phys. Plasmas*, **1**, 968–989.

Borghesi, M., MacKinnon, A. J., Barringer, L., Gaillard, R., Gizzi, L. A., Meyer, C., Willi, O., Pukhov, A., and Meyer-ter-Vehn, J. (1997). Relativistic channeling of a picosecond laser pulse in a near-critical preformed plasma. *Phys. Rev. Lett.*, **78**, 879–882.

Borghesi, M., Mackinnon, A. J., Gaillard, R., Willi, O., Pukhov, A., and Meyer-ter-Vehn, J. (1998). Large quasistatic magnetic fields generated by a relativistically intense laser pulse propagating in a preionized plasma. *Phys. Rev. Lett.*, **80**, 5137–5140.

Borghesi, M., Mackinnon, A. J., Campbell, D. H., Hicks, D. G., Kar, S., Patel, P. K., Price, D., Romagnani, L., Schiavi, A., and Willi, O. (2004). Multi-MeV proton source investigations in ultraintense laser-foil interactions. *Phys. Rev. Lett.*, **92**, 055003.

Borisov, A. B., Borovskiy, A. V., Korobkin, V. V., Prokhorov, A. M., Rhodes,

C. K., and Shiryaev, O. B. (1990). Stabilization of relativistic self-focusing of intense subpicosecond ultraviolet pulses in plasmas. *Phys. Rev. Lett.*, **65**, 1753–1756.

Borisov, A. B., Borovskiy, A. V., Korobkin, V. V., Prokhorov, A. M., Shiryaev, O. B., Shi, X. M., Luk, T. S., McPherson, A, Solem, J. C., Boyer, K., and Rhodes, C. K. (1992a). Observation of relativistic and charge-displacement self-channeling of intense subpicosecond ultraviolet (248nm) radiation in plasmas. *Phys. Rev. Lett.*, **68**, 2309–2312.

Borisov, A. B., Borovskiy, A. V., Shiryaev, O. B., Korobkin, V. V., Prokhorov, A. M., Solem, J. C., Luk, T., Boyer, K., and Rhodes, C. K. (1992b). Relativistic and charge-displacement self-channeling of intense ultrashort laser pulses in plasmas. *Phys. Rev. A*, **45**, 5830–5845.

Borisov, A. B., Song, X. Y., Frigeni, F., Koshman, Y., Dai, Y., Boyer, K., and Rhodes, C. K. (2003). Ultrabright multikilovolt coherent tunable x-ray source at lambda similar to 2.71–2.93 Angstrom. *J. Phys. B*, **36**, 3433–3455.

Born, M., and Wolf, E. (1980). *Principles of Optics*. 6th edn. Pergamon Press, Oxford.

Bourdier, A. (1983). Oblique incidence of a strong electromagnetic wave on a cold inhomogeneous electron plasma. Relativistic effects. *Phys. Fluids*, **26**, 1804–1807.

Boyer, K., Luk, T. S., and Rhodes, C. K. (1988). Possibility of optically induced nuclear fission. *Phys. Rev. Lett.*, **60**, 557–560.

Brown, L. S., and Kibble, T. W. B. (1964). Interaction of intense laser beams with electrons. *Phys. Rev.*, **133**, A705–A719.

Brueckner, K. A., and Jorna, S. (1974). Laser-driven fusion. *Rev. Mod. Phys.*, **46**, 325–367.

Brunel, F. (1987). Not-so-resonant, resonant absorption. *Phys. Rev. Lett.*, **59**, 52–55.

Brunel, F. (1988). Anomalous absorption of high intensity subpicosecond laser pulses. *Phys. Fluids*, **31**, 2714–2719.

Bula, C., McDonald, K. T., Prebys, E. J., Bamber, C., Boege, S., Kotseroglou, T., Melissinos, A. C., Meyerhofer, D. D., Ragg, W., Burke, D. L., Field, R. C., Horton-Smith, G., Odian, A. C., Spencer, J. E., Walz, D., Berridge, S. C., Bugg, W. M., Shmakov, K., and Wiedemann, A. W. (1996). Observation of nonlinear effects in Compton scattering. *Phys. Rev. Lett.*, **76**, 3116–3119.

Bulanov, S. V., Kirsanov, V. I., and Sakharov, A. S. (1989). Excitation of ultrarelativistic langmuir waves by electromagnetic pulses. *Phys. Scripta*, **T30**, 208–209.

Bulanov, S. V., Naumova, N. M., and Pegoraro, F. (1994). Interaction of an ultrashort, relativistically strong laser pulse with an overdense plasma. *Phys. Plasmas*, **1**, 745–757.

Bulanov, S. V., Pegoraro, F., and Pukhov, A. (1995). Two-dimensional regimes of self-focusing, wake-field generation and induced focusing of a short intense laser pulse in an underdense plasma. *Phys. Rev. Lett.*, **74**, 710–713.

Bulanov, S. V., Esirkepov, T. Zh., Khoroshkov, V. S., Kuznetsov, A. V., and Pegoraro, F. (2002). Oncological hadron therapy with laser ion accelerators.

Phys. Lett. A, **299**, 240–247.

Bulanov, S. V., Esirkepov, T., and Tajima, T. (2003). Light intensification towards the Schwinger limit. *Phys. Rev. Lett.*, **91**, 085001.

Buneman, O. (1973). Inversion of the Helmholtz (or Laplace-Poisson) operator for slab geometry. *J. Comp. Phys.*, **12**, 124–130.

Burnett, K., Reed, V. C., and Knight, P. L. (1993). Atoms in ultra-intense laser fields. *J. Phys. B–At. Mol. Opt. Phys.*, **26**, 561–598.

Burnett, N. H., and Corkum, P. B. (1987). Cold-plasma production for recombination extreme-ultraviolet lasers by optical-field-induced ionization. *J. Opt. Soc. Am. B*, **6**, 1195–1199.

Burnett, N. H., and Enright, G. D. (1990). Population inversion in the recombination of optically-ionized plasmas. *IEEE J. Quant. Elec.*, **26**, 1797–1808.

Burnett, N. H., Baldis, H. A., Richardson, M. C., and Enright, G. D. (1977). Harmonic generation in CO_2 laser target interaction. *Appl. Phys. Lett.*, **31**, 172–174.

Cairns, R. A., Rau, B., and Airila, M. (2000). Enhanced transmission of laser light through thin slabs of overdense plasmas. *Phys. Plasmas*, **7**, 3736–3742.

Carman, R. L., Rhodes, C. K., and Benjamin, R. F. (1981a). Observation of harmonics in the visible and ultraviolet created in CO_2-laser-produced plasmas. *Phys. Rev. A*, **24**, 2649–2663.

Carman, R. L., Forslund, D. W., and Kindel, J. M. (1981b). Visible harmonic emission as a way of measuring profile steepening. *Phys. Rev. Lett.*, **46**, 29–32.

Castillo-Herrera, C. I., and Johnston, T. W. (1993). Incoherent harmonic emission from strong electromagnetic waves in plasmas. *IEEE Trans. Plasma Sci.*, **21**, 125–135.

Catto, P. J., and More, R. M. (1977). Sheath inverse bremsstrahlung in laser produced plasmas. *Phys. Fluids*, **20**, 704–705.

Cavalleri, A., Siders, C. W., Brown, F. L. H., Leitner, D. M., Tóth, C., Squier, J. A., Barty, C. P. J., Wilson, K. R., Sokolowski-Tinten, K., von Hoegen, M. Horn, von der Linde, D., and Kammler, M. (2000). Anharmonic lattice dynamics in germanium measured with ultrafast x-ray diffraction. *Phys. Rev. Lett.*, **85**, 586–589.

Cavalleri, A., Tóth, Cs., Squier, C. W. Siders J. A., Ráksi, F., Forget, P., and Kieffer, J. C. (2001). Femtosecond structural dynamics in VO_2 during an ultrafast solid-solid phase transition. *Phys. Rev. Lett.*, **87**, 237401.

Chaker, M., Boily, S., Lafontaine, B., Keffer, J. C., Pépin, H., Toubhans, I., and Fabbro, R. (1990). X-ray wavelength optimization of the laser plasma X-ray lithography source. *Microelectronic Engineering*, **10**, 91–105.

Chaker, M., Kieffer, J. C., Matte, J. P., Pépin, H, Audebert, P., Maine, P., Strickland, D., Bado, P., and Mourou, G. (1991). Short-pulse laser absorption in very steep plasma density gradients. *Phys. Fluids B*, **3**, 167–175.

Chen, F. F. (1974). *Introduction to Plasma Physics*. Plenum Press, New York.

Chen, H., Soom, B., Yaakobi, B., Uchida, S., and Meyerhofer, D. D. (1993). Hot-electron characterization from $K\alpha$ measurements in high-contrast, p-polarized, picosecond laser-plasma interactions. *Phys. Rev. Lett.*, **70**, 3431–3434.

Chen, P., and Tajima, T. (1999). Testing Unruh radiation with ultraintense lasers. *Phys. Rev. Lett.*, **83**, 256–259.

Chen, S. Y., Sarkisov, G. S., Maksimchuk, A., Wagner, R., and Umstadter, D. (1998a). Evolution of a plasma waveguide created during relativistic-ponderomotive self-channeling of an intense laser pulse. *Phys. Rev. Lett.*, **80**, 2610–2613.

Chen, S.-Y., Maksimchuk, A., and Umstadter, D. (1998b). Experimental observation of relativistic nonlinear Thomson scattering. *Nature*, **396**, 653–655.

Chen, S.-Y., Krishnan, M., Maksimchuk, A., Wagner, R., and Umstadter, D. (1999). Detailed dynamics of electron beams self-trapped and accelerated in a self-modulated laser wakefield. *Phys. Plasmas*, **6**, 4739–4749.

Chen, S.-Y., Maksimchuk, A., Esarey, E., and Umstadter, D. (2000). Observation of phase-matched relativistic harmonic generation. *Phys. Rev. Lett.*, **84**, 5528–5531.

Chen, X. L., and Sudan, R. N. (1993a). Necessary and sufficient conditions for self-focusing of short ultraintense laser pulses in underdense plasmas. *Phys. Rev. Lett.*, **70**, 2082–2085.

Chen, X. L., and Sudan, R. N. (1993b). Two-dimensional self-focusing of short intense laser pulse in underdense plasma. *Phys. Fluids B*, **5**, 1336–1348.

Chernikov, A. A., Schmidt, G., and Neishtadt, A. I. (1992). Unlimited particle acceleration by waves in a magnetic field. *Phys. Rev. Lett.*, **68**, 1507–1510.

Christov, I. P., Murnane, M. M., and Kapteyn, H. C. (1997). High-harmonic generation of attosecond pulses in the single-cycle regime. *Phys. Rev. Lett.*, **78**, 1251–1254.

Clark, E. L. 2001. *Measurements of energetic particles from ultra intense laser plasma interactions*. Ph.D. thesis, Physics, University of London.

Clark, E. L., Krushelnick, K., Davies, J. R., Zepf, M., Tatarakis, M., Beg, F. N., Machacek, A., Norreys, P. A., Santala, M. I. K., Watts, I., and Dangor, A. E. (2000). Measurements of energetic proton transport through magnetized plasma from intense laser interactions with solids. *Phys. Rev. Lett.*, **84**, 670–673.

Clayton, C. E., Joshi, C., Darrow, C., and Umstadter, D. (1985). Relativistic plasma-wave excitation by collinear optical mixing. *Phys. Rev. Lett.*, **54**, 2343–2346.

Clayton, C. E., March, K. A., Dyson, A., Everett, M., Lal, A., Leemans, W. P., Williams, R., and Joshi, C. (1993). Ultrahigh-gradient acceleration of injected electrons by laser-excited relativistic electron-plasma waves. *Phys. Rev. Lett.*, **70**, 37–40.

Coffey, T. P. (1971). Breaking of large amplitude plasma oscillations. *Phys. Fluids*, **14**, 1402–1406.

Corkum, P. B. (1993). Plasma perspective on strong-field multiphoton ionization. *Phys. Rev. Lett.*, **71**, 1994–1997.

Coverdale, C. A., Darrow, C. B., Decker, C. D., Mori, W. B., Tzeng, K.-C., Marsh, K. A., Clayton, C. E., and Joshi, C. (1995). Propagation of intense subpicosecond laser pulses through underdense plasmas. *Phys. Rev. Lett.*, **74**, 4659–4662.

Cowan, T. E., Hunt, W. W., Phillips, T. W., Wilks, S. C., Perry, M. D., Brown,

C., Fountain, W., Hatchett, S., Johnson, J., Key, M. H., Parnell, T., Pennington, D. M., Snavely, R. A., and Takahasi, Y. (2000). Photonuclear Fission
from high energy electrons from ultraintense laser-solid interactions. *Phys.
Rev. Lett.*, **84**, 903–906.

Cowan, T. E., Fuchs, J., Ruhl, H., Kemp, A., Audebert, P., Roth, M., Stephens,
R., Barton, I., Blazevic, A., Brambrink, E., Cobble, J., Fernandez, J., Gauthier, J.-C., Geissel, M., Hegelich, M., Kaae, J., Karsch, S., Le Sage, G. P., Letzring, S., Manclossi, M., Meyroneinc, S., Newkirk, A., Pépin, H., and Renard-
LeGalloudec, N. (2004). Ultralow emittance, multi-MeV proton beams froma
laser virtual-cathode plasma accelerator. *Phys. Rev. Lett.*, **92**, 204801.

Dalla, S., and Lontano, M. (1993). On the maximum longitudinal electric field
of a large amplitude electron plasma wave excited by a short electromagnetic
radiation pulse. *Phys. Lett. A*, **173**, 456–461.

Dangor, A. E., Bradshaw, A. K. L., and Dyson, A. E. (1990). Observation of
relativistic plasma waves generated by the beat-wave with 1 μm lasers. *Phys.
Scripta*, **T30**, 107–109.

Darrow, C. B., Coverdate, C., Perry, M. D., Mori, W. B., Clayton, C., Marsh,
K., and Joshi, C. (1992). Strongly coupled stimulated Raman backscatter from
subpicosecond laser-plasma interactions. *Phys. Rev. Lett.*, **69**, 442–445.

Davidson, R. C. (1972). *Nonlinear Plasma Theory*. Academic Press, New York.

Davies, J. R. (2004). Alfvén limit in fast ignition. *Phys. Rev. E*, **69**(6), 065402.

Davies, J. R., Bell, A. R., Haines, M. G., and Guérin, S. M. (1997). Short-pulse
hight-intensity laser-generated fast electron transport into thick solid targets.
Phys. Rev. E, **56**, 7193–7203.

Davis, J., Clark, R., and Giuliani, J. (1995). Ultrashort pulse laser-produced
Al/Si plasma. *Laser and Part. Beams*, **13**, 3–18.

Dawson, J. M. (1959). Nonlinear electron oscillations in a cold plasma. *Phys.
Rev.*, **113**, 383–387.

Dawson, J. M. (1962). One-dimensional plasma model. *Phys. Fluids*, **5**, 445–459.

Dawson, J. M., and Oberman, C. (1962). High-frequency conductivity and the
emission and absorption coefficients of a fully ionized plasma. *Phys. Fluids*, **5**,
517.

Decker, C. D., Mori, W. B., and Katsouleas, T. (1994). Particle-in-cell simulations
of Raman forward scattering from short-pulse high-intensity lasers. *Phys. Rev.
E*, **50**, R3338–R3341.

Decoster, A. (1978). Nonlinear travelling waves in a homogeneous cold collisionless plasma. *Phys. Rep.*, **47**, 285–422.

Delettrez, J. (1988). Simulations of the interaction of sub-ps laser pulses with
solid targets. *Bull. Amer. Phys. Soc.*, **33**, 2060.

Denavit, J. (1979). Collisionless plasma expansion into a vacuum. *Phys. Fluids*,
22, 1384–1392.

Denavit, J. (1992). Absorption of high intensity subpicosecond lasers on solid
density targets. *Phys. Rev. Lett.*, **69**, 3052–3055.

Dendy, R. (ed) (1993). *Plasma Physics: An Introductory Course*. Cambridge
University Press, Cambridge.

Denisov, N. G. (1957). On a singularity of the field of an electromagnetic wave propagated in an inhomogeneous plasma. *Sov. Phys. JETP*, **4**, 544–553.

Deutsch, C., Furukawa, H., Mima, K., Murakami, M., and Nishihara, K. (1996). Interaction physics of the fast ignitor concept. *Phys. Rev. Lett.*, **77**, 2483–2486.

Disdier, L., Garconnet, J.-P., Malka, G., and Miquel, J.-L. (1999). Fast neutron emission from a high-energy ion beam produced by a high-intensity subpicosecond laser pulse. *Phys. Rev. Lett.*, **82**, 1454–1457.

Ditmire, T., Donnelly, T., Falcone, R. W., and Perry, M. D. (1995). Strong x-ray emission from high-temperature plasmas produced by intense irradiation of clusters. *Phys. Rev. Lett.*, **75**, 3122–3125.

Ditmire, T., Zweiback, J., Yanovsky, V. P., Cowan, T. E., Hays, G., and Wharton, K. B. (1999). Nuclear fusion from explosions of femtosecond laser-heated deuterium clusters. *Nature*, **398**, 489–492.

Djaoui, A. (1999). Time-dependent hydrogenic ionization model for non-LTE mixtures. *J. Quant. Spectrosc. Radiat. Transfer*, **62**, 303–320.

Doumy, G., Dobosz, S., D'Oliveira, P., Monot, P., Perdrix, M., Quere, F., Reau, F., Martin, P., Audebert, P., Gauthier, J.-C., and Geindre, J.-P. (2004). High order harmonic generation by nonlinear reflection of a pedestal-free intense laser pulse on a plasma. *Appl. Phys. B*, **78**, 901–904.

Drake, J. F., Kaw, P. K., Lee, Y. C., Schmidt, G., Liu, C. S., and Rosenbluth, M. N. (1974). Parametric instabilities of electromagnetic waves in plasmas. *Phys. Fluids*, **17**, 778–785.

Drell, S. D. 1994 (May). *High Energy Physics Advisory Panel's Subpanel on Vision for the Future of High Energy Physics*. Tech. rept. DOE/ER-0614P. U. S. Department of Energy, Washington DC.

Drude, P. (1900). Zur Elektronentheorie. I. *Annalen der Physik*, **1**, 566–613.

Dunne, M., Afshar-Rad, T., Edwards, J., MacKinnon, A. J., Viana, S. M., Willi, O., and Pert, G. (1994). Experimental observations of the expansion of an optical-field- induced ionization channel in a gas jet target. *Phys. Rev. Lett.*, **72**, 1024–1027.

Durfee III, C. G., and Milchberg, H. M. (1993). Light pipe for high intensity laser pulses. *Phys. Rev. Lett.*, **71**, 2409–2412.

Düsterer, S., Schwoerer, H., Ziegler, W., Ziener, C., and Sauerbrey, R. (2001). Optimization of EUV radiation yield from laser-produced plasma. *Appl. Phys. B*, **73**, 693–698.

Eason, R. W., Bradley, D. K, Kilkenny, J. D., and Greaves, G. N. (1984). Improved laser-EXAFS studies of aluminium foil. *J. Phys. C*, **17**, 5067–5074.

Eberly, J. H., and Sleeper, A. (1968). Trajectory and mass shift of a classical electron in a radiation pulse. *Phys. Rev.*, **176**, 1570–1573.

Ebrahim, N. A., Lavigne, P., and Aithal, S. (1985). Experiments on the plasma beat-wave accelerator. *IEEE Trans. Nuc. Sci.*, **NS232**.

Eder, D. C., Amendt, P., DaSilva, L. B., London, R. A., MacGowan, B. J., Matthews, D. L., Penetrante, B. M., Rosen, M. D., Wilks, S. C., Donnelly, T. D., Falcone, R. W., and Strobel, G. L. (1994). Tabletop x-ray lasers. *Phys. Plasmas*, **1**, 1744–1752.

Eder, D. C., Pretzler, G., Fill, E., Eidmann, K., and Saemann, A. (2000). Spatial characteristics of K_α radiation from weakly relativistic laser plasmas. *Appl.*

Phys. B, **70**, 211–217.

Edwards, J., and Rose, S. J. (1993). Ionization timescales in hot dense plasmas. *J. Phys. B: At. Mol. Opt. Phys.*, **26**, L523–L527.

Einstein, A. (1905). Über einen die Erzeugung und Verwandlung des Lichtes betreffenden heuristischen Gesichtspunkt. *Annalen der Physik*, **17**, 132–148.

Elliot, J. A. (1993). Plasma Kinetic Theory. *In:* Dendy, R. O. (ed), *Plasma Physics: An Introductory Course, Chapter 2*. Cambridge University Press, Cambridge.

Elton, R. C. (1990). *X-Ray Lasers*. Academic Press, Boston, MA.

Englert, T. J., and Rinehart, E. A. (1983). Second-harmonic photons from the interaction of free electrons with intense laser radiation. *Phys. Rev. A*, **28**, 1539–1545.

Esarey, E., Ting, A., and Sprangle, P. (1988). Relativistic focusing and beat wave phase velocity control in the plasma beat wave accelerator. *Appl. Phys. Lett.*, **53**, 1266–1268.

Esarey, E., Ting, A., Sprangle, P., and Joyce, G. (1989). The laser wakefield accelerator. *Comments Plasma Phys. Contr. Fusion*, **12**, 191–204.

Esarey, E., Ting, A., Sprangle, P., Umstadter, D., and Liu, X. (1993). Nonlinear analysis of relativistic harmonic generation by intense lasers in plasmas. *IEEE Trans. Plasma Sci.*, **21**, 95–104.

Esarey, E., Krall, J., and Sprangle, P. (1994). Envelope analysis of intense laser pulse self-modulation in plasmas. *Phys. Rev. Lett.*, **72**, 2887–2890.

Esarey, E., Sprangle, P., and Krall, J. (1995). Laser acceleration of electrons in vacuum. *Phys. Rev. E*, **52**, 5443–5453.

Esarey, E., Sprangle, P., Krall, J., and Ting, A. (1997). Self-focusing and guiding of short laser pulses in ionizing gases and plasmas. *IEEE J. Quant. Elec.*, **33**, 1879–1914.

Esirkepov, T., Bulanov, S. V., Nishihara, K., and Tajima, T. (2004). Soliton synchrotron afterglow in a laser plasma. *Phys. Rev. Lett.*, **92**, 255001.

Esirkepov, T. Zh. (2001). Exact charge conservation scheme for particle-in-cell simulation with an arbitrary form-factor. *Comp. Phys. Commun.*, **135**, 144–153.

Estabrook, K. G., and Kruer, W. L. (1978). Properties of resonantly heated electron distributions. *Phys. Rev. Lett*, **40**, 42–45.

Estabrook, K. G., and Kruer, W. L. 1986. *Resonant absorption in very steep density gradients*. Tech. rept. Lawrence Livermore National Laboratory.

Estabrook, K. G., Valeo, E. J., and Kruer, W. L. (1975). Two-dimensional relativistic simulations of resonance absorption. *Phys. Fluids*, **18**, 1151–1159.

Evans, R. G. (1985). The development of fluid codes for the laser compression of plasma. *Pages 247–278 of:* Hooper, M. B. (ed), *Laser-Plasma Interactions*, vol. 3. SUSSP, Edinburgh.

Everett, M., Lal, A., Gordon, D., Clayton, C. E., Marsh, K. A., and Joshi, C. (1994). Trapped electron acceleration by a laser-driven relativistic plasma wave. *Nature*, **368**, 527–529.

Everett, M. J., Lal, A., Gordon, D., Wharton, K., Clayton, C. E., Mori, W. B., and Joshi, C. (1995). Evolution of stimulated Raman into stimulated Compton-scattering of laser light via wave breaking of plasma waves. *Phys. Rev. Lett.*, **74**, 1355–1358.

Ewald, F., Schwoerer, H., and Sauerbrey, R. (2002). K$_\alpha$ radiation from relativistic laser-produced plasmas. *Europhys. Lett.*, **60**, 710–716.

Faure, J., Clinec, Y., Pukhov, A., Kiselev, S., Gordienko, S., Lefebvre, E., Rousseau, J.-P., Burgy, F., and Malka, V. (2004). A laser-plasma accelerator producing monoenergetic electron beams. *Nature*, **431**, 541–544.

Fedosejevs, R., Ottman, R., Sigel, R., Kühnle, G., Szatmari, S., and Schäfer, F. P. (1990a). Absorption of femtosecond laser pulses in high-density plasmas. *Phys. Rev. Lett.*, **64**, 1250–1253.

Fedosejevs, R., Ottman, R., Sigel, R., Kühnle, G., Szatmari, S., and Schäfer, F. P. (1990b). Absorption of subpicosecond ultraviolet laser pulses in high-density plasma. *Appl. Phys. B*, **50**, 79–99.

Feit, M. D., and Fleck Jr., J. A. (1988). Beam nonparaxiality, filament formation, and beam breakup in the self-focusing of optical beams. *J. Opt. Soc. Am. B*, **5**, 633–640.

Ferray, M., L'Huillier, A., Li, X. F., Lompre, L. A., Mainfray, G., and Manus, C. (1988). Multiple-harmonic conversion of 1064-nm radiation in rare gases. *J. Phys. B*, **21**, L31–L35.

Feurer, T., Theobald, W., Sauerbrey, R., Uschmann, I., Altenbernd, D., Teubner, U., Gibbon, P., Förster, E., Malka, G., and Miquel, J. L. (1997). Onset of diffuse reflectivity and fast electron flux inhibition in 528-nm-laser-solid interactions at ultrahigh intensity. *Phys. Rev. E*, **56**, 4608–4614.

Feurer, T., Morak, A., Uschmann, I., Ziener, C., Schwoerer, H., Förster, E., and Sauerbrey, R. (2001). An incoherent sub-picosecond x-ray source for time-resolved x-ray diffraction experiments. *App. Phys. B*, **72**, 15–20.

Fews, A. P., Norreys, P. A., Beg, F. N., Bell, A. R., Dangor, A. E., Danson, C. N., Lee, P., and Rose, S. J. (1994). Plasma ion emission from high intensity picosecond laser pulse interactions with solid targets. *Phys. Rev. Lett.*, **73**, 1801–1804.

Fonesca, R. A., Silva, L. O., Tsung, F. S., Decyk, V. K., Lu, W., Ren, C., Mori, W. B., Deng, S., Lee, S., Katsouleas, T., and Adam, J. C. (2002). OSIRIS: A three-dimensional, fully relativistic particle-in-cell code for modeling plasma based accelerators. *Lecture Notes in Computer Science*, **2331**, 342–351.

Forget, P., Dorchies, F., Kieffer, J.-C., and Peyrusse, O. (2004). Ultrafast broadband laser plasma x-ray source for femtosecond time-resolved EXAFS. *Chem. Phys.*, **299**, 259–263.

Forslund, D. W., Kindel, J. M., Lee, K., Lindman, E. L., and Morse, R. L. (1975a). Theory and simulation of resonance absorption in a hot plasma. *Phys. Rev. A*, **11**, 679–683.

Forslund, D. W., Kindel, J. M., and Lindman, E. L. (1975b). Theory of stimulated scattering processes in laser-irradiated plasmas. *Phys. Fluids*, **18**, 1002–1016.

Forslund, D. W., Kindel, J. M., and Lee, K. (1977). Theory of hot-electron spectra at high laser intensity. *Phys. Rev. Lett*, **39**, 284–288.

Forslund, D. W., Kindel, J. M, Mori, W. B., Joshi, C., and Dawson, J. M. (1985). Two-dimensional simulations of single-frequency and beat-wave laser-plasma heating. *Phys. Rev. Lett.*, **54**, 558–561.

Förster, E., Fill, E. E., Gäbel, K., He, H., Missalla, Th., Renner, O., Uschmann, I., and Wark, J. (1994). X-Ray emission spectroscopy. *J. Quant. Spectrosc. Radiat. Trans.*, **51**, 101–111.

Freeman, R. R., and Bucksbaum, P. H. (1991). Investigations of above-threshold ionization using subpicosecond laser-pulses. *J. Phys. B–At. Mol. Opt.*, **24**, 325–437.

Fritzler, S., Najmudin, Z., Malka, V., Krushelnick, K., Marle, C., Walton, B., Wei, M. S., Clarke, R. J., and Dangor, A. E. (2002). Ion heating and thermonuclear neutron production from high-intensity subpicosecond laser pulses interacting with underdense plasmas. *Phys. Rev. Lett.*, **89**, 165004.

Fromy, P., Deutsch, C., and Maynard, G. (1992). Thomas-Fermi-like and average atom model equations of state for highly compressed matter at any temperature. *Laser and Part. Beams*, **10**, 263–275.

Fuchs, J., Malka, G., Adam, J. C., Amiranoff, F., Baton, S. D., Blanchot, N., Héron, A., Laval, G., Miquel, J. L., Mora, P., Pépin, H., and Rousseaux, C. (1998a). Dynamics of subpicosecond relativistic laser pulse self-channeling in an underdense preformed plasma. *Phys. Rev. Lett.*, **80**, 1658–1661.

Fuchs, J., Adam, J. C., Amiranoff, F., Baton, S. D., Gallant, P., Gremillet, L., Héron, A., Kieffer, J. C., Laval, G., Malka, G., Miquel, J. L., Mora, P., Pépin, H., and Rousseaux, C. (1998b). Transmission through highly overdense plasma slabs with a subpicosecond relativistic laser pulse. *Phys. Rev. Lett.*, **80**, 2326–2329.

Gahn, C., Tsakiris, G. D., Pukhov, A., ter Vehn, J. Meyer, Pretzler, G., Thirolf, P., Habs, D., and Witte, K. J. (1999). Multi-MeV electron beam generation by direct laser acceleration in high-density plasma channels. *Phys. Rev. Lett.*, **83**, 4772–4775.

Gamaliy, E. G. (1994). Ultrashort powerful laser matter interaction: Physical problems, models and computations. *Laser and Part. Beams*, **12**, 185–208.

Gamaliy, E. G., and Dragila, R. (1990). Interaction of ultrashort laser pulses at relativistic intensities with solid targets: relativistic skin effect. *Phys. Rev. A*, **42**, 929–935.

Gamaliy, E. G., and Tikhonchuk, V. T. (1988). Anomalous skin effect. *JETP Lett.*, **48**, 453–455.

Gauthier, J.-C. (1988). Spectroscopy and atomic physics. *In:* Hooper, M. B. (ed), *Laser-Plasma Interactions*, vol. 4. SUSSP, Edinburgh.

Geddes, C. G. R., Toth, Cs., van Tilborg, J., Esarey, E., Schroeder, C. B., Bruhwiller, D., Nieter, C., Cary, J., and Leemans, W. P. (2004). High-quality electron beams from a laser wakefield acclerator using plasma-channel guiding. *Nature*, **431**, 538–541.

Gibbon, P. (1990). The self-trapping of light waves by beat-wave excitation. *Phys. Fluids B*, **2**, 2196–2208.

Gibbon, P. (1992). A numerical model of the plasma beat-wave accelerator. *Comp. Phys. Commun.*, **69**, 299–305.

Gibbon, P. (1994). Efficient production of fast electrons from femtosecond laser interaction with solid targets. *Phys. Rev. Lett.*, **73**, 664–667.

Gibbon, P. (1996). Harmonic generation by femtosecond laser-solid interaction: A coherent water-window light source? *Phys. Rev. Lett.*, **76**, 50–53.

Gibbon, P. (1997). High order harmonic generation in plasmas. *IEEE J. Quant. Elec.*, **33**, 1915–1924.

Gibbon, P., and Bell, A. R. (1988). Cascade focusing in the beat-wave accelerator. *Phys. Rev. Lett.*, **61**, 1599–1602.

Gibbon, P., and Bell, A. R. 1990. Resonant absorption of picosecond laser pulses in steep density gradients. *In: 20th ECLIM*. Schliersee, January 22–26.

Gibbon, P., and Bell, A. R. (1992). Collisionless absorption in sharp-edged plasmas. *Phys. Rev. Lett.*, **68**, 1535–1538.

Gibbon, P., Monot, P., Auguste, T., and Mainfray, G. (1995). Measurable signatures of relativistic self-focusing in underdense plasmas. *Phys. Plasmas*, **2**, 1305–1310.

Gibbon, P., Jakober, F., Monot, P., and Auguste, T. (1996). Experimental study of relativistic self-focusing and self-channeling of an intense laser pulse in an underdense plasma. *IEEE Trans. Plasma Sci.*, **24**, 343–350.

Gibbon, P., Altenbernd, D., Teubner, U., Förster, E., Audebert, P., Geindre, J.-P., Gauthier, J. C., and Mysyrowicz, A. (1997). Plasma density determination by transmission of laser-generated surface harmonics. *Phys. Rev. E*, **55**, R6352–R6355.

Gibbon, P., Andreev, A. A., Lefebvre, E., Bonnaud, G., Ruhl, H., Delettrez, J., and Bell, A.R. (1999). Calibration of 1D boosted kinetic codes for modeling high-intensity laser-solid interactions. *Phys. Plasmas*, **6**, 947–953.

Gibbon, P., Beg, F. N., Evans, R. G., Clark, E. L., , and Zepf, M. (2004). Tree code simulations of proton acceleration from laser-irradiated wire targets. *Phys. Plasmas*, **11**, 4032–4040.

Ginzburg, V.L. (1964). *The Propagation of Electromagnetic Waves in Plasmas*. Pergamon, New York.

Gitomer, S. J., Jones, R. D., Begay, F., Ehler, A. W., Kephart, J. F., and Kristal, R. (1986). Fast ions and hot electrons in the laser-plasma interaction. *Phys. Fluids*, **29**, 2679–2688.

Giulietti, D., Gizzi, L. A., Giulietti, A., Macchi, A., Teychenné, D., Chessa, P., Rousse, A., Cheriaux, G., Chambaret, J. P., and Darpentigny, G. (1997). Observation of solid-density laminar plasma transparency to intense 20 femtosecond laser pulses. *Phys. Rev. Lett.*, **79**, 3194–3197.

Glinsky, M. E. (1995). Regimes of suprathermal electron transport. *Phys. Plasmas*, 2796–2806.

Godwin, R. P. (1972). Optical mechanism for enhanced absorption of laser energy incident on solid targets. *Phys. Rev. Lett.*, **28**, 85–87.

Godwin, R. P. (1994). Fresnel absorption, resonance absorption and x-rays in laser-produced plasmas. *Appl. Optics*, **33**, 1063.

Gontier, Y., Poirier, M., and Trahin, M. (1980). Multiphoton absorption above the ionization threshold. *J. Phys. B–At. Mol. Opt.*, **13**, 1381–1387.

Gorbunov, L., Mora, P., and Antonsen, Jr., T. M. (1996). Magnetic field of a plasma wake driven by a laser pulse. *Phys. Rev. Lett.*, **76**, 2495–2498.

Gorbunov, L., Mora, P., and Antonsen, Jr., T. M. (1997). Quasistatic magnetic field generated by a short laser pulse in an underdense plasma. *Phys. Plasmas*, **4**, 4358–4368.

Gorbunov, L. M., and Kirsanov, V. I. (1987). Excitation of plasma waves by an electromagnetic wave packet. *Sov. Phys. JETP*, **66**, 290–294.

Gordon, D., Tzeng, K. C., Clayton, C. E., Dangor, A. E., Malka, V., Marsh, K. A., Modena, A., Mori, W. B., Muggli, P., Najmudin, Z., Neely, D., Danson, C., and Joshi, C. (1998). Observation of electron energies beyond the linear dephasing limit from a laser-excited relativistic plasma wave. *Phys. Rev. Lett.*, **80**, 2133–2136.

Gordon III, C. L., Yin, G. Y., Lemoff, B. E., Bell, P. M., and Barty, C. P. J. (1995). Time-gated imaging with an ultrashort-pulse, laser-produced-plasma x-ray source. *Op. Lett.*, **20**, 1056–1058.

Grebogi, C., Tripathi, V. K., and Chen, H-H. (1983). Harmonic generation of radiation in a steep density profile. *Phys. Fluids*, **26**, 1904–1908.

Greschik, F., Dimou, L., and Kull, H.-J. (2000). Electrostatic model of laser pulse absorption by thin foils. *Laser and Part. Beams*, **18**, 367–373.

Grillon, G., Balcou, Ph., Chambaret, J.-P., Hulin, D., Martino, J., Moustaizis, S., Notebaert, L., Pittman, M., Pussieux, Th., Rousse, A., Rousseau, J.-Ph., Sebban, S., Sublemontier, O., and Schmidt, M. (2002). Deuterium-deuterium fusion dynamics in low-density molecular-cluster jets irradiated by intense ultrafast laser pulses. *Phys. Rev. Lett.*, **89**, 065005.

Grimes, M. K., Rundquist, A. R., Lee, Y.-S., and Downer, M. C. (1999). Experimental identification of vacuum heating at femtosecond-laser-irradiated metal surfaces. *Phys. Rev. Lett.*, **82**, 4010–4013.

Guérin, S., Laval, G., Mora, P., Adam, J. C., Héron, A., and Bendib, A. (1995). Modulational and Raman instabilities in the relativistic regime. *Phys. Plasmas*, **2**, 2807–2814.

Guérin, S., Mora, P., Adam, J. C., Héron, A., and Laval, G. (1996). Propagation of ultra-intense laser pulses through overdense plasma layers. *Phys. Plasmas*, **3**, 2693–2701.

Gunn, J. E., and Ostriker, J. P. (1971). On the motion and radiation of charged particles in strong electromagnetic waves. I. Motion in plane and spherical waves. *Astrophys. J.*, **165**, 523–541.

Hain, S., and Mulser, P. (2001). Fast ignition without hole boring. *Phys. Rev. Lett.*, **86**(6), 1015–1018.

Haines, M. G. (1979). Physics of the coronal region. *In:* Cairns, R. A., and Sanderson, J. J. (eds), *Laser-Plasma Interactions*, vol. 1. SUSSP, Edinburgh.

Haines, M. G. (1997). Saturation mechanisms for the generated magnetic field in nonuniform laser-matter irradiation. *Phys. Rev. Lett.*, **78**, 254–257.

Hamster, H., Sullivan, A., Gordon, S., White, W., and Falcone, F. W. (1993). Subpicosecond, electromagnetic pulses from intense laser-plasma interaction. *Phys. Rev. Lett.*, **71**, 2715–2728.

Hares, J. D., Kilkenny, J. D., Key, M. H., and Lunney, J. G. (1979). Measurement of fast-electron energy spectra and preheating in laser-irradiated targets. *Phys. Rev. Lett.*, **42**, 1216–1219.

Harrach, R. J., and Kidder, R. E. (1981). Simple model of energy deposition by suprathermal electrons in laser-irradiated targets. *Phys. Rev. A*, **23**, 887–896.

Harris, S. E., and Kmetec, J. D. (1988). Mixed-species targets for femtosecond-time-scale x-ray generation. *Phys. Rev. Lett.*, **61**, 62–65.

Hartemann, F. V. (1998). High-intensity scattering processes of relativistic electrons in vacuum. *Phys. Plasmas*, **5**, 2037–2047.

Hartemann, F. V., Fochs, S. N., Sage, G. P. Le, Luhmann, N. C., Woodworth, J. G., Perry, M. D., Chen, Y. J., and Kerman, A. K. (1995). Nonlinear ponderomotive scattering of relativistic electrons by an intense laser field at focus. *Phys. Rev. E*, **51**, 4833–4843.

He, F., Lau, Y. Y., Umstadter, D. P., and Kowalczyk, R. (2003). Backscattering of an intense laser beam by an electron. *Phys. Rev. Lett.*, **90**, 055002.

Heisenberg, W., and Euler, H. (1936). *Z. Physik.*

Hentschel, M., Kienberger, R., Spielmann, Ch., Reider, G. A., Milosevic, N., Brabec, T., Corkum, P., Heinzmann, U., Drescher, M., and Krausz, F. (2001). Attosecond metrology. *Nature*, **414**, 509–513.

Herrlin, K., Svahn, G., Olsson, C., Pettersson, H., Tillman, C., Persson, A., Wahlström, C.-G., and Svanberg, S. (1993). Generation of X-rays for medical imaging by high-power lasers: Preliminary results. *Radiology*, **189**, 65–68.

Hertz, H. M., Johansson, G. A., Stollberg, H., de Groot, J., Hemberg, O., Holmberg, A., Rehbein, S., Jansson, P., Eriksson, F., and Birch, J. (2003). Table-top x-ray microscopy: Sources, optics and applications. *J. Phys. IV*, **104**, 115–119.

Hockney, R.L., and Eastwood, J.W. (1981). *Computer Simulation Using Particles*. McGraw-Hill, New York.

Honda, M., Meyer-ter-Vehn, J., and Pukhov, A. (2000). Collective stopping and ion heating in relativistic-electron-beam transport for fast ignition. *Phys. Rev. Lett.*, **85**(10), 2128–2131.

Hora, H. (1981). *Physics of Laser Driven Plasmas*. Wiley, New York.

Hu, S. X., and Starace, A. F. (2002). GeV electrons from ultraintense laser interaction with highly charged ions. *Phys. Rev. Lett.*, **88**, 245003.

Huba, J. D. (1994). *NRL Plasma Formulary*. Naval Research Laboratory, Washington DC.

Hutchinson, I. H. (1987). *Principles of Plasma Diagnostics*. Cambridge University Press, Cambridge.

Jackson, J. D. (1975). *Classical Electrodynamics*. 2nd edn. Wiley, New York.

Jiang, Z., Kieffer, J. C., Matte, J. P., Chaker, M., Peyrusse, O., Gilles, D., Korn, G., Maksimchuk, A., Coe, S., and Mourou, G. (1995). X-ray spectroscopy of hot solid density plasmas produced by subpicosecond high contrast laser pulses at 10^{18}–10^{19} Wcm^{-2}. *Phys. Plasmas*, **2**, 1702–1711.

Joshi, C., and Katsouleas, T. (2003). Plasma accelerators at the energy frontier and on tabletops. *Physics Today*, **56**, 47–53.

Kalashnikov, M. P., Nickles, P. V., Schlegel, Th., Schnürer, M., Billhardt, F., Will, I., Sandner, W., and Demchenko, N. N. (1994). Dynamics of laser-plasma interaction at 10^{18} Wcm^{-2}. *Phys. Rev. Lett.*, **73**, 260–263.

Kapteyn, H. C. (1992). Photoionization-pumped x-ray lasers using ultrashort-pulse excitation. *Appl. Opt.*, **31**, 4931–4939.

Karttunen, S. J., and Salomaa, R. R. E. (1986). Electromagnetic field cascading in the beat-wave generation of plasma waves. *Phys. Rev. Lett.*, **56**, 604–607.

Karttunen, S. J., and Salomaa, R. R. E. (1989). Role of plasmon detuning in beat wave acceleration. *Phys. Scr.*, **39**, 741–748.

Kasparian, J., Rodriguez, M., Méjean, G., Yu, J., Salmon, E., Wille, H., Bourayou, R., Frey, S., André, Y.-B., Mysyrowicz, A., Sauerbrey, R., Wolf, J.-P., and Wöste, L. (2003). White-light filaments for atmospheric analysis. *Science*, **301**, 61–64.

Kato, S., Bhattacharyya, B, Nishiguchi, A., and Mima, K. (1993). Wave breaking and absorption efficiency for short pulse p-polarized laser light in a very steep density gradient. *Phys. Fluids B*, **5**, 564–570.

Kato, Y. 2001 (May 28–30). *Workshop on compact high-intensity short-pulse lasers: future directions and applications.* Tech. rept. OECD Global Science Forum, Paris. http://www.oecd.org/dataoecd/8/34/2451461.pdf.

Katsouleas, T., and Dawson, J. M. (1983). Laser surfatron accelerator. *Phys. Rev. Lett.*, **51**, 392–395.

Katsouleas, T., and Mori, W. B. (1988). Wave-breaking amplitude of relativistic oscillations in a thermal plasma. *Phys. Rev. Lett.*, **61**, 90–93.

Katsouleas, T., Bingham, R., and (Eds.) (1996). Special issue on 2nd generation plasma accelerators. *IEEE Trans. Plas. Sci.*

Kaw, P., and Dawson, J. (1970). Relativistic nonlinear propagation of laser beams in cold overdense plasmas. *Phys. Fluids*, **13**, 472–481.

Keldysh, L. V. (1965). Ionization in the field of a strong electromagnetic wave. *Sov. Phys. JETP*, **20**, 1307.

Kieffer, J. C., Matte, J. P., Bélair, S., Chaker, M., Audebert, P., Pépin, H., Maine, P., Strickland, D., Bado, P., and Mourou, G. (1989a). Absorption of an ultrashort laser pulse in very steep plasma density gradients. *IEEE J. Quantum Electronics*, **25**, 2640–2647.

Kieffer, J. C., Audebert, P., Chaker, M., Matte, J.P., Johnston, T. W., Maine, P., Delettrez, J., Strickland, D., Bado, P., and Mourou, G. (1989b). Short-pulse laser absorption in very steep plasma density gradients. *Phys. Rev. Lett.*, **62**, 760–763.

Kieffer, J. C., Chaker, M., Matte, J. P., Pépin, H., Côté, C. Y., Beaudoin, Y., Johnston, T. W., Cien, C. Y., Coe, S., Mourou, G., and Peyrusse, O. (1993). Ultrafast x-ray sources. *Phys. Fluids B*, **5**, 2676–2681.

Kieffer, J. C., Krol, A., Jiang, Z., Chamberlain, C. C., Scalzetti, E., and Ichalalene, Z. (2002). Future of laser-based x-ray sources for medical imaging. *Appl. Phys. B*, **74**, S75–S81.

Kingham, R., and Bell, A. R. (1997). Enhanced wakefields for the 1D laser wakefield accelerator. *Phys. Rev. Lett.*, **79**, 4810–4813.

Kingham, R. J., Gibbon, P., Theobald, W., Veisz, L., and Sauerbrey, R. (2001). Phase modulation ultra-short pulses reflected from high density plasmas. *Phys. Rev. Lett.*, **86**, 810–813.

Kishimoto, Y., Mima, K., Watanabe, T., and Nishikawa, K. (1983). Analysis of fast-ion velocity distributions in laser plasmas with a truncated Maxwellian velocity distrbution of hot electrons. *Phys. Fluids*, **26**, 2308–2315.

Kitagawa, Y., Matsumoto, T., Minamihata, T., Sawai, K., Matsuo, K., Mima,

K., Nishihara, K., Azechi, H., Tanaka, K. A., Takabe, H., and Nakai, S. (1992). Beat-wave excitation of plasma wave and observation of accelerated electrons. *Phys. Rev. Lett.*, **68**, 48–51.

Klem, D. E., Darrow, C., Lane, S., and Perry, M. D. (1993). Absorption and x-ray measurements from ultra-intense laser-plasma interactions. *Pages 98–102 of:* Baldis, H. A. (ed), *Short-pulse high-intensity lasers and applications II*, vol. 1860. SPIE, Bellingham, USA.

Kmetec, J. D., Gordon, C. L., Macklin, J. J., Lemoff, B. E., Brown, G. S., and Harris, S. E. (1992). MeV X-ray generation with a femtosecond laser. *Phys. Rev. Lett.*, **68**, 1527–1530.

Kodama, R. (2002). Fast heating scalable to laser fusion ignition. *Nature*, **418**, 933–934.

Kodama, R., Norreys, P. A., Mima, K., Dangor, A. E., Evans, R. G., Fujita, H., Kitagawa, Y., Krushelnick, K., Miyakoshi, T., Miyanaga, N., Norimatsu, T., Rose, S. J., Shozaki, T., Shigemori, K., Sunahara, A., Tampo, M., Tanaka, K. A., Toyama, Y., Yamanaka, T., and Zepf, M. (2001). Fast heating of ultrahigh-density plasma as a step towards laser fusion ignition. *Nature*, **412**, 798–802.

Kohlweyer, S., Tsakiris, G. D., Wahlström, C.-G., Tillman, C., and Mercer, I. (1995). Harmonic generation from solid-vacuum interface irradiated at high laser intensities. *Op. Commun.*, **117**, 431–438.

Kong, Q., Ho, Y. K., Wang, J. X., Wang, P. X., Feng, L., and Yuan, Z. S. (2000). Conditions for electron capture by an ultraintense stationary laser beam. *Phys. Rev. E*, **61**, 1981–1984.

Krall, N. A., and Trivelpiece, A. W. (1986). *Principles of Plasma Physics*. McGraw-Hill, New York.

Krause, J. L., Schafer, K. J., and Kulander, K. C. (1992). High-order harmonic generation from atoms and ions in the high intensity regime. *Phys. Rev. Lett.*, **68**, 3535–3538.

Kruer, W. L. (1988). *The Physics of Laser Plasma Interactions*. Addison-Wesley, New York.

Kruer, W. L., and Estabrook, K. (1985). J x B heating by very intense laser light. *Phys. Fluids*, **28**, 430–432.

Kruit, P., Kimman, J., Muller, H. G., and Vanderwiel, M. J. (1983). Electron spectra from multiphoton ionization of xenon at 1064, 532 and 355 nm. *Phys. Rev. A*, **28**, 248–255.

Krushelnick, K., Ting, A., Moore, C. I., Burris, H. R., Esarey, E., Sprangle, P., and Baine, M. (1997). Plasma channel formation and guiding during high intensity short pulse laser plasma experiments. *Phys. Rev. Lett.*, **78**, 4047–4050.

Krushelnick, K., Clark, E. L., Najmudin, Z., Salvati, M., Santala, M. I. K., Tatarakis, M., Dangor, A. E., Malka, V., Neely, D., Allot, R., and Danson, C. (1999). Multi-MeV ion production from high-intensity laser interactions with underdense plasmas. *Phys. Rev. Lett.*, **83**, 737–740.

Kuhlke, D., Herpes, U., and von der Linde, D. (1987). Soft X-ray emission from subpicosecond laser-produced plasmas. *Appl. Phys. Lett*, **50**, 1785.

Kull, H. (1983). Linear mode conversion in laser plasmas. *Phys. Fluids*, **26**,

1881–1887.

Lambropoulos, P. (1985). Mechanisms for multiple ionization of atoms by strong pulsed lasers. *Phys. Rev. Lett.*, **55**, 2141–2144.

Landau, L. D., and Lifshitz, L. M. (1962). *Classical Theory of Fields*. 2nd edn. Addison-Wesley, Reading, Mass.

Landen, O. L., Stearns, D. G., and Campbell, E. M. (1989). Measurement of the expansion of picosecond laser-produced plasmas using resonance absorption profile spectroscopy. *Phys. Rev. Lett.*, **63**, 1475.

Last, I., and Jortner, J. (2001). Nuclear fusion induced by Coulomb explosion of heteronuclear clusters. *Phys. Rev. Lett.*, **87**, 033401.

Lawson, J. D. (1979). Lasers and accelerators. *IEEE Trans. Nuc. Sci.*, **26**, 4217–4219.

Le Blanc, S. P., Sauerbrey, R., Rae, S. C., and Burnett, K. (1993). Spectral blue shifting of a femtosecond laser pulse propagating through a high-pressure gas. *J. Opt. Soc. Am. B*, **10**, 1801–1809.

Ledingham, K. W. D., Spencer, I., McCanny, T., Singhal, R. P., Santala, M. I. K., Clark, E., Watts, I., Beg, F. N., Zepf, M., Krushelnick, K., Tatarakis, M., Dangor, A. E., Norreys, P. A., Allot, R., Neely, D., Clarke, R. J., Machacek, A. C., Wark, J. S., Cresswell, A. J., Sanderson, D. C. W., and Magill, J. (2000). Photonuclear physics when a multiterawatt laser pulse interacts with solid targets. *Phys. Rev. Lett.*, **84**, 899–902.

Lee, K., Forslund, D. W., Kindel, J. M., and Lindman, E. L. (1976). Theoretical derivation of laser induced plasma profiles. *Phys. Fluids*, **20**, 51–54.

Lee, Y. T. (1987). A model for ionization balance and L-shell spectroscopy of non-LTE plasmas. *J. Quant. Spectrosc. Radiat. Transfer*, **38**, 131–145.

Leemans, W., Clayton, C. E., Mori, W. B., Marsh, K. A., Kaw, P. K., Dyson, A., Joshi, C., and Wallace, J. M (1992). Experiments and simulations of tunnel-ionized plasmas. *Phys. Rev. A*, **46**, 1091–1105.

Leemans, W. P., Shoenlein, R. W., Volfbeyn, P., Chin, A. H., Glover, T. E., Balling, P., Zolotorev, M., Kim, K. J., Chattopadhyay, S., and Shank, C. V. (1997). Interaction of relativistic electrons with ultrashort laser pulses: Generation of femtosecond X-rays and microprobing of electron beams. *IEEE J. Quant. Elec.*, **33**, 1925–1934.

Lefebvre, E., and Bonnaud, G. (1995). Transparence/opacity of a solid target illuminated by an ultrahigh-intensity laser pulse. *Phys. Rev. Lett.*, **74**, 2002–2005.

L'Huillier, A., and Balcou, Ph. (1993). High-order harmonic generation in rare gases with a 1-ps 1053-nm laser. *Phys. Rev. Lett.*, **70**, 774–777.

Lichters, R., Meyer-ter-Vehn, J., and Pukhov, A. (1996). Short-pulse laser harmonics from oscillating plasma surfaces driven at relativistic intensity. *Physics of Plasmas*, **3**, 3425–3437.

Limpouch, J., Drska, L., and Liska, R. (1994). Fokker-Planck simulations of interactions of femtosecond laser pulses with dense plasmas. *Laser and Part. Beams*, **12**, 101–110.

Lindl, J. (1995). Development of the indirect-drive approach to inertial confinement fusion and the target physics basis for ignition and gain. *Phys. Plasmas*,

2, 3933–4024.

Lindl, J. (1997). *Inertial Confinement Fusion: The Quest for Ignition and Energy Gain Using Indirect Drive*. AIP Press, New York.

Litvak, A. G. (1970). Finite-amplitude wave beams in a magnetoactive plasma. *Sov. Phys. JETP*, **30**, 344–347.

Liu, X., Umstadter, D., Esarey, E., and Ting, A. (1993). Harmonic generation by an intense laser pulse in neutral and ionized gases. *IEEE Trans. Plas. Sci.*, **21**, 90–93.

London, R. A., Rosen, M. D., and Trebes, J. E. (1989). Wavelength choice for soft x-ray laser holography of biological samples. *Appl. Opt.*, **28**, 3397–3404.

Luther-Davies, B., Perry, A., and Nugent, K. A. (1987). K_α measurements and superthermal electron transport in layered laser-irradiated disk targets. *Phys. Rev. A*, **35**, 4306–4313.

Mackinnon, A. J., Allfrey, S., Borghesi, M., Iwase, A., Willi, O., and Walsh, F. (1995). Optical guiding of 10 TW picosecond laser pulses in preformed density channels at high intensities and densities. *Pages 337–340 of:* Rose, S. J. (ed), *Laser interaction with matter*, vol. 140. Inst. Phys. Conf. Ser., Bristol.

Magill, J., Schwoerer, H., Ewald, F., Galy, J., Schenkel, R., and Sauerbrey, R. (2003). Laser transmutation of iodine-129. *Appl. Phys. B*, **77**, 387–390.

Maine, P., Strickland, D., Bado, P., Pessot, M., and Mourou, G. (1988). Generation of ultrahigh peak power pulses by chirped-pulse amplification. *IEEE J. Quantum Electron.*, **24**, 398–403.

Mainfray, G., and Manus, C. (1991). Multiphoton ionization of atoms. *Rep. Mod. Phys.*, **54**, 1333–1372.

Maksimchuk, A., Gu, S., Flippo, K., Umstadter, D., and Bychenkov, V. Yu. (2000). Forward ion acceleration in thin films driven by a high-intensity laser. *Phys. Rev. Lett.*, **84**, 4108–4111.

Maldonado, J. R. (1995). Prospects for granular X-ray lithography sources. *Pages 2–22 of:* Richardson, M. C., and Kyrala, G. A. (eds), *Applications of Laser Plasma Radiation II*, vol. 2523. SPIE, Bellingham, USA.

Malka, G., Fuchs, J., Amiranoff, F., Baton, S. D., Gaillard, R., Miquel, J. L., Pépin, H., Rousseaux, C., Bonnaud, G., Busquet, M., and Lours, L. (1997). Suprathermal electron generation and channel formation by an ultrarelativistic laser pulse in an underdense plasma. *Phys. Rev. Lett.*, **79**, 2053–2056.

Malka, V., Fritzler, S., Lefebvre, E., Aleonard, M.-M., Burgy, F., Chambaret, J.-P., Chemin, J.-F., Krushelnick, K., Malka, G., Mangles, S. P. D., Najmudin, Z., Pittman, M., Rousseau, J.-P., Scheurer, J.-N., Walton, B., and Dangor, A. E. (2002). Electron acceleration by a wake field forced by an intense ultrashort laser pulse. *Science*, **298**, 1596–1600.

Malkin, V. M., and Fisch, N. J. (2002). Collective deceleration of relativistic electrons precisely in the core of an inertial-fusion target. *Phys. Rev. Lett.*, **89**, 125004.

Malkin, V. M., Shvets, G., and Fisch, N. J. (2000). Ultra-powerful compact amplifiers for short laser pulses. *Phys. Plasmas*, **7**, 2232–2240.

Mangles, S. P. D., Murphy, C. D., Najmudin, Z., Thomas, A. G. R., Collier, J. L., Dangor, A. E., Divall, E. J., Forster, P. S., Gallacher, J. G., Hooker,

C. J., Jaroszynski, D. A., Langley, A. J., Mori, W. B., Norreys, P. A., Tsung, F. S., Viskup, R., Walton, B. R., and Krushelnick, K. (2004). Monoenergetic beams of relativistic electrons from intense laser-plasma interactions. *Nature*, **431**, 535–538.

Marchand, R., Rankin, R., Capjack, C. E., and Birnboim, A. (1987). Diffraction, self-focusing, and the geometrical optics limit in laser-produced plasmas. *Phys. Fluids*, **30**, 1521–1525.

Mason, R. J., and Tabak, M. (1998). Magnetic field generation in high-intensity laser-matter interactions. *Phys. Rev. Lett.*, **80**, 524–527.

Mathews, J., and Walker, R. L. (1970). *Mathematical Methods of Physics*. 2nd edn. Addison-Wesley, Redwood City, CA.

Matte, J. P., and Aguenaou, K. (1992). Numerical studies of the anomalous skin effect. *Phys. Rev. A*, **45**, 2558–2566.

Matte, J. P., Kieffer, J. C., Ethier, S., Chaker, M., and Peyrusse, O. (1994). Spectroscopic signature of non-maxwellian and nonstationary effects in plasmas heated by intense, ultrashort laser pulses. *Phys. Rev. Lett.*, **72**, 1208–1211.

Max, C. E., Arons, J., and Langdon, A. B. (1974). Self-modulation and self-focusing of electromagnetic waves in plasmas. *Phys. Rev. Lett.*, **33**, 209–212.

Maxon, S., Estabrook, K. G., Prasad, M. K., Osterheld, A. L., London, R. A., and Eder, D. C. (1993). High gain X-ray lasers at the water window. *Phys. Rev. Lett.*, **70**, 2285–2288.

McDonald, K. T. (1998). Comment on Malka et al., PRL 78, 3314 (1997). *Phys. Rev. Lett.*, **80**, 1350.

McKenna, P., Ledingham, K. W. D., McCanny, T., Singhal, R. P., Spencer, I., Santala, M. I. K., Beg, F. N., Krushelnick, K., Tatarakis, M., Wei, M. S., Clark, E. L., Clarke, R. J., Lancaster, K. L., Norreys, P. A., Spohr, K., Chapman, R., and Zepf, M. (2003). Demonstration of fusion-evaporation and direct-interaction nuclear reactions using high-intensity laser-plasma-accelerated ion beams. *Phys. Rev. Lett.*, **91**, 075006.

McKinstrie, C., and Forslund, D. W. (1986). The detuning of relativistic Langmuir waves in the beat-wave accelerator. *Phys. Fluids*, **30**, 904–908.

McKinstrie, C. J., and Bingham, R. (1992). Stimulated Raman forward scattering and the relativistic modulational instability of light waves in rarefied plasma. *Phys. Fluids B*, **4**, 2626–2633.

McKinstrie, C. J., and DuBois, D. F. (1988). A covariant formalism for wave propagation applied to stimulated Raman scattering. *Phys. Fluids*, **31**, 278–287.

McKinstrie, C. J., and Russell, D. A. (1988). Nonlinear focusing of coupled waves. *Phys. Rev. Lett.*, **61**, 2929–2932.

McLean, E. A., Stamper, J. A., Ripin, B. H., Griem, H. R., McMahon, J. M., and Bodner, S. E. (1978). Harmonic generation in Nd:laser-produced plasmas. *Appl. Phys. Lett.*, **31**, 825–827.

McPherson, A., Gibson, G., Jara, H., Johann, U., Luk, T. S., McIntyre, I. A., Boyer, K., and Rhodes, C. K. (1987). Studies of multiphoton production of vacuum ultraviolet-radiation in the rare gases. *J. Op. Soc. Am. B*, **4**, 595–601.

McPherson, A., Thompson, B. D., Borisov, A. B., Boyer, K., and Rhodes, C. K.

(1994). Multiphoton-induced x-ray emission at 4–5 keV from xenon atoms with multiple core vacancies. *Nature*, **370**, 631–634.

Meyer-ter-Vehn, J. (1982). On energy gain of fusion targets: the model of Kidder and Bodner improved. *Nuc. Fusion*, **22**, 561–565.

Meyerhofer, D. D. (1997). High-intensity laser-electron scattering. *IEEE J. Quantum Elec.*, **33**, 1935–1941.

Meyerhofer, D. D., Chen, H., Delettrez, J. A., Soom, B., Uchida, S., and Yaakobi, B. (1993). Resonance absorption in high-intensity contrast, picosecond laser-plasma interactions. *Phys. Fluids B*, **5**, 2584–2588.

Miano, G., and de Angelis, U. (1989). Saturation and cross-field coupling of beat-driven 3-D plasma waves. *Plas. Phys. Contr. Fus.*, **31**, 1383–1389.

Milchberg, H. M., and Freeman, R. R. (1989). Light absorption in ultrashort scale length plasmas. *J. Opt. Soc. Am. B*, **6**, 1351–1355.

Milchberg, H. M., Freeman, R. R., Davey, S. C., and More, R. M. (1988). Resistivity of a simple metal from room temperature to 1000000K. *Phys. Rev. Lett.*, **61**, 2364–2367.

Milchberg, H. M., Lyubomirsky, I., and Durfee III, C. G. (1991). Factors controlling the x-ray pulse emission from an intense femtosecond laser heated solid. *Phys. Rev. Lett.*, **67**, 2654–2657.

Millat, T., Selchow, A., Wierling, A., Reinholz, H., Redmer, R., and Röpke, G. (2003). Dynamic collsion frequency for a two-component plasma. *J. Phys. A*, **36**, 6259–6264.

Millikan, R. A. (1916). Einstein's photoelectric equation and contact electromotive force. *Physical Review*, **7**, 18–32.

Modena, A., Najmudin, Z., Dangor, A. E., Clayton, C. E., Marsh, K. A., Joshi, C., Malka, V., Darrow, C. B., Danson, C., Neely, D., and Walsh, F. N. (1995). Electron acceleration from the breaking of relativistic plasma waves. *Nature*, **377**, 606–608.

Mohideen, U., Tom, H. W. K., Freeman, R. R., Bokor, J., and Bucksbaum, P. H. (1992). Interaction of free electrons with an intense focused laser pulse in Gaussian and conical axicon geometries. *J. Opt. Soc. Am. B*, **9**, 2190–2195.

Monot, P., Auguste, T., Gibbon, P., Jackober, F., and Mainfray, G. (1995a). Collimation of an intense laser beam by a weakly relativistic plasma. *Phys. Rev. E*, **52**, R5780–R5783.

Monot, P., Auguste, T., Gibbon, P., Jackober, F., Mainfray, G., Dulieu, A., Louis-Jacquet, M., Malka, G., and Miquel, J. L. (1995b). Experimental demonstration of relativistic self-channeling of a multi-terawatt laser pulse in an underdense plasma. *Phys. Rev. Lett.*, **74**, 2953–2956.

Montgomery, D., and Tidman, P. (1964). Secular and nonsecular behavior for the cold plasma equations. *Phys. Fluids*, **7**, 242–249.

Moore, C. I., Knauer, J. P., and Meyerhofer, D. D. (1995). Observation of the transition from Thomson to Compton scattering in multiphoton interactions with low-energy electrons. *Phys. Rev. Lett.*, **74**, 2439–2442.

Moore, C. I., Ting, A., Krushelnick, K., Esarey, E., Hubbard, R. F., Hafizi, B., Burris, H. R., Manka, C., and Sprangle, P. (1997). Electron trapping in self-modulated laser wakefields by Raman backscatter. *Phys. Rev. Lett.*, **79**,

3909–3912.

Mora, P. (1988). Utilisation des plasmas pour l'accélération de particules à très haute énergie. Revue Phys. Appl., 23, 1489–1494.

Mora, P. (2003). Plasma expansion into a vacuum. Phys. Rev. Lett., 90, 185002.

Mora, P., and Quesnel, B. (1998). Comment on Malka et al., PRL 78, 3314 (1997). Phys. Rev. Lett., 80, 1352.

Mora, P., Pesme, D., Héron, A., Laval, G., and Silvestre, N. (1988). Modulational instability and its consequences for the beat-wave accelerator. Phys. Rev. Lett., 61, 1611–1614.

More, R. M. (1979). Quantum-statistical model for high-density matter. Phys. Rev. A, 19, 1234–1246.

More, R. M., Zinamon, Z., Warren, K. H., Falcone, R., and Murnane, M. (1988a). Heating of solids with ultra-short laser pulses. J. de Physique, 49 C7-, 43–51.

More, R. M., Warren, K. H., Young, D. A., and Zimmerman, G. B. (1988b). A new quotidian equation of state (QEOS) for hot dense matter. Phys. Fluids, 31, 3059–3078.

Mori, W. B. (1994). Overview of laboratory plasma radiation sources. Physica Scripta, T52, 28–35.

Mori, W. B. (1997). The physics of the nonlinear optics of plasmas at relativistic intensities for short-pulse lasers. IEEE J. Quan. Elec., 33, 1942–1953.

Mori, W. B., Joshi, C., Dawson, J. M., Forslund, D. W., and Kindel, J. M. (1988). Evolution of self-focusing of intense electromagnetic waves in plasma. Phys. Rev. Lett., 60, 1298–1301.

Mori, W. B., Decker, C. D., and Leemans, W. P. (1993). Relativistic harmonic content of nonlinear electromagnetic waves in underdense plasmas. IEEE Trans. Plasma Sci., 21, 110–119.

Mori, W. B., Decker, C. D., Hinkel, D. E., and Katsouleas, T. (1994). Raman forward scattering of short-pulse high-intensity lasers. Phys. Rev. Lett., 72, 1482–1485.

Mulser, P. (1979). Laser radiation transport and ponderomotive force. Pages 91–144 of: Cairns, R. A., and Sanderson, J. J. (eds), Laser-Plasma Interactions, vol. 1. SUSSP, Edinburgh.

Mulser, P., Pfalzner, S., and Cornolti, F. (1989). Plasma production with intense femtosecond laser pulses. Pages 142–145 of: Velarde, G., Minguez, E., and Perlado, J. M. (eds), Laser Interaction with Matter. World Scientific, Singapore.

Murnane, M. M., Kapteyn, H. C., and Falcone, R. W. (1989). High-density plasmas produced by ultrafast laser pulses. Phys. Rev. Lett., 62, 155–158.

Murnane, M. M., Kapteyn, H. C., Rosen, M. D., and Falcone, R. W. (1991). Ultrafast X-ray pulses from laser-produced plasmas. Science, 251, 531–536.

Nakajima, K., Fisher, D., Kawakubo, T., Nakanishi, H, Ogata, A., Kato, Y., Kitagawa, Y., Kodama, R., Mima, K., Shiraga, H., Suzuki, K., Yamakawa, K., Zhang, T., Sakawa, Y., Shoji, T., Nishida, Y., Yugami, N., Downer, M., and Tajima, T. (1995). Observation of ultrahigh gradient electron acceleration by a self-modulated intense short laser pulse. Phys. Rev. Lett., 74, 4428–4431.

Naumova, N. M., Bulanov, S. V., Esirkepov, T. Zh., Farina, D., Nishihara, K., Pegoraro, F., Ruhl, H., and Sakharov, A. S. (2001). Formation of electromagnetic postsolitons in plasmas. *Phys. Rev. Lett.*, **87**, 185004.

Nemoto, K., Maksimchuk, A., Banerjee, S., Flippo, K., Mourou, G., Umstadter, D., and Bychenkov, V. Yu. (2001). Laser-triggered ion acceleration and table top isotope production. *Appl. Phys. Lett.*, **78**, 595–597.

Ng, A., Celliers, P., Forsman, A., More, R. M., Lee, Y. T., Perrot, F., Dharma-wardana, M. W. C., and Rinker, G. A. (1994). Reflectivity of intense femtosecond laser pulses from a simple metal. *Phys. Rev. Lett.*, **72**, 3351–3354.

Nicholas, D. J. (1982). The development of fluid codes for the laser compression of plasma. *Pages 129–183 of:* Cairns, R. A. (ed), *Laser-Plasma Interactions*, vol. 2. SUSSP, Edinburgh.

Nickles, P. V., v. N. Shlyaptsev, Kalachnikov, M., Schnürer, M., Will, I., and Sandner, W. (1997). Short pulse x-ray laser at 32.6 nm based on transient gain in Ne-like titanium. *Phys. Rev. Lett.*, **78**, 2748–2751.

Noble, R. J. (1984). Plasma wave generation in the beat-wave accelerator. *Phys. Rev. A*, **32**, 460–471.

Norreys, P. A., Zepf, M., Moustaizis, S., Fews, A. P., Zhang, J., Lee, P., Bakarezos, M., Danson, C. N., Dyson, A., Gibbon, P., Loukakos, P., Neely, D., Walsh, F. N., Wark, J. S., and Dangor, A. E. (1996). Efficient XUV harmonics generated from picosecond laser pulse interactions with solid targets. *Phys. Rev. Lett.*, **76**, 1832–1835.

Norreys, P. A., Santala, M., Clark, E., Zepf, M., Watts, I., Beg, F. N., Krushelnick, K., Tatarakis, M., Dangor, A. E., Fang, X., Graham, P., McCanny, T., Singhal, R. P., Ledingham, K. W. D., Creswell, A., Sanderson, D. C. W., Magill, J., Machacek, A., Wark, J. S., Allott, R., Kennedy, B., and Neely, D. (1999). Observation of a highly directional γ-ray beam from ultrashort, ultraintense laser pulse interactions with solids. *Phys. Plasmas*, **6**, 2150–2156.

Norreys, P. A., Lancaster, K. L., Murphy, C. D., Habara, H., Karsch, S., Clarke, R. J., Collier, J., Heathcote, R., Hemandez-Gomez, C., Hawkes, S., Neely, D., Hutchinson, M. H. R., Evans, R. G., Borghesi, M., Romagnani, L., Zepf, M., Akli, K., King, J. A., Zhang, B., Freeman, R. R., MacKinnon, A. J., Hatchett, S. P., Patel, P., Snavely, R., Key, M. H., Nikroo, A., Stephens, R., Stoeckl, C., Tanaka, K. A., Norimatsu, T., Toyama, Y., and Kodama, R. (2004). Integrated implosion/heating studies for advanced fast ignition. *Phys. Plasmas*, **11**(5), 2746–2753.

Nuckolls, J. H., Wood, L., Thiessen, A., and Zimmerman, G. B. (1972). Laser compression of matter to super-high densities: thermonuclear (CTR) applications. *Nature*, **239**, 129.

Offenberger, A. A., Blyth, W., Dangor, A. E., Djaoui, A., Key, M. H., Najmudin, Z., and Wark, J. S. (1993). Electron temperature of optically ionized gases produced by high intensity 268nm radiation. *Phys. Rev. Lett.*, **71**, 3983–3986.

Ostriker, J. P., and Gunn, J. E. (1969). On the nature of pulsars. I. Theory. *Astrophys. J.*, **157**, 1395–1417.

Patel, P. K., Mackinnon, A. J., Key, M. H., Cowan, T. E., Foord, M. E., Allen, M., Price, D. F., Ruhl, H., Springer, P. T., and Stephens, R. (2004). Isochoric heating of solid-density matter with an ultrafast proton beam. *Phys. Rev. Lett.*,

91, 125004.

Pearlman, J. S., and Morse, R. L. (1978). Maximum expansion velocities of laser-produced plasmas. *Phys. Rev. Lett.*, **40**, 1652–1655.

Pegoraro, F., Bulanov, S. V., Califano, F., and Lontano, M. (1996). Nonlinear development of the Weibel instability and magnetic field generation in collisionless plasmas. *Phys. Scr.*, **T63**, 262–265.

Penetrante, B. M., and Bardsley, J. N. (1991). Residual energy in plasmas produced by intense subpicosecond lasers. *Phys. Rev. A*, **43**, 3100.

Perelomov, A. M., Popov, V. S., and Terentev, M. V. (1966). Ionization of atoms in an alternating electric field. *Sov. Phys. JETP*, **23**, 924.

Pert, G. J. (1979). Fluid Codes. *Pages 323–386 of:* Cairns, R. A., and Sanderson, J. J. (eds), *Laser-Plasma Interactions*, vol. 1. SUSSP, Edinburgh.

Pert, G. J. (1988). Computer simulation of laser-produced plasmas. *Pages 193–211 of:* Hooper, M. B. (ed), *Laser-Plasma Interactions*, vol. 4. SUSSP, Edinburgh.

Pert, G. J. (1995). Inverse bremsstrahlung in strong radiation fields at low temperatures. *Phys. Rev. E*, **51**, 4778–4789.

Pesme, D., Karttunen, S. J., Salomaa, R. R. E., Laval, G., and Silvestre, N. (1988). Modulational instability in the beat wave accelerator. *Laser Part. Beams*, **6**, 199–210.

Petsko, G. A., and Ringe, D. (1984). Fluctuations in protein structure from x-ray diffraction. *Ann. Rev. Biophys. Bioeng.*, **13**, 331–371.

Pfalzner, S. (1991). Statistical description of ionization potentials in dense plasmas. *Appl. Phys. B*, **53**, 203–206.

Pfalzner, S. (1992). Field ionization in dense plasmas. *J. Phys. B: At. Mol. Opt. Phys.*, **25**, L545–L549.

Pfalzner, S., and Gibbon, P. (1998). Direct calculation of inverse-bremsstrahlung absorption in strongly coupled, nonlinearly driven laser plasmas. *Phys. Rev. E*, **57**, 4698–4705.

Pickwell, E., Cole, B. E., Fitzgerald, A. J., Pepper, M., and Wallace, V. P. (2004). *In vivo* study of human skin using pulse terahertz radiation. *Phys. Med. Biol.*, **49**, 1595–1607.

Pisani, F., Bernardinello, A., Batani, D., Antonicci, A., Martinolli, E., andL. Gremillet, M. Koenig, Amiranoff, F., Baton, S., Davis, J., Hall, T. A., Scott, D., Norreys, P., Djaoui, A., Rousseaux, C., Fews, P., Bandulet, H., and Pepin, H. (2000). Experimental evidence of electric inihibition in fast electron penetration and of electric-field-limited fast electron transport in dense matter. *Phys. Rev. E*, **62**.

Playa, L., Roso, L., Rzazewski, K., and Lewenstein, M. (1998). Generation of attosecond pulse trains during the reflection of a very intense laser on a solid surface. *J. Op. Soc. Am. B*, **15**, 1904–1911.

Popov, V. S. (2002). The Schwinger effect and possibilities for its observation using optical and x-ray lasers. *J. Exp. Theor. Phys.*, **94**, 1057–1069.

Potter, D. (1972). *Computational Physics*. John Wiley, New York.

Press, W. H., Flannery, B. P., Teukolsky, S. A., and Vetterling, W. T. (1989). *Numerical Recipes: The Art of Acientific Computing*. Cambridge University

Press, Cambridge.

Preston, S. G., Sanpera, A., Zepf, M., Blyth, W. J., Smith, C. G., Wark, J. S., Key, M. H., Burnett, K., Nakai, M., Neely, D., and Offenberger, A. A. (1996). High-order harmonics of 248.6-nm KrF laser from helium and neon ions. *Phys.Rev. A*, **53**(1), R31–R34.

Pretzler, G., Saemann, A., Pukhov, A., Rudolph, D., Schätz, T., Schramm, U., Thirolf, P., Habs, D., Eidmann, K., Tsakiris, G. D., Meyer-ter-Vehn, J., and Witte, K. J. (1998). Neutron production by 200mJ ultrashort laser pulses. *Phys. Rev. E*, **58**, 1165–1168.

Pretzler, G., Schlegel, T., Fill, E., and Eder, D. (2000). Hot electron generation in copper and photopumping of cobalt. *Phys. Rev. E*, **62**, 5618–5623.

Price, D. F., More, R. M., Walling, R. S., Guethlein, G., Shepherd, R. L., Stewart, R. E., and White, W. E. (1995). Absorption of ultrashort laser pulses by solid targets heated rapidly to temperatures 1–1000eV. *Phys. Rev. Lett.*, **75**, 252–255.

Priedhorsky, W., Lier, D., Day, R., and Gerke, D. (1981). Hard-X-ray measurements of 10.6-μm laser-irradiated targets. *Phys. Rev. Lett.*, **47**, 1661–1664.

Protopapas, M., Lappas, D. G., Keitel, C. H., and Knight, P. L. (1996). Recollisions, bremsstrahlung, and attosecond pulses from intense laser fields. *Phys. Rev. A*, **53**(5), R2933–R2936.

Protopapas, M., Keitel, C. H., and Knight, P. L. (1997). Atomic physics with super-high intensity lasers. *Rep. Prog. Phys.*, **60**, 389–486.

Pukhov, A. (2001). Three-dimensional simulations of ion acceleration from a foil irradiated by a short-pulse laser. *Phys. Rev. Lett.*, **86**, 3562–3565.

Pukhov, A. (2003). Strong field interaction of laser radiation. *Rep. Prog. Phys.*, **66**, 47–101.

Pukhov, A., and Meyer-ter-Vehn, J. (1996). Relativistic magnetic self-channeling of light in plasma. Three-dimensional PIC simulation. *Phys. Rev. Lett.*, **76**, 3975–3978.

Pukhov, A., and Meyer-ter-Vehn, J. (1997). Laser hole boring into overdense plasma and relativistic electron currents for fast ignition of ICF targets. *Phys. Rev. Lett.*, **79**, 2686–2689.

Quesnel, B., Mora, P., Adam, J. C., Gu'erin, S., H'eron, A., and Laval, G. (1997). Electron parametric instabilities of ultraintense short laser pulses propagating in plasmas. *Phys. Rev. Lett.*, **78**, 2132–2135.

Rae, S. C. (1993). Ionization-induced defocusing of intense laser pulses in high-pressure gases. *Optics Commun.*, **97**, 25–28.

Rae, S. C., and Burnett, K. (1991). Reflectivity of steep-gradient plasmas in intense subpicosecond laser pulses. *Phys. Rev. A*, **44**, 3835–3840.

Rajeev, P. P., Taneja, P., Ayyub, P., Sandhu, A. S., and Kumar, G. R. (2003). Metal nanoplasmas as bright sources of hard x-ray pulses. *Phys. Rev. Lett.*, **90**, 115002.

Rax, J. M., and Fisch, N. J. (1992). Third harmonic generation with ultrahigh-intensity laser pulses. *Phys. Rev. Lett.*, **69**, 772–775.

Rax, J. M., and Fisch, N. J. (1993). Phase-matched third harmonic generation in a plasma. *IEEE Trans. Plasma Sci.*, **21**, 105–109.

Reich, Ch., Gibbon, P., Uschmann, I., and Förster, E. (2000). Yield optimisation and time-structure of femtosecond laser plasma $K\alpha$ sources. *Phys. Rev. Lett.*, **84**, 4846–4849.

Reich, Ch., Uschmann, I., Ewald, F., Düsterer, S., Lübcke, A., Schwoerer, H., Sauerbrey, R., Förster, E., and Gibbon, P. (2003). Spatial characteristics of x-ray emission from relativistic laser plasmas. *Phys. Rev. E*, **68**, 056408.

Ress, D., Da Silva, L. B., London, R. A., Trebes, J. E., Mrowka, S., Barbee Jr., T. W., and Lehr, D. E. (1994). Measurement of laser-plasma electron density with a soft x-ray deflectometer. *Science*, **265**, 514–517.

Richardson, M. C., McKenty, P. W., Marshall, F. J., Verdon, C. P., Soures, J. M., McCrory, R. L., Barnouin, O., Craxton, R. S., Delettrez, J., Hutchison, R. L., Jannimagi, P. A., Keck, R., Kessler, T., Kim, H., Letzring, S. A., Roback, D. M., Seka, W., Skupsky, S., Yaakobi, B., Lane, S. M., and Prussin, S. (1986). Ablatively driven targets imploded with the 24 UV beam OMEGA system. *Pages 421–448 of:* Schwartz, H., and Hora, H. (eds), *Laser Interaction and Related Plasma Phenomena*, vol. 7. Plenum Press, New York.

Richtmyer, R. D., and Morton, K. W. (1967). *Difference Methods for Initial Value Problems*. 2nd edn. Interscience, New York.

Ride, S.K., Esarey, E., and Baine, M. (1995). Thomson scattering of intense lasers from electron-beams at arbitrary interaction angles. *Phys. Rev. E*, **52**, 5425–5442.

Riley, D., Weaver, I., McSherry, D., Dunne, M., Neely, D., Notley, M., and Nardi, E. (2002). Direct observation of strong coupling in a dense plasma. *Phys. Rev. E*, **66**, 046408.

Rischel, C., Rousse, A., Uschmann, I., Albouy, P.-A., Geindre, J.-P., Audebert, P., Gauthier, J.-C., Förster, E., Martin, J.-L., and Antonetti, A. (1997). Femtosecond time-resolved X-ray diffraction from laser-heated organic films. *Nature*, **390**, 490–492.

Rochow, T. G., and Tucker, P. A. (1994). *Introduction to Microscopy by Means of Light, Electrons, X Rays or Acoustics*. 2nd edn. Plenum Press, New York.

Rose, S. J. (1988). Dense Plasma Physics. *Pages 171–192 of:* Hooper, M. B. (ed), *Laser-Plasma Interactions 4*. SUSSP, Edinburgh.

Rose-Petruck, C., Jimenez, R., Guo, T., Cavalleri, A., Siders, C. W., Raksi, F., Squier, J. A., Walker, B. C., Wilson, K. R., and Barty, C. P. J. (1999). Picosecond-milliangstrom lattice dynamics measured by ultrafast x-ray diffraction. *Nature*, **398**, 310–312.

Rosen, M. D. 1990. Scaling laws for femtosecond laser plasma interactions. *Pages 160–170 of:* Campbell, E. M. (ed), *Femtosecond to Nanosecond High-Intensity Lasers and Applications*, vol. 1229.

Rosenbluth, M. N., and Liu, C. S. (1972). Excitation of plasma waves by two laser beams. *Phys. Rev. Lett.*, **29**, 701–705.

Ross, I. N., Matousek, P., Towrie, M., Langley, A. J., and Collier, J. L. (1997). The prospects for ultrashort pulse duration and ultrahigh intensity using optical parametric chirped pulse amplifiers. *Opt. Commun.*, **144**, 125–133.

Roth, M., Cowan, T. E., Key, M. H., Hatchett, S. P., Brown, C., Fountain, W., Johnson, J., Pennington, D. M., Snavely, R. A., Wilks, S. C., Yasuike, K.,

Ruhl, H., Pegoraro, F., Bulanov, S. V., Campbell, E. M., Perry, M. D., and Powell, H. (2001). Fast ignition by intense laser-accelerated proton beams. *Phys. Rev. Lett.*, **86**, 436–439.

Rousse, A., Audebert, P., Geindre, J. P., Falliès, F., Gauthier, J. C., Mysyrowicz, A., Grillon, G., and Antonetti, A. (1994). Efficient K_α x-ray source from femtosecond laser-produced plasmas. *Phys. Rev. E*, **50**, 2200–2207.

Rousse, A., Rischel, C., and Gauthier, J.-C. (2001a). Colloquium: Femtosecond x-ray crystallography. *Rev. Mod. Phys.*, **73**, 17–31.

Rousse, A., Rishcel, C., Fourmaux, S., Uschmann, I., Sebban, S., Grillon, G., Balcou, P., Förster, E., Geindre, J.-P., Audebert, P., Gauthier, J.-C., and Hulin, D. (2001b). Non-thermal melting in semiconductors measured at femtosecond resolution. *Nature*, **410**, 65–68.

Rozmus, W., and Tikhonchuk, V. T. (1990). Skin effect and interaction of short laser pulses with dense plasmas. *Phys. Rev. A*, **42**, 7401.

Ruhl, H., and Mulser, P. (1995). Relativistic Vlasov simulation of intense fs laser pulse-matter interaction. *Physics Letters A*, **205**, 388–392.

Ruhl, H., Sentoku, Y., Mima, K., Tanaka, K. A., and Kodama, R. (1999). Collimated electron jets by intense laser-beam-plasma surface interaction under oblique incidence. *Phys. Rev. Lett.*, **82**, 743–746.

Ruhl, H., Bulanov, S. V., Cowan, T. E., Liseikina, T. V., Nickles, P., Pegoraro, F., Roth, M., and Sandner, W. (2001). Computer simulation of the three-dimensional regime of proton acceleration in the interaction of laser radiation with a thin spherical target. *Plasma Phys. Rep.*, **27**, 387–396.

Rymell, L., Berglund, M., and Hertz, H. M. (1995). Debris-free single-line laser-plasma x-ray source for microscopy. *Appl. Phys. Lett.*, **66**, 2625–2627.

Sakharov, A. S., and Kirsanov, V. I. (1994). Theory of Raman scattering for a short ultrastrong lser pulse in a rarefied plasma. *Phys. Rev. E*, **49**, 3274–3282.

Salamin, Y. I., and Keitel, C. H. (2000a). Analysis of electron acceleration in a vacuum beat wave. *J. Phys. B.*, **33**, 5057–5076.

Salamin, Y. I., and Keitel, C. H. (2000b). Subcycle high electron acceleration by crossed laser beams. *Appl. Phys. Lett.*, **77**, 1082–1084.

Salamin, Y. I., and Keitel, C. H. (2002). Electron acceleration by a tightly focused laser beam. *Phys. Rev. Lett.*, **88**, 095005.

Salieres, P., and Lewenstein, M. (2001). Generation of ultrashort coherent XUV pulses by harmonic conversion of intense laser pulses in gases: Towards attosecond pulses. *Meas. Sci. Tech.*, **12**, 1818–1827.

Salzmann, D. (1998). *Atomic Physics in Hot Plasmas*. Oxford University Press, New York.

San Roman, J., Roso, L., and Reiss, H. R. (2000). Evolution of a relativistic wavepacket describing a free electron in a very intense laser field. *J. Phys. B–At. Mol. Opt.*, **33**, 1869–1880.

Santala, M. I. K., Zepf, M., Watts, I., Beg, F. N., Clark, E., Tatarakis, M., Krushelnick, K., Dangor, A. E., McCanny, T., Spencer, I., Singhal, R. P., Ledingham, K. W. D., Wilks, S. C., Machacek, A. C., Wark, J. S., Allott, R., Clarke, R. J., and Norreys, P. A. (2000). Effect of the plasma density scale-length on the direction of fast electrons in relativistic laser-solid interactions.

Phys. Rev. Lett., **84**, 1459–1462.

Sarachik, E. S., and Schappert, G. T. (1970). Classical theory of the scattering of intense laser radiation by free electrons. Phys. Rev. D, 1, 2738–2753.

Sauerbrey, R., Fure, J., LeBlanc, S. P., van Wonterghem, B., Teubner, U., and Schäfer, F. P. (1994). Reflectivity of laser-produced plasmas generated by a high-intensity ultrashort pulse. Phys. Plasmas, 1, 1635–1642.

Schlanges, M., Bornath, T., Kremp, D., and Hilse, P. (2003). Quantum kinetic approach to transport processes in dense laser plasmas. Contr. Plas. Phys., **43**, 360–362.

Schnürer, M., Kalashnikov, M.P., Nickles, P.V., Schlegel, Th., Sandner, W., Demchenko, N., Nolte, R., and Ambrosi, P. (1995). Hard X-ray emission above from intense short pulse laser plasmas. Phys. Plasmas, **2**, 3106–3110.

Schnürer, M., Nolte, R., Schegel, T., Kalachnikov, M. P., Nickles, P. V., Ambrosi, P., and Sandner, W. (1997). On the distribution of hot electrons produced in short-pulse laser-plasma interaction. J. Phys. B, **30**, 4653–4661.

Schnürer, M., Spielmann, C., Wobrauschek, P., Streli, C., Burnett, N. H., Kan, C., Ferencz, K., Koppitsch, R., Cheng, Z., Brabec, T., and Krausz, F. (1998). Coherent 0.5-keV X-ray emission from helium driven by a sub-10-fs laser. Phys. Rev. Lett., **80**, 3236–3238.

Schnürer, M., Nolte, R., Rousse, A., Grillon, G., Cheriaux, G., Kalachnikov, M. P., Nickles, P. V., and Sandner, W. (2000). Dosimetric measurements of electron and photon yields from solid targets irradiated with 30fs pulses from a 14 TW laser. Phys. Rev. E, **61**, 4394–4401.

Schwinger, J. (1949). On the classical radiation of accelerated electrons. Physical Review, **75**, 1912–1925.

Schwinger, J. (1951). On gauge invariance and vacuum polarization. Phys. Rev., **82**, 664–679.

Schwoerer, H., Gibbon, P., Düsterer, S., Behrens, R., Ziener, C., Reich, C., and Sauerbrey, R. (2001). MeV x-rays and photoneutrons from femtosecond laser-produced plasmas. Phys. Rev. Lett., **86**, 2317–2320.

Schwoerer, H., Ewald, F., Sauerbrey, R., Galy, J., Magill, J., Rondinella, V., Schenkel, R., and Butz, T. (2003). Fission of actinides using a tabletop laser. Europhys. Lett., **61**, 47–52.

Sentoku, Y., Mima, K., Kaw, P., and Nishsikawa, K. (2003). Anomalous resistivity resulting from MeV-electron transport in overdense plasma. Phys. Rev. Lett., **90**, 155001.

Shen, B., and Yu, M. Y. (2002). High-intensity laser-field amplification between two foils. Phys. Rev. Lett., **89**, 275004.

Shen, Y. R. (1984). The Principles of Nonlinear Optics. John Wiley, New York.

Sheng, Z. M., Meyer-ter-Vehn, J., and Pukhov, A. (1998). Analytic and numerical study of magnetic fields in the plasma wake of an intense laser pulse. Phys. Plasmas, **5**, 3764–3773.

Sheng, Z.-M., Sentoku, Y., Mima, K., Zhang, J., Yu, W., and Meyer-ter-Vehn, J. (2000). Angular distributions of fast electrons, ions and bremsstrahlung x-/γ-rays in intense laser interaction with solid targets. Phys. Rev. Lett., **85**, 5340–5343.

Shepherd, R., Booth, R., Price, D., Bowers, M, Swan, D., Bonlie, J., Young, B., Dunn, J., White, B., and Stewart, R. (1995). Ultrafast x-ray streak camera for use in ultrashort laser-produced plasma research. *Rev. Sci. Instr.*, **66**, 719–721.

Shepherd, R. L., Kania, D. R., and Jones, L. A. (1988). Measurement of resistivity in a partially degenerate, strongly coupled plasma. *Phys. Rev. Lett.*, **61**, 1278–1281.

Shore, B. W., and Knight, P. L. (1987). Enhancement of high optical harmonics by excess-photon ionization. *J. Phys. B*, **20**, 413–423.

Shukla, P. K., Rao, N. N., Yu, M. Y., and Tsintsadze, N. L. (1986). Relativistic nonlinear effects in plasmas. *Phys. Rep.*, **138**, 1–149.

Shvets, G., and Wurtele, J. S. (1994). Instabilities of short-pulse laser propagation through plasma channels. *Phys. Rev. Lett.*, **73**, 3540–3543.

Siders, C. W., Cavalleri, A., Sokolowski-Tinten, K., Tóth, Cs., Guo, T., Kammler, M., Horn von Hoegen, M., Wilson, K. R., von der Linde, D., and Barty, C. P. J. (1999). Detection of nonthermal melting by ultrafast X-ray diffraction. *Science*, **286**, 1340–1342.

Siegman, A. E. (1986). *Lasers*. University Science Books, Mill Valley, CA.

Silin, V. P. (1965). Nonlinear high-frequency plasma conductivity. *Sov. Phys. JETP*, **20**, 1510.

Snavely, R. A., Key, M. H., Hatchett, S. P., Cowan, T. E., Roth, M., Phillips, T. W., Stoyer, M. A., Henry, E. A., Sangster, T. C., Singh, M. S., Wilks, S. C., MacKinnon, A., Offenberger, A., Pennington, D. M., Yasuike, K., Langdon, A. B., Lasinski, B. F., Johnson, J., Perry, M. D., and Campbell, E. M. (2000). Intense high-energy proton beams from Petawatt-laser irradiation of solids. *Phys. Rev. Lett.*, **85**, 2945–2948.

Sokolowski-Tinten, K., Blome, C., Blums, J., Cavalleri, A., Dietrich, C., Tarasevitch, A., Uschmann, I., Förster, E., Kammler, M., Horn-von-Hoegen, M., and von der Linde, D. (2003). Femtosecond x-ray measurement of coherent lattice vibrations near the Lindemann stability limit. *Nature*, **422**, 287–289.

Solem, J. C., and Chapline, G. F. (1984). X-ray biomicroholography. *Opt. Eng.*, **23**, 193–203.

Soom, B., Chen, H., Fisher, Y., and Meyerhofer, D. D. (1993). Strong $K\alpha$ emission in picosecond laser-plasma interactions. *J. Appl. Phys.*, **74**, 5372–5377.

Spitzer Jr., L. (1962). *Physics of Fully Ionized Gases*. John Wiley, New York.

Spitzer, Jr., L., and Härm, R. (1952). Transport phenomena in a completely ionized gas. *Phys. Rev.*, **89**, 977–981.

Sprangle, P., and Esarey, E. (1991). Stimulated backscattered harmonic generation from intense laser interactions with beams and plasmas. *Phys. Rev. Lett.*, **67**, 2021–2024.

Sprangle, P., Tang, C.-M., and Esarey, E. (1987). Relativistic self-focusing of short-pulse radiation beams in plasmas. *IEEE Trans. Plasma Sci.*, **15**, 145–153.

Sprangle, P., Esarey, E., Ting, A., and Joyce, G. (1988). Laser wakefield acceleration and relativistic optical guiding. *Appl. Phys. Lett.*, **53**, 2146–2148.

Sprangle, P., Esarey, E., and Ting, A. (1990a). Nonlinear interaction of intense laser pulses in plasmas. *Phys. Rev. A*, **41**, 4463–4469.

Sprangle, P., Esarey, E., and Ting, A. (1990b). Nonlinear theory of intense laser-plasma interactions. *Phys. Rev. Lett.*, **64**, 2011–2014.

Sprangle, P., Esarey, E., Krall, J., and Joyce, G. (1992). Propagation and guiding of intense laser pulses in plasmas. *Phys. Rev. Lett.*, **69**, 2200–2203.

Stamper, J., Papadapoulos, K., Sudan, R. N., McLean, E., Dean, S., and Dawson, J. M. (1971). Spontaneous magnetic fields in laser-produced plasmas. *Phys. Rev. Lett.*, **26**, 1012–1015.

Stamper, J. A., Lehmberg, R. H., Schmitt, A., Herbst, M. J., Young, F. C., Gardner, J. H., and Obenshain, S. P. (1985). Evidence in the second-harmonic emission for self-focusing of a laser pulse in a plasma. *Phys. Fluids*, **28**, 2563–2569.

Startsev, E. A., and McKinstrie, C. J. (1997). Multiple scale derivation of the relativistic ponderomotive force. *Phys. Rev. E*, **55**, 7527–7535.

Stearns, D. G., Landen, O. L., Campbell, E. M., and Scofield, J. H. (1988). Generation of ultrashort x-ray pulses. *Phys. Rev. A*, **37**, 1684–1690.

Strickland, D., and Mourou, G. (1985). Compression of amplified chirped optical pulses. *Op. Commun.*, **56**, 219–211.

Stupakov, G. V., and Zolotorev, M. S. (2001). Ponderomotive laser acceleration and focusing in vacuum for generation of attosecond electron bunches. *Phys. Rev. Lett.*, **86**, 5274–5277.

Sudan, R. (1993). Mechanism for the generation of 10^9 G magnetic fields in the interaction of ultraintense short laser pulse with an overdense plasma target. *Phys. Rev. Lett*, **70**, 3075–3078.

Sullivan, A., Hamster, H., Gordon, S. P., Nathel, H., and Falcone, R. W. (1994). Propagation of intense, ultrashort laser pulses in plasmas. *Optics Lett.*, **19**, 1544–1546.

Sun, G.-Z., Ott, E., Lee, Y. C., and Guzdar, P. (1987). Self-focusing of short intense pulses in plasmas. *Phys. Fluids*, **30**, 526–532.

Svanberg, S. (2001). Some applications of ultrashort laser pulses in biology and medicine. *Meas. Sci. Tech.*, **12**, 1777–1783.

Tabak, M., Hammer, J., Glinsky, M. E., Kruer, W. L., Wilks, S. C., Woodworth, J., Campbell, E. M., Perry, M. D., and Mason, R. J. (1994). Ignition and high gain with ultrapowerful lasers. *Phys. Plasmas*, **1**, 1626–1634.

Tajima, T. (1988). *Computational Plasma Physics with Applications to Fusion and Astrophysics*. Addison-Wesley, New York.

Tajima, T., and Dawson, J. M. (1979). Laser electron accelerator. *Phys. Rev. Lett.*, **43**, 267–270.

Tallents, G. J., Key, M. H., Ridgeley, A., Shaikh, W., Lewis, C. L. S., O'Neill, D., Davidson, S. J., Freeman, N. J., and Perkins, D. (1990). An investigation of the X-ray point-source brightness for a short-pulse laser plasma. *J. Quant. Spec. Rad. Trans.*, **43**, 53–60.

Tanaka, K. A., Kodama, R., Mima, K., Kitagawa, Y., Fujita, H., Miyanaga, N., Nagai, K., Norimatsu, T., Sato, T., Sentoku, Y., Shigemori, K., Sunahara, A., Shozaki, T., Tanpo, M., Tohyama, S., Yabuuchi, T., Zheng, J., Yamanaka, T., Norreys, P. A., Evans, R., Zepf, M., Krushelnick, K., Dangor, A., Stephens, R.,

Hatchett, S., Tabak, M., and Turner, R. (2003). Basic and integrated studies for fast ignition. *Phys. Plasmas*, **10**(5), 1925–1930.

Tang, C. M., Sprangle, P., and Sudan, R. N. (1985). Dynamics of space-charge waves in the laser beat-wave accelerator. *Phys. Fluids*, **28**, 1974–1983.

Tarasevitch, A., Orisch, A., von der Linde, D., Balcou, Ph., Rey, G., Chambaret, J.-P., Teubner, U., Klöpfel, D., and Theobald, W. (2000). Generation of high-order spatially coherent harmonics from solid targets by femtosecond laser pulses. *Phys. Rev. A*, **62**, 3816.

Tatarakis, M., Davies, J. R., Lee, P., Norreys, P. A., Kassapakis, N. G., Beg, F. N., Bell, A. R., Haines, M. G., and Dangor, A. E. (1998). Plasma formation on the front and rear of plastic targets due to high-intensity laser-generated fast electrons. *Phys. Rev. Lett.*, **81**, 999–1002.

Tatarakis, M., Gopal, A., Watts, I., Beg, F. N., Dangor, A. E., Krushelnick, K., Wagner, U., Norreys, P. A., Clark, E. L., Zepf, M., and Evans, R. G. (2002). Measurements of ultrastrong magnetic fields during relativistic laser-plasma interactions. *Phys. Plasmas*, **9**, 2244–2250.

Teubner, U., Bergmann, J., van Wonterghem, B., F. P. Schäfer, and Sauerbrey, R. (1993). Angle-dependent X-ray emission and resonance absorption in a laser-produced plasma generated by a high intensity ultrashort pulse. *Phys. Rev. Lett.*, **70**, 794–797.

Teubner, U., Wülker, C., and E. Förster, W. Theobald (1995). X-ray spectra from high-intensity subpicosecond laser produced plasmas. *Phys. Plasmas*, **2**, 972–981.

Teubner, U., Uschmann, I., Gibbon, P., Altenbernd, D., Förster, E., Feurer, T., Theobald, W., Sauerbrey, R., Hirst, G., Key, M. H., Lister, J., and Neely, D. (1996a). Absorption and hot electron production by high intensity femtosecond UV-Laser pulses in solid targets. *Phys. Rev. E*, **54**, 4167–4177.

Teubner, U., Gibbon, P., Förster, E., Falliès, F., Audebert, P., Geindre, J. P., and Gauthier, J. C. (1996b). Subpicosecond KrF laser plasma interaction at intensities between 10^{14} Wcm^{-2} and 10^{17} Wcm^{-2}. *Phys. Plasmas*, **3**, 2679–2685.

Teubner, U., Eidmann, K., Wagner, U., Andiel, U., Pisani, F., Tsakiris, G. D., Witte, K., Meyer-ter-Vehn, J., Schlegel, T., and Förster, E. (2004). Harmonic emission from the rear side of thin overdense foils irradiated with intense ultrashort laser pulses. *Phys. Rev. Lett.*, **92**, 185001.

Teychenné, D., Giulietti, A., Giulietti, D., and Gizzi, L. A. (1998). Magnetically induced optical transparency of overdense plasmas due to ultrafast ionization. *Phys. Rev. E*, **58**, R1245–R1247.

Theobald, W., Häßner, R., Kingham, R., Sauerbrey, R., Fehr, R., Gericke, D. O., Schlanges, M., Kraeft, W. D., and Ishikawa, K. (1999). Electron densities, temperatures and the dielectric function of femtosecond laser-produced plasmas. *Phys. Rev. E*, **59**, 3544–3553.

Tillman, C., Persson, A., Wahlström, C.-G., Svanberg, S., and Herrlin, K. (1995). Imaging using hard X-rays from a laser-produced plasma. *Appl. Phys. B*, **61**, 333–338.

Tillman, C., Mercer, I., Svanberg, S., and Herrlin, K. (1996). Elemental biological imaging by differential absorption with a laser-produced x-ray source. *J. Opt. Soc. Am. B*, **13**, 1–7.

Ting, A., Krushelnick, K., Moore, C. I., Burris, H. R., Esarey, E., Krall, J., and Sprangle, P. (1996). Temporal evolution of self-modulated laser wakefields measured by coherent Thomson scattering. *Phys. Rev. Lett.*, **77**, 5377–5380.

Tisch, J. W. G., Smith, R. A., Muffett, J. E., Ciarrocca, M., Marangos, J. P., and Hutchinson, M. H. R. (1994). Angularly resolved high-order harmonic generation in helium. *Phys. Rev. A*, **49**(1), R28–31.

Tomov, I. V., Chen, P., and Rentzepis, P. M. (1995). Picosecond time-resolved X-ray diffraction during laser-pulse heating of an Au(111) crystal. *J. Appl. Cryst.*, **28**, 358–362.

Town, R. P. J., Bell, A. R., and Rose, S. (1994). Fokker-Planck simulations of short-pulse laser-solid experiments. *Phys. Rev. E*, **50**, 1413–1421.

Town, R. P. J., Bell, A. R., and Rose, S. (1995). Fokker-Planck calculations with ionization dynamics of short-pulse laser-solid interactions. *Phys. Rev. Lett.*, **74**, 924–927.

Tripathi, V. K., and Liu, C. S. (1994). Self-generated magnetic field in an amplitude modulated laser filament in a plasma. *Phys. Plasmas*, **1**, 990–992.

Tsytovich, V. N., De Angelis, U., and Bingham, R. (1989). Beat-wave-wake acceleration by short high intensity laser pulses. *Comments Plasma Phys. Contr. Fusion*, **12**, 249–256.

Tzeng, K.-C., and Mori, W. B. (1998). Suppression of electron ponderomotive blowout and relativistic self-focusing by the occurrence of Raman scattering and plasma heating. *Phys. Rev. Lett.*, **81**, 104–107.

Tzeng, K.-C., Mori, W. B., and Decker, C. D. (1996). The anomalous absorption and scattering of short-pulse high-intensity lasers in underdense plasmas. *Phys. Rev. Lett.*, **76**, 3332–3335.

Umstadter, D., Chen, S. Y., Maksimchuk, A., Mourou, G., and Wagner, R. (1996). Nonlinear optics in relativistic plasmas and laser wakefield acceleration. *Science*, **273**, 472–475.

Uschmann, I., Gibbon, P., Klöpfel, D., Feurer, Th., Förster, E., Audebert, P., Geindre, J.-P., Gauthier, J.-C., Rousse, A., and Rischel, Ch. (1999). X-ray emission produced by hot electrons from fs-laser produced plasma – diagnostic and application. *Laser and Part. Beams,*, **17**, 671–679.

Vachaspati (1962). Harmonics in the scattering of light by free electrons. *Physical Review*, **128**, 664–666.

Veisz, L., Theobald, W., Schillinger, H., Gibbon, P., Sauerbrey, R., and Jovanovic, M. S. (2002). Three-halves harmonic emission from femtosecond laser produced plasmas. *Phys. Plasmas*, **9**, 3197–3200.

Veisz, L., Theobald, W., Feurer, T., Schwoerer, H., Uschmann, I., Renner, O., and Sauerbrey, R. (2004). Three-halves harmonic emission from femtosecond laser-produced plasmas with steep density gradients. *Phys. Plasmas*, **11**, 3311–3323.

Vidal, F., and Johnston, T. W. (1996). Electromagnetic beam breakup: Multiple filaments, single beam equilibria and radiation. *Phys. Rev. Lett.*, **77**, 1282–1285.

Volkov, D. M. (1935). Über eine Klasse von Lösungen der Dirac'schen Gleichung. *Z. Phys.*, **94**, 250–260.

von der Linde, D., and Rzazewski, K. (1996). High-order optical harmonic generation from solid surfaces. *Appl. Phys. B*, **63**, 499–506.

von der Linde, D., Engers, T., Jenke, G., Agostini, P., Grillon, G., Nibbering, E., Mysyrowicz, A., and Antonetti, A. (1995). Generation of high-order harmonics from solid surfaces by intense femtosecond laser pulses. *Phys. Rev. A*, **52**, R25–R27.

Voronov, G. S., and Delone, N. B. (1965). Ionization of xenon atom by electric field of ruby laser emission. *Sov. Phys. JETP Lett.*, **1**, 66.

Wagner, R., Chen, S. Y., Maksimchuk, A., and Umstadter, D. (1997). Electron acceleration by a laser wakefield in a relativistic self-guided channel. *Phys. Rev. Lett.*, **78**, 3125–3128.

Wallace, J. M. (1985). Nonlocal energy deposition in high-intensity laser-plasma interactions. *Phys. Rev. Lett.*, **55**, 707–710.

Walser, M. W., Keitel, C. H., Scrinzi, A., and Brabec, T. (2000). High harmonic generation beyond the electric dipole approximation. *Phys. Rev. Lett.*, **85**, 5082–5085.

Wang, P. X., Ho, Y. K., Yuan, X. Q., Kong, Q., Cao, N., Sessler, A. M., Esarey, E., and Nishida, Y. (2001). Vacuum electron acceleration by an intense laser. *Appl. Phys. Lett.*, **78**, 2253–2255.

Ward, J. F., and New, G. H. C. (1969). Optical third harmonic generation in gases by a focused laser beam. *Phys. Rev.*, **185**, 57–72.

Watts, I., Zepf, M., Clark, E. L., Tatarakis, M., Krushelnick, K., Dangor, A. E., Allott, R. M., Clarke, R. J., Neely, D., and Norreys, P. A. (2002). Dynamics of the critical surface in high-intensity laser-solid interactions: modulation of the XUV harmonic spectra. *Phys. Rev. Lett.*, **88**, 155001.

Weibel, E. S. (1967). Anomalous skin effect in a plasma. *Phys. Fluids*, **10**, 741–748.

Wharton, K. B., Hatchett, S. P., Wilks, S. C., Key, M. H., Moody, J. D., Yanovsky, V., Offenberger, A. A., Hammel, B. A., Perry, M. D., and Joshi, C. (1998). Experimental measurements of hot electrons generated by ultraintense laser-plasma interactions on solid-density targets. *Phys. Rev. Lett.*, **81**, 822–825.

Wickens, L. M., Allen, J. E., and Rumsby, P. T. (1978). Ion emission from laser-produced plasmas with two electron temperatures. *Phys. Rev. Lett.*, **41**, 243–246.

Wilks, S. C. (1993). Simulations of ultraintense laser-plasma interactions. *Phys. Fluids B*, **5**, 2603–2608.

Wilks, S. C., and Kruer, W. L. (1997). Absorption of ultrashort, ultra-intense laser light by solids and overdense plasmas. *IEEE J. Quant. Elec.*, **33**, 1954–1968.

Wilks, S. C., Kruer, W. L., Tabak, M., and Langdon, A. B. (1992). Absorption of ultra-intense laser pulses. *Phys. Rev. Lett.*, **69**, 1383–1386.

Wilks, S. C., Kruer, W. L., and Mori, W. B. (1993). Odd harmonic generation of ultra-intense laser pulses reflected from an overdense plasma. *IEEE Trans. Plasma Sci.*, **21**, 120–124.

Wilks, S. C., Langdon, A. B., Cowan, T. E., Roth, M., Singh, M., Hatchett, S., Key, M. H., Pennington, D., MacKinnon, A., and Snavely, R. A. (2001). Energetic proton generation in ultra-intense laser-solid interactions. *Phys. Plasmas*, **8**, 542–549.

Winick, H. (1994). *Synchrotron Radiation Sources: A Primer.* World Scientific, Singapore.

Woodward, P. M. (1947). A method of calculating the field over a plane. *J. Inst. Electr. Eng.*, **93**, 1554–1558.

Workman, J., Maksimchuk, A., Liu, X., Ellenberger, U., Coe, J. S., Chien, C.-Y., and Umstadter, D. (1995). Control of bright picosecond X-ray emission from intense subpicosecond laser-plasma interactions. *Phys. Rev. Lett.*, **75**, 2324–2327.

Yang, T.-Y. B., Kruer, W. L., More, R. M., and Langdon, A. B. (1995). Absorption of laser light in overdense plasma by sheath inverse bremsstrahlung. *Phys. Plasmas*, **2**, 3146–3154.

Yang, T.-Y. B., Kruer, W. L., Langdon, A. B., and Johnston, T. W. (1996). Mechanisms for collisionless absorption of light waves obliquely incident on overdens plasmas with steep density gradients. *Phys. Plasmas*, **3**, 2702–2709.

Young, P. E., Foord, M. E., Hammer, J. H., Kruer, W. L., Tabak, M., and Wilks, S. C. (1995). Time-dependent channel formation in a laser-produced plasma. *Phys. Rev. Lett.*, **75**(6), 1082–1085.

Yu, J., Jiang, Z., Kieffer, J. C., and Krol, A. (1999a). Hard x-ray emission in high-intensity femtosecond laser-target interaction. *Phys. Plasmas*, **6**, 1318–1322.

Yu, J., Jiang, Z., Kieffer, J. C., and Krol, A. (1999b). High magnification imaging with a laser-based hard x-ray source. *IEEE Trans. Quant. Elec.*, **5**, 911–915.

Yu, W., Yu, M. Y., Sheng, Z. M, and Zhang, J. (1998). Model for fast electrons in ultrashort-pulse laser interaction with solid targets. *Phys. Rev. E*, **58**, 2456–2460.

Yu, W., Sheng, Z. M, Yu, M. Y., Zhang, J., Jiang, Z. M., and Xu, Z. (1999c). Model for transmission of ultrastrong laser pulses through thin foil targets. *Phys. Rev. E*, **59**, 3583–3587.

Zakharov, V. E. (1972). Collapse of langmuir waves. *JETP*, **35**, 908.

Zeitoun, Ph., Faivre, G., Sebban, S., Mocek, T., Hallou, A., Fajardo, M., Aubert, D., Balcou, Ph., Burgy, F., Douillet, D., Kazamias, S., de Lacheze-Murel, G., Lefrou, T., le Pape, S., Mercere, P., Merdji, H., Morlens, A. S., Rousseau, J. P., and Valentin, C. (2004). A high-intensity highly coherent soft x-ray femtosecond laser seeded by a high harmonic beam. *Nature*, **431**, 426–429.

Zel'dovich, Ya. B., and Raizer, Yu. P. (1966). *Physics of Shock Waves and High-Temperature Phenomena.* Academic Press, New York.

Zeng, G., Shen, B., Yu, W., and Xu, Z. (1996). Relativistic harmonic generation excited in the ultrashort laser pulse regime. *Phys. Plasmas*, **3**, 4220–4224.

Zepf, M., Castro-Colin, M., Chambers, D., Preston, S.G., Wark, J.S., Zhang, J., Danson, C.N., Neely, D., Norreys, P.A., Dangor, A.E., Dyson, A., Lee, P., Fews,

A.P., Gibbon, P., Moustaizis, S., and Key, M.H. (1996). Measurements of hole boring velocities and inferred energy absorption fractions for laser interaction with solid targets at 10^{19}Wcm$^{-2}\mu m^2$. *Phys. Plasmas*, **3**, 3242–3244.

Zepf, M., Tsakiris, G. D., Pretzler, G., Watts, I., Chambers, D. M., Norreys, P. A., Andiel, U., Dangor, A. E., Eidmann, K., Gahn, C., Machacek, A., Wark, J. S., and Witte, K. (1998). Role of the plasma scale length in the harmonic generation from solid targets. *Phys. Rev. E*, **58**, R5253–R5256.

Zepf, M., Clark, E. L., Beg, F. N., Clarke, R. J., Dangor, A. E., Gopal, A., Krushelnick, K., Norreys, P. A., Tatarakis, M., Wagner, U., and Wei, M. S. (2003). Proton acceleration from high-intensity laser interactions with thin foil targets. *Phys. Rev. Lett.*, **90**, 064801–1.

Zhang, J., Zepf, M., Norreys, P. A., Dangor, A. E., Bakarezos, M., Danson, C. N., Dyson, A., Fews, A. P., Gibbon, P., Lee, P., Loukakos, P., Moustaizis, S., Neely, D., Walsh, F. N., and Wark, J. S. (1996). Coherence and bandwidth measurements of harmonics generated from solid surfaces irradiated by intense picosecond laser pulses. *Phys. Rev. A*, **54**, 1597–1603.

Zhidkov, A., Koga, J., Sasaki, A., and Uesaka, M. (2002). Radiation damping effects on the interaction of ultraintense laser pulses with an overdense plasma. *Phys. Rev. Lett.*, **88**, 185002.

Zhou, J., Peatross, J., Murnane, M. M., and Kapteyn, H. C. (1996). Enhanced high-harmonic generation using 25 fs laser pulses. *Phys. Rev. Lett.*, **76**, 752–755.

Zigler, A., Burkhalter, P. G., and Nagel, D. J. (1991). High intensity generation of 9–13 Å x-rays from BaF$_2$ targets. *Appl. Phys. Lett.*, **59**, 777–778.

Ziman, J. M. (1969). *Principle of the Theory of Solids*. Cambridge University Press, Cambridge, England.

Index

Printed in the United States
By Bookmasters